Trends in Communication Technologies and Engineering Science

Lecture Notes in Electrical Engineering

Volume 33

For further volumes:
http://www.springer.com/series/7818

Sio-Iong Ao · Xu Huang
Ping-Kong Alexander Wai
Editors

Trends in Communication Technologies and Engineering Science

Springer

Editors
Dr. Sio-Iong Ao
Harvard School of Engineering
and Applied Sciences
Harvard University
60 Oxford Street
Cambridge MA 02138
USA
siao@seas.harvard.edu

Prof. Xu Huang
University of Canberra
School of Information
Sciences & Engineering
Canberra ACT 2601
Australia
Xu.Huang@canberra.edu.au

Prof. Ping-Kong Alexander Wai
Hong Kong Polytechnic
University
Faculty of Engineering
Hung Hom, Kowloon
Hong Kong/PR China
enwai@polyu.edu.hk

ISSN 1876-1100 e-ISSN 1876-1119
ISBN 978-90-481-8139-1 e-ISBN 978-1-4020-9532-0
DOI 10.1007/978-1-4020-9532-0
Springer Dordrecht Heidelberg London New York

Printed on acid-free paper

Springer is part of Springer Science+Business Media (www.springer.com)

Preface

A large international conference on Advances in Communication Technologies and Engineering Science was held in Hong Kong, March 19–21, 2008, under the International MultiConference of Engineers and Computer Scientists (IMECS 2008). The IMECS 2008 is organized by the International Association of Engineers (IAENG). IAENG is a non-profit international association for the engineers and the computer scientists, which was found originally in 1968 and has been undergoing rapid expansions in recent few years. The IMECS conferences serve as good platforms for the engineering community to meet with each other and to exchange ideas. The conferences have also stroke a balance between theoretical and application development. The conference committees have been formed with over two hundred committee members who are mainly research center heads, faculty deans, department heads, professors, and research scientists from over 30 countries. The conferences are truly international meetings with a high level of participation from many countries. The response that we have received for the congress is excellent. There have been more than five hundred manuscript submissions for the IMECS 2008. All submitted papers have gone through the peer review process and the overall acceptance rate is 56.03%.

This volume contains revised and extended research articles written by prominent researchers participating in the conference. Topics covered include communications theory, communications protocols, network management, wireless networks, telecommunication, electronics, power engineering, control engineering, signal processing, and industrial applications. The book will offer the states of arts of tremendous advances in communication systems and engineering science and also serve as an excellent reference work for researchers and graduate students working with/on communication technologies and engineering science.

Cambridge, Massachusetts — Sio-Iong Ao
Canberra, ACT — Xu Huang
Hong Kong, P.R. China — Ping-Kong Alexander Wai

Contents

Contributors

Titik Khawa Abdul Rahman Academic Affairs Department, Universiti Teknologi MARA Malaysia, takitik@streamyx.com

Witold Abramowicz Department of Information Systems, Poznań University of Economics, Al. Niepodległości 10, 60-967 Poznan, Poland, W.Abramowicz@kie.ae.poznan.pl

Mohd Rafi Adzman Faculty of Electrical System, Universiti Malaysia Perlis (UniMAP), mohdrafi@unimap.edu.my

Zekeriya Aliyazicioglu California State Polytechnic University, Pomona, CA, USA

Ebrahim Ansari Chelche Department of Computer Science and Engineering, Shiraz University, Shiraz, Iran

Farhat Anwar Department of ECE, Faculty of Engineering, IIUM, Kuala Lumpur, Malaysia

Md. Saiful Azad Department of ECE, Faculty of Engineering, IIUM, Kuala Lumpur, Malaysia

Tao Ban National Institute of Information and Communications Technology, Tokyo, Japan, bantao@nict.go.jp

Uwe Baumgarten Institut für Informatik, Technische Universität München, Boltzmannstr. 3, 85748 Garching b., München, Germany, baumgaru@in.tum.de

Chun-Che Chang Department of Computer Engineering and Science, Yuan-Ze University, Chung-Li, Taiwan 32026, ROC

Yanan Chang Institute of Computer Network and Communication, Huazhong Normal University, Wuhan, Hubei 430079, P.R. China

Limiao Chen Institute of Computer Network and Communication, Huazhong Normal University, Wuhan, Hubei 430079, P.R. China

Chung-Man Cheung Department of Electrical and Electronic Engineering, The University of Hong Kong, Pokfulam Road, Hong Kong, edmondcheung@eee.hku.hk

G.H. Dastghaibyfard Department of Computer Science and Engineering, Shiraz University, Shiraz, Iran

Nur Hazima Faezaa Ismail Aviation Engineering Programme, Malaysian Airline System (MAS), Malaysia, mkhayat@salam.uitm.edu.my

K.M. Gangaraju Center for Artificial Intelligence and Robotics, Bangalore, India

Viji Gopal R.M.K Engineering College, Chennai, India

N.S. Gowri Ganesh Centre for Development of Advanced Computing, Chennai, India

Wolfgang Haberl Institut für Informatik, Technische Universität München, Boltzmannstr. 3, 85748 Garching b. München, Germany, haberl@in.tum.de

Konstanty Haniewicz Department of Information Systems, Poznań University of Economics, Al. Niepodległości 10, 60-967 Poznan, Poland, K.Haniewicz@kie.ae.poznan.pl

Genta Hashimoto Yamaguchi University, Graduate School of Medicine, Ube, Yamaguchi 753-8512, Japan

Kenji Hatano Faculty of Culture and Information Science, Doshisha University, Kyoto, Japan

Takafumi Hiro Yamaguchi University, Graduate School of Medicine, Ube, Yamaguchi 753-8512, Japan

Radosław Hofman Department of Information Systems, Poznań University of Economics, Al. Niepodległości 10, 60-967 Poznan, Poland, radekh@teycom.pl

Hsi-Chin Hsin Department of Computer Science and Information Engineering, National United University, Miaoli, Taiwan, ROC

H.K. Hwang California State Polytechnic University, Pomona, CA, USA

I-Shyan Hwang Department of Computer Engineering and Science, Yuan-Ze University, Chung-Li, Taiwan 32026, ROC

Shohei Ichiyama Yamaguchi University, Graduate School of Science and Engineering, Yoshida, Yamaguchi 753-8512, Japan

Maria Indrawan Monash University, Melbourne, Australia

Masumi Ishikawa Kyushu Institute of Technology, 2-4 Hibikino, Wakamatsu, Kitakyushu 808-0196, Japan, ishikawa@brain.kyutech.ac.jp

A. Jangwanitlert Electrical Engineering Department, Faculty of Engineering, King Mongkut's Institute of Technology Ladkrabang, Bangkok 10520, Thailand

Chang-Sung Jeong Department Electronics Engineering, Korea University, Seoul, South Korea

Hani Kaabi Department of Computer Science and Engineering, Shiraz University, Shiraz, Iran

Monika Kaczmarek Department of Information Systems, Poznań University of Economics, Al. Niepodległości 10, 60-967 Poznan, Poland, M.Kaczmarek@kie.ae.poznan.pl

Youki Kadobayashi National Institute of Information and Communications Technology, Tokyo, Japan

S. Kaitwanidvilai Electrical Engineering Department, Faculty of Engineering, King Mongkut's Institute of Technology Ladkrabang, Bangkok 10520, Thailand

Mira Kajko-Mattsson Department of Computer and Systems Sciences, Stockholm University/Royal Institute of Technology, SE-164 40 Kista, Sweden, mira@dsv.su.se

Mohd. Rozely Kalil Politeknik Ungku Omar, Ipoh Malaysia, rozleymsc@yahoo.com.

S. Karnprachar Electrical and Computer Engineering Department, Faculty of Engineering, Naresuan University, Phitsanulok 65000, Thailand

Liang-Yu Ke Department of Computer Engineering and Science, Yuan-Ze University, Chung-Li, Taiwan 32026, ROC

Morteza Keshtakaran Department of Computer Science and Engineering, Shiraz University, Shiraz, Iran

W. Khanngern Electrical Engineering Department, Faculty of Engineering, King Mongkut's Institute of Technology Ladkrabang, Bangkok 10520, Thailand

Mohamad Khayat Idris Faculty of Electrical Engineering, Universiti Teknologi MARA Malaysia, mkhayat@salam.utim.edu.my

Fuminori Kimura Department of Media Technology, College of Information Science and Engineering, Ritsumeikan University, Kyoto, Japan

Ryosuke Kubota Ube National College of Technology, Department of Intelligent System Engineering, Ube, Yamaguchi 755-8555, JAPAN, kubota@ube-k.ac.jp

Masahito Kurihara Graduate School of Information Science and Technology, Hokkaido University, Hokkaido, Japan

Hing-Kit Kwan Department of Electrical and Electronic Engineering, The University of Hong Kong, Pokfulam Road, Hong Kong, kwanhk@eee.hku.hk

Tae-Dong Lee School of Electronics Engineering, Korea University, Seoul, South Korea

Chi-Un Lei Department of Electrical and Electronic Engineering, The University of Hong Kong, Pokfulam Road, Hong Kong, culei@eee.hku.hk

Dongyang Long Department of Computer Science, School of Information Science and Technology, Sun Yat-Sen University, Guangzhou, China

Akira Maeda Department of Media Technology, College of Information Science and Engineering, Ritsumeikan University, Kyoto, Japan

Masunori Matsuzaki Yamaguchi University, Graduate School of Medicine, Ube, Yamaguchi 753-8512, Japan

Ismail Musirin Faculty of Electrical Engineering, Universiti Teknologi MARA Malaysia, i_musirin@yahoo.co.uk

Jun Miyazaki Graduate School of Information Science, Nara Institute of Science and Technology, Nara, Japan

I. Ngarmroo Electrical Engineering Department, Faculty of Engineering, King Mongkut's Institute of Technology Ladkrabang, Bangkok 10520, Thailand

Jaana Nyfjord Department of Computer and Systems Sciences, Stockholm University/Royal Institute of Technology, SE-164 40 Kista, Sweden, jaana@dsv.su.se

Md. Arafatur Rahman Department of ECE, Faculty of Engineering, IIUM, Kuala Lumpur, Malaysia

M.H. Sadreddini Department of Computer Science and Engineering, Shiraz University, Shiraz, Iran

G.V.K. Sasirekha Center for Artificial Intelligence and Robotics, Bangalore, India

Haruhiko Sato Graduate School of Information Science and Technology, Hokkaido University, Hokkaido, Japan

Dong-Bum Seo Department of Information and Communication Engineering, Korea University, Seoul, South Korea

Zen-Der Shyu Department of General Studies, Army Academy, Chung-Li, Taiwan 32092, ROC

Alex Tze Hiang Sim Monash University, Melbourne, Australia

Bala Srinivasan Monash University, Melbourne, Australia

Noriaki Suetake Yamaguchi University, Graduate School of Science and Engineering, Yoshida, Yamaguchi 753-8512, Japan

Tze-Yun Sung Department of Microelectronics Engineering, Chung Hua University, Hisnchu, Taiwan, ROC

Su-Kit Tang Department of Computer Science, School of Information Science and Technology, Sun Yat-Sen University, Guangzhou, China

Michael Tautschnig Institut für Informatik, Technische Universität Darmstadt, Hochschulstr. 10, 64289 Darmstadt, Germany, tautschnig@cs.tu-darmstadt.de

Hidekazu Terai Ritsumeikan University, Kyoto, Japan

Eiji Uchino Yamaguchi University, Graduate School of Science and Engineering, Yoshida, Yamaguchi 753-8512, Japan, uchino@yamaguchi-u.ac.jp

Shunsuke Uemura Faculty of Informatics, Nara Sangyo University, Nara, Japan

Ngai Wong Department of Electrical and Electronic Engineering, The University of Hong Kong, Pokfulam Road, Hong Kong, nwong@eee.hku.hk

Xiaoqiong Wu Institute of Computer Network and Communication, Huazhong Normal University, Wuhan, Hubei 430079, P.R. China

Debao Xiao Institute of Computer Network and Communication, Huazhong Normal University, Wuhan, Hubei 430079, P.R. China

Hui Xu Institute of Computer Network and Communication, Huazhong Normal University, Wuhan, Hubei 430079, P.R. China

Hironori Yamauchi Ritsumeikan University, Kyoto, Japan

Masaya Yoshikawa Meijo university, Nagoya, Japan

Hong Zhang Kyushu Institute of Technology, 2-4 Hibikino, Wakamatsu, Kitakyushu 808-0196, Japan, zhang@brain.kyutech.ac.jp

Samar Zutshi Swinburne University, Melbourne, Australia.

Dominik Zyskowski Department of Information Systems, Poznań University of Economics, Al. Niepodległości 10, 60-967 Poznan, Poland, D.Zyskowski@kie.ae.poznan.pl

Chapter 1
Survivable Architecture with Dynamic Wavelength and Bandwidth Allocation Scheme in WDM-EPON

I-Shyan Hwang, Zen-Der Shyu and Chun-Che Chang

Abstract This study proposes a novel fault-tolerant architecture in WDM-EPON, Cost-based Fault-tolerant WDM-EPON (CFT-WDM-EPON), to provide overall protection. The CFT-WDM-EPON only equips a backup feeder fiber to recover the system failure. Additionally, a prediction-based fair wavelength and bandwidth allocation (PFWBA) scheme is also proposed to enhance the differentiated services for WDM-EPON based on the Dynamic Wavelength Allocation (DWA) and Prediction-based Fair Excessive Bandwidth Reallocation (PFEBR). The PFEBR involves an Early-DBA mechanism, which improves prediction accuracy by delaying report messages of unstable traffic ONUs, and assigns linear estimation credit to predict the arrival of traffic during waiting time. The DWA can operate in coordination with an unstable degree list to allocate the available time of wavelength precisely. Simulation results show that the proposed PFWBA scheme outperforms the WDM IPACT-ST with a single polling table and the Dynamic wavelength and bandwidth 3 (DWBA3) in terms of end-to-end delay and jitter performance.

Keywords Fault tolerance · WDM-EPON · Differentiated services · PFEBR

1.1 Introduction

Compared with the current access network technologies, passive optical network (PON) technologies are a promising solution for the full service access network. The ITU-T recommends a series of ATM-based Broadband PON systems (i.e. ATM-PON, BPON and GPON). Furthermore, Ethernet PON (EPON) has been discussed in IEEE 802.3ah as an extension of Gigabit-Ethernet [1]. Although the EPON or ATM-based PON provides higher bandwidth than traditional copper-based access

I.-S. Hwang (✉)
Department of Computer Engineering and Science, Yuan-Ze University, Chung-Li, Taiwan, 32026, ROC

S.-I. Ao et al. (eds.), *Trends in Communication Technologies and Engineering Science*, Lecture Notes in Electrical Engineering 33, DOI 10.1007/978-1-4020-9532-0_1,

networks, the bandwidth of the PON needs to be increased further. The WDM-PON architecture can adopt wavelength-division multiple access (WDMA) to support multiple wavelengths in either or both upstream and downstream directions [2]. For employing the WDMA technology, the passive arrayed waveguide grating (AWG) is deployed in the WDM-PON architecture. The AWG allows for spatial reuse of the wavelength channels; thus, a multi-wavelength source at the optical line terminal (OLT) is used to transmit multiple wavelengths to the various optical network units (ONUs) [3]. In the WDM-PON, such a framework provides guaranteed Quality-of-Service (QoS), high security and privacy. However, the limitations of WDM-PON are lack of mature device technologies, lack of suitable network protocols and software to support the architecture, and high overall cost of deploying optical modules [4]. To integrate the advantages of EPON and WDM-PON to provide high link capacity, and to lower the overall system cost, a smooth migration to WDMA from EPON is expected a promising solution for optical access network technology.

The WDM-EPON is the expected solution which employs EPON and WDM-PON systems to provide additional link capacity and lower the cost of optical units. The WDM-EPON manages different wavelength channels in order to increase the available bandwidth of the EPON, but not to increase the cost of the system [5, 6]. In the WDM-EPON architecture, the OLT is upgraded as an array of fixed-tuned transceivers, and reserves one control wavelength channel to forward broadcast frames to each ONU. For the ONU node structure, the WDM-EPON adds tunable transceivers which employ different tuning times and tuning ranges. In the downstream direction, WDM-EPON broadcasts control messages from the OLT to each ONU through the entire bandwidth of one wavelength. Each ONU discards or accepts the incoming frames depending on the packet header addressing. In the upstream direction, WDM-EPON adopts time-division multiple access (TDMA) coupled with multi-point control protocol data unit (MPCPDU) mechanism to avoid collision. The MPCPDU involves both GATE MPCPDU and REPORT messages. The OLT allocates upstream bandwidth to each ONU by sending GATE MPCPDU messages which contains a timestamp, granted timeslots and wavelengths indicating the periods during which the ONU can transmit data. Each ONU can send REPORT messages concerning the queue state to the OLT, enabling the OLT to allocate the appropriate upstream bandwidth, wavelengths and timeslots to each ONU. With multiple ONUs sharing the same upstream bandwidth and wavelengths to transmit data on the WDM-EPON, any data collision lengthens the end-to-end delay and degrades the system performance. Hence, the bandwidth and wavelength allocation is a major concern of research in the WDM-EPON, especially with the large demand for bandwidth and critical applications [5, 6].

Wavelength and bandwidth allocation schemes can be divided into two categories, *Static Wavelength Dynamic Time* (SWDT) and *Dynamic Wavelength Dynamic Time* (DWDT) which is also called *Dynamic Wavelength and Bandwidth Allocation* (DWBA) [7]. In the SWDT, the OLT allocates wavelengths statically and timeslots dynamically. The ONUs are divided into different groups according to the number of wavelengths, and each group of ONUs shares a pre-defined wavelength. However, the number of ONUs on each wavelength is identified in the SWDT, which does not exploit the inter-channel statistical multiplexing, thus lowering utilization. The DWBA assigns the bandwidth and wavelength based on the requested bandwidth, wavelength loading and QoS requirement by each ONU [6, 7]. The DWBA can also exploit both inter-channel and intra-channel statistical multiplexing. Therefore, the DWBA scheme provides more efficient bandwidth allocation than the SWDT scheme, allowing each ONU to share the network resources, and improving QoS for end-users.

Another significant issue in the WDM-EPON is how to protect and recover the optical failure in the WDM-EPON system. Since optical passive networks transmit aggregated high-speed data from several hundreds of end-users, failure in network units or links results in serious data loss. This study proposes a novel fault-tolerant architecture for WDM-EPON, *Cost-based Fault-tolerant WDM-EPON* (*CFT-WDM-EPON*), to lower the cost of conventional protection architecture, and in the short term to recover the functionality of the failed equipment. The CFT-WDM-EPON can recover the optical failure by fast wavelength switching between control and data channel, and only equips a backup feeder fiber to connect the adjacent PON system. Additionally, this study also proposes a robust prediction-based fair wavelength and bandwidth allocation (PFWBA) scheme, which includes the *dynamic wavelength allocation* (DWA) [7] and *Early DBA* (E-DBA) *mechanism* [8]. The E-DBA reduces the idle period and waiting time in the conventional DBA scheme, and obtains fresh queue information for unstable traffic ONUs to improve the accurate prediction. The DWA can operate in coordination with the unstable degree list to allocate the wavelength available time precisely. Furthermore, to improve the system performance, this study also considers the fairness of excessive bandwidth reallocation among ONUs for differentiated traffic classes.

The rest of this paper is organized as follows. Section 1.2 describes related work of DWBA and existing protection scheme in WDM-EPON. Section 1.3 then proposes a novel fault-tolerant architecture, CFT-WDM-EPON, capable of providing overall protection for optical nodes and fibers. Next, Section 1.4 presents the PFWBA scheme, which incorporates the DWA and the E-DBA mechanism for dealing with prediction and fairness allocation of wavelength and bandwidth. Section 1.5 presents the simulation results of the proposed system and other well-known methodologies. Conclusions are finally drawn in Section 1.6.

1.2 Related Work

In WDM-EPON, allocating the bandwidth and wavelength efficiently is the key factor to satisfying various QoS requirements for end-users. Recent studies on wavelength and bandwidth allocation in WDM-EPON can be classified as dynamic wavelength allocation (DWA) and dynamic bandwidth allocation (DBA). The WDM-EPON DWA concerns how the OLT allocates suitable wavelength from multiple wavelengths to ONUs. The WDM-EPON DBA is applied, after the OLT assigns wavelengths to ONUs, in order to allocate bandwidth for each ONU efficiently according to the QoS requirement and network traffic.

Previously proposed DWA systems include the sequential scheduling algorithm [9], which emulates a virtual global first-in-first-out queuing for all incoming requests, and assigns a suitable wavelength for each request. This scheduling algorithm may suffer from wasted bandwidth and poor fairness guarantee if some ONUs have large Round Trip Times (RTTs). To overcome the wasted bandwidth problem, K.S. Kim et al. [10] presents a batch scheduling system that provides priority queuing by scheduling over more than one frame. The batch scheduling system stores the bandwidth requests arriving at OLT during the batch period in queues, and schedules them at the end of batch period. The scheduling delay of the batch scheduling system may increase when the system load is very low and the batch period is short [11]. M. McGarry et al. investigated another scheduling algorithm for REPORT messages such as *online scheduling* and *offline scheduling* [5]. In online scheduling, the OLT follows a grant-on-the-fly manner to allocation timeslots. The OLT allocates a transmission window for each ONU as soon as the OLT receives REPORT message from each ONU for the next cycle. Unlike the online scheduling, the OLT follows a wait-for-all manner in offline scheduling. The OLT allocates transmission windows for all ONUs in a round-robin manner after having received all REPORT messages from all ONUs for the next cycle. The offline scheduling with wait-for-all leads to a long waiting time and idle period because of the long inter-scheduling cycle gap.

In terms of WDM-EPON DBA, K.H. Kwong et al. [12] proposed the WDM IPACT-ST scheme based on the interleaved polling with adaptive cycle time (IPACT), which is proposed for EPON access network [1]. The WDM IPACT-ST applies IPACT to multi-channel PON, where ONUs are equipped with fixed transceivers. Nonetheless, the WDM IPACT-ST lacks the ability to handle the excessive bandwidth, which is collected from lightly-loaded ONUs. As an extension of the WDM IPACT-ST, the excessive bandwidth reallocation (EBR) [7, 13] redistributes the available excessive bandwidth to heavily-loaded ONUs according to the proportion of each request, and improves the performance in terms of packet delay. However, EBR has some drawbacks, namely unfairness and excessive bandwidth allocated to ONUs over that requested. This is termed the *redundant bandwidth problem* [12]. A.R. Dhaini et al. [7] proposed the DWBA3 scheme, the extension of the EBR in the WDM-EPON, which allocates the bandwidth for two steps. The DWBA3 allocates first the guaranteed bandwidth for heavily-loaded ONUs, and the requested bandwidth for lightly-loaded ONUs. Finally, upon receiving all REPORT messages, the DWBA3 redistributes the available excessive bandwidth to heavily-loaded ONUs based on the proportion of each request in next cycle. The upstream in different transmission cycle for heavily-loaded ONUs increases the number of guard time, which decreases the available bandwidth, and increases the end-to-end delay. The PFEBR [8] executes the DBA scheme after the REPORT messages from unstable traffic ONUs are received at the end of ONU_{N-1}, instead of at the end of ONU_N in the standard DBA scheme. The operation reduces the idle period in the standard DBA scheme, and obtains more fresh information of unstable traffic ONUs to enhance the accuracy of prediction in the following cycle. Additionally, the bandwidth is allocated to each ONU in the next cycle according to the unstable degree list. The unstable degree list is calculated using variance of historical traffic, and sorted in decreasing order of all ONUs. The PFEBR alleviates traffic variance by shortening the waiting time before transmitting data for unstable traffic ONUs, and thus improves prediction accuracy.

The fault tolerance of WDM-based PON topology has been discussed recently [9–16]. The fault-tolerant architecture considers two types of network failure, namely link failure and node failure. The node protection of PON concentrates on the most important optical unit, namely OLT, and the link protection focuses on the feeder fibers that connect the OLT node with the remote node (RN) in the PON system. The protection mechanism for the PON topology constructs the backup links or nodes to recover the failure. The authors of [9, 16] provide recovery mechanism against fiber-cut of feeder fiber which connects RN and OLT. H. Nakamura et al. [16] presented the protection fiber for feeder fibers to avoid the fiber-cut situation. The protection fiber may recover fiber failure occurring on the working fiber, but cannot recover from failure occurring on OLT. F.T. An et al. [9] proposed a new Hybrid TDM/WDM-PON architecture named Stanford University aCCESS Hybrid WDM/TDM Passive Optical Network (SUCCESS-HPON). The SUCCESS-HPON is based on a ring-plus-distribution-trees topology, which can provide users with better protection and restoration capabilities than conventional PONs. The SUCCESS provides bi-directional transmission on the same wavelength and fiber. Transmission occurs in reverse direction to recover the failure if the feeder fiber fails. However, the SUCCESS-HPON cannot provide protection for OLT and ONUs. Furthermore, the ring-plus-distribution-trees topology has a long RTT due to the very large number of ONUs. The authors of [14, 15] proposed redundant fibers topology to avoid the failure occurs. However, the protection for feeder and distribution fiber is provided by duplicate deployment of fiber in [14], which is not a cost-effective protection scheme. Furthermore, this protection architecture does not provide any recovery system for node failure. X. F. Sun et al. presented a ring-plus-tree architecture, which not only constructs protection links for feeder fibers, but also constructs protection links between ONUs [15]. The ONU still can transmit data by protection fiber connecting neighboring ONUs when the fiber-cut occurs on the distribution fibers. However, the protection architecture still cannot prevent node failure on optical nodes, and is not cost-effective due to duplicating deployment of fibers. The above models focus on the fiber-cut situation on optical fibers, but do not consider optical node failures, even if the node failure may damage the PON system. If optical node failure occurs on the OLT, then the dynamic wavelength and bandwidth allocation cannot be processed. Additionally, the bandwidth and wavelength requested by all ONUs cannot be granted without leading to significant damage. Therefore, a good fault-tolerant scheme should protect the most important optical nodes, the OLT and the feeder fiber to prevent the failures that degrade the system performance.

1.3 The Proposed CFT-WDM-EPON Architecture

The proposed protection architecture, CFT-WDM-EPON, based on WDM-EPON, comprises a centralized OLT and AWG, and connects a group of associated ONUs to provide complete protection for overall PON architecture, illustrated in Fig. 1.1. Each PON system connects adjacent PON systems by protection feeder fiber, and each AWG connects two OLTs with many ONUs. When no failures occur on the CFT-WDM-EPON, ONUs transmit the REPORT message to OLTs through wavelength, $\lambda_{\mathrm{REPORT}}$, which is reserved for transmitting control messages for each ONU. The OLT considers takes the loading balancing between different wavelengths when the OLT assigns wavelengths and bandwidth to each ONU by GATE MPCPDU message. The ONUs transmit data by different wavelengths, λ_α and λ_β, which are routed and transmitted through different feeder fibers by AWG. When the failure occurs on the CFT-WDM-EPON, OLT #1 cannot receive the REPORT messages from the ONUs in PON #1, and OLT #2 will starts the recovery scheme to take over the REPORT message from PON #1, and allocate wavelengths, $\lambda_{\mathrm{REPORT}}$ and λ_β, for upstream through protection feeder fiber to PON #1. The protection architectures in papers [9, 14–16] do not provide any recovery scheme for the OLT node, shown in Table 1.1. To prevent optical node failure and fiber failure from causing significant damage to the PON system, the CFT-WDM-EPON only equips backup optical fiber to provide the overall protection scheme.

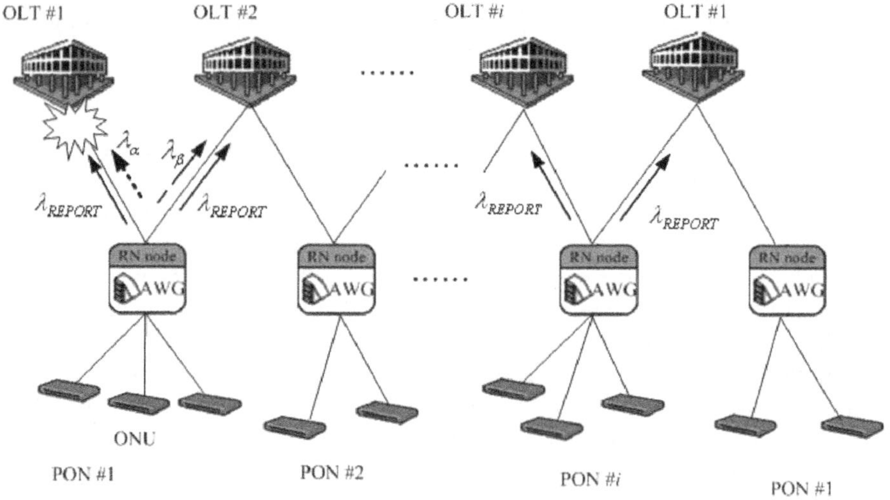

Fig. 1.1 The CFT-WDM-EPON architecture

Table 1.1 The comparisons of the protection architecture

Architecture	Protected OLT	Protected feeder fiber	# PON	# OLT	# Feeder fiber
CFT-WDM-EPON	Yes	Yes	N	N	$2N$
F.T. An et al. [89]	No	Yes	N	1	$N+1$
H. Nakamura et al. [1316]	No	Yes	N	1	$2N$
E.S. Son et al. [1114]	No	Yes	N	N	$2N$
X.F. Sun et al. [1215]	No	Yes	N	N	$2N$

1.4 The Proposed Dynamic Wavelength and Bandwidth Allocation

This study proposes a robust *prediction-based fair wavelength and bandwidth allocation* (*PFWBA*) scheme, which includes the dynamic wavelength allocation (DWA) and E-DBA mechanism. The E-DBA mechanism allocates the bandwidth to each ONU according to the decreasing order of unstable degree list and improves the prediction accuracy. Additionally, the DWA mechanism selects the wavelength with the least available time for each ONU to reduce the average delay time. To reduce the prediction inaccuracy resulting from a long waiting time, the DWA divides all ONUs into three groups based on the unstable degree list. The DWA can cooperate with the PFWBA scheme to select a suitable wavelength, and reduce the delay time for each ONU.

1.4.1 PFEBR Scheme with Early DBA Mechanism

1.4.1.1 The Operation of Early DBA Mechanism

The E-DBA mechanism arranges the sequence of transmitting REPORT messages to OLT by delaying some unstable traffic ONUs of β_V. The E-DBA mechanism consists of two operations. First, the OLT executes the DBA scheme after the REPORT messages from β_V are received at the end of ONU_{N-1}, instead of ONU_N in the standard DBA scheme. The operation reduces the idle period in the standard DBA scheme, and obtains the fresh queue information for unstable traffic ONUs to improve the prediction accuracy. Second, the bandwidth request for each ONU is allocated based on the traffic variation of all ONUs in decreasing order, and β_V is updated by assigning some unstable traffic ONUs with higher variations. This operation alleviates variance by shortening the waiting time for unstable traffic ONUs, to enhance the prediction accuracy.

1.4.1.2 PFEBR Scheme

- *Unstable degree list*

The PFEBR [8] calculates the variance of each ONU from the historical traffic required, and sorts the variances in decreasing order to obtain the unstable degree list. The variance of ONU_i, V_i, can be expressed as follows:

$$V_i = \frac{1}{N_H} \sum_{n \in \text{historical cycle}} (B_{i,n}^{Total} - \overline{B}_i)^2 \quad \text{and} \quad \overline{B}_i = \frac{1}{N_H} \sum_{n=1}^{N_H} B_{i,n}^{Total}, \tag{1.1}$$

where $B_{i,n}^{Total}$ represents the sum of differentiated traffic classes of ONU_i in the n*th* cycle, $\overline{B}_i$ is the mean of the $B_{i,n}^{Total}$, and N_H represents the number of historical REPORT messages piggybacked. β_V denotes a set of ONUs in unstable degree list with a high variance which is greater than the mean variance $\overline{V}$, where

$$\overline{V} = \frac{1}{N} \sum_{i=1}^{N} V_i.$$

The bandwidth prediction of each ONU after obtaining the unstable degree list is described as follows. Unlike the mechanism that piggybacks all REPORT messages in the data timeslots, the E-DBA mechanism shifts the REPORT messages of β_V between the $(N-1)$th and Nth ONU. The PFEBR requires the fresh queue information of unstable traffic ONUs to avoid prediction inaccuracy, which degrades the system performance.

- *Prediction based on unstable degree list*

After the sequence of all ONUs from the unstable degree list is uploaded, the PFEBR predicts the traffic bandwidth required according to the unstable degree list. The predicted request, $R_{i,n+1}^{c}$, for differentiated traffic classes of all ONUs is defined as follows:

$$\begin{cases} R_{i,n}^{EF} = B_{i,n}^{EF} \\ R_{i,n+1}^{c} = (1+\alpha) B_{i,n}^{c}, \quad c \in \{AF, BE\}, \end{cases} \tag{1.2}$$

where $B_{i,n}^{c}$ represents the requested bandwidth of ONU_i in the n*th* cycle, for differentiated traffic classes $c \in \{AF, BE\}$, and α denotes the linear estimation credit modified from the PFEBR. To achieve a better performance for a time-critical application, such as EF traffic, the constant bit rate (CBR) bandwidth should be assigned to the ONUs according to the rate of these applications. Therefore, this study assigns the CBR bandwidth to EF traffic.

- *Excessive bandwidth allocation*

The PFEBR executes the EBR to assign uplink bandwidth to each ONU after it has finished predicting the bandwidth needed for each ONU. The proposed PFEBR scheme can provide fairness for excessive bandwidth allocation according to the guaranteed bandwidth rather than requested bandwidth [13], with no partiality or increase in bandwidth utilization. The operation of fair EBR in the PFEBR is described as follows. First, calculate the $R_{i,n}^{Total}$ of all ONUs. The available bandwidth, $B_{available}$, can be expressed as

$$B_{available} = C_{capacity} \times (T_{cycle} - N \cdot g - N_v \cdot g) - N \times 512, \tag{1.3}$$

where $C_{capacity}$ represents the OLT link capacity (bits/sec), T_{cycle} denotes the maximum cycle time; g is the guard time; N is the number of ONUs, and N_V is the number of ONUs in β_V. The ONU$_i$ with the maximal residue bandwidth, i.e. $\max(S_i - R_{i,n}^{Total})$, is then selected from unassigned ONUs. S_i is the minimum guaranteed time slots for the differential traffics of ONU$_i$ determined by service level agreement (SLA).The granted bandwidth for ONU$_i$, $G_{i,n+1}^{Total}$, is given as follows:

$$G_{i,n+1}^{Total} = \min\left(B_{available} \times \frac{S_i}{\sum\limits_{k \in unassigned} S_k}, R_{i,n}^{Total} \right), \tag{1.4}$$

where $R_{i,n}^{Total}$ represents the sum of the differentiated traffic load after being predicted from ONU$_i$ in the nth cycle. The granted bandwidth for EF, AF and BE classes are described as follows:

$$\begin{cases} G_{i,n+1}^{EF} = R_{i,n}^{EF} \\ G_{i,n+1}^{AF} = \min\left(G_{i,n+1}^{Total} - G_{i,n+1}^{EF}, R_{i,n}^{AF}\right). \\ G_{i,n+1}^{BE} = G_{i,n+1}^{Total} - G_{i,n+1}^{EF} - G_{i,n+1}^{AF} \end{cases} \tag{1.5}$$

The process $B_{available} = B_{available} - G_{i,n+1}^{Total}$ continues until all ONUs has been assigned. Finally, the PFEBR arranges the upload sequence of each ONU by unstable degree list.

1.4.2 Dynamic Wavelength Allocation

The PFWBA defines the following global status variables used in the scheme description:

1. CAT: Channel available times. CAT[i] $= t$ indicates that the wavelength λ_i is available for transmission after time t, where $i = 1, 2, \ldots, w$, and w is the number of wavelengths.
2. RTT: RTT[i] *represents the round trip time (RTTs) between the OLT and the* ith ONU.

The PFWBA considers the unstable degree list when scheduling the upload sequence after collecting all REPORT messages from the ONUs in order to improve the prediction accuracy. First, the PFWBA divides all ONUs into three levels based on the variance of all ONUs and allocates the wavelength for each ONU group by group, which is determined as follows:

$$\begin{cases} Group\ 1, & if\ \mathrm{ONU}_i \in \beta_V \\ Group\ 2, & V_i > \overline{V}\ and\ \mathrm{ONU}_i \notin \beta_V \\ Group\ 3, & otherwise \end{cases},$$

The wavelength scheduling process is described as follows.

1. Schedule the PFWBA from Group 1 to Group 3 according to the unstable degree list.
2. Select the requested frame in the same Group with the minimum transmission time, and schedule its transmission.

$$Transmission\ time = RTT[i] + g + transmission\ timeslots,$$

where g is the guard time, and the transmission timeslots are obtained from the PFEBR.
3. Choose the earliest available wavelength transmission time CAT[i].
4. Update the CAT[i] is as CAT[i] = *transmission time* + CAT[i].
5. Repeat the above operation until the requested frames in the same Group are scheduled, and schedule the requested frames in the following Group.

1.5 Performance Analysis

The performance of the proposed PFWBA was compared with that of the WDM IPACT-ST [12] and DWBA3 [7] in terms of end-to-end delay and jitter performance. The performance evaluation was examined by the OPNET simulation tool. Two wavelength channels were adopted, and the link capacity was 1 Gb/s. The distance from one ONU to the OLT was assumed to be 20 km, and each ONU had an infinite buffer. The service policy was first-in-first-out. For the traffic model considered here, an extensive study shows that most network traffic can be characterized according to self-similarity and long-range dependence (LRD) [17]. This model was adopted to generate highly bursty BE and AF traffic classes with the Hurst parameter of 0.7. The packet sizes were uniformly distributed between 64 and 1518 bytes. Additionally, high-priority traffic (e.g. voice applications) was modeled by a Poisson distribution, and the packet size was fixed to 70 bytes [18]. The traffic profile was as follows: 20% of the total generated traffic was considered as high priority, and the remaining 80% was equally distributed between low- and medium-priority traffic [19].

1.5.1 End-to-End Delay

Figure 1.2 compares the average end-to-end packet delay of the PFWBA for EF, AF and BE traffics classes with different numbers of wavelengths and ONUs vs. traffic loads. The EF traffics class in the PFWBA with two wavelengths and 64 ONUs had the longest packet delay when the traffic load exceeded 90%, as shown in Fig. 1.2(b). The possible reason is that the wavelength scheduling mechanism initially chose the wavelength with the first available time for transmitting the ONUs. The prior selection of wavelength for transmission may result in prediction inaccuracy. The ITU-T recommendation G.114 specifies the delay for voice traffic in access network at 1.5 ms [20]. Although the PFWBA scheme with two wavelengths and 64 ONUs had the longest delay, the EF end-to-end delay was still less than 1.5 ms. Figure 1.3 compares the end-to-end packet delays among the PFWBA, WDM IPACT-ST and DWBA3 of all EF, AF and BE traffics classes with two channels and 64 ONUs for different traffic loads. Simulation results show that the proposed PFWBA outperformed the other two schemes.

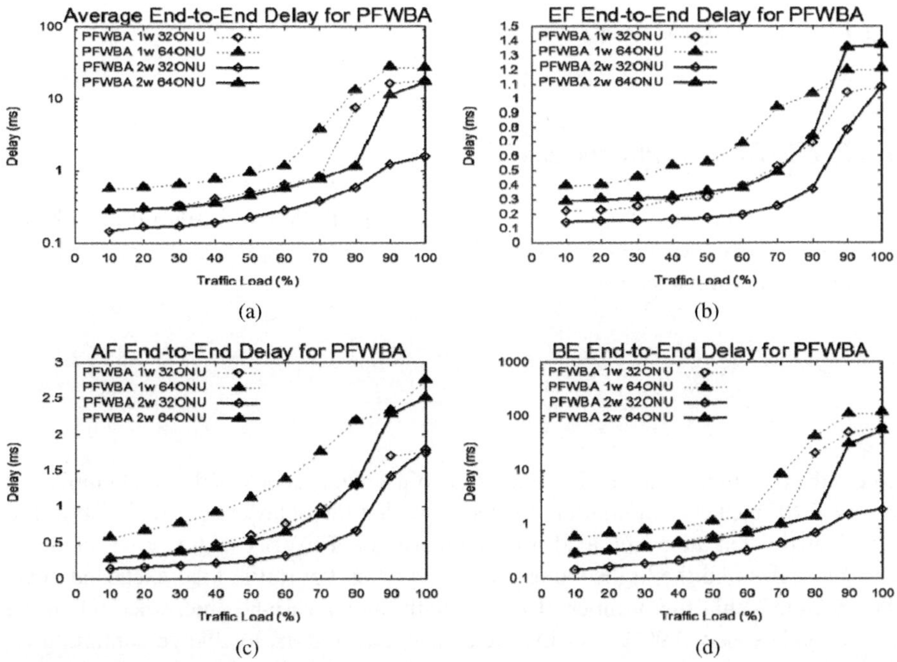

Fig. 1.2 End-to-end delay for PFWBA

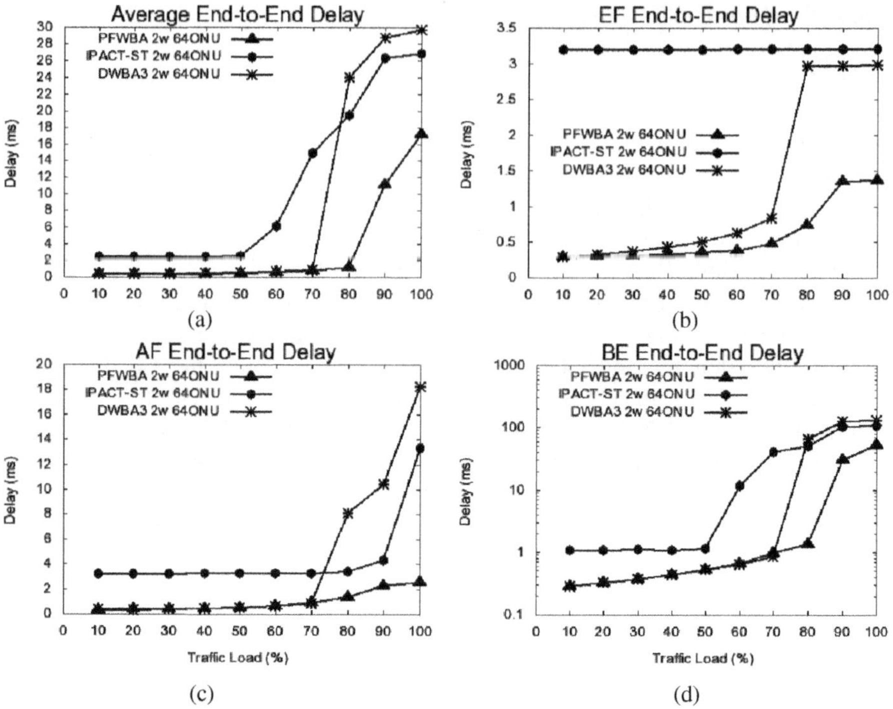

Fig. 1.3 End-to-end delay comparison

1.5.2 EF Jitter Performance

Figure 1.4(a) compares the jitter performance of the PFWBA for EF traffic with different numbers of wavelengths and ONUs vs. traffic loads, respectively. The delay variance σ^2 is calculated as

$$\sigma^2 = \frac{\sum_{i=1}^{N} (d_i^{EF} - \overline{d})^2}{N},$$

where the d_i^{EF} represents the EF delay time of packet i, and N is the total number of received EF packets. Simulation results show that the delay variance for EF traffic lengthened when the traffic load rises. However, the PFWBA with two wavelengths and 64 ONUs had the highest delay variance when the traffic load exceeded 90%. The reason is that the number of ONUs with higher variance increased when the traffic load exceeded 90%, causing the unstable degree list to change continuously. Therefore, the wavelength scheduling mechanism and the upload sequence of ONUs also changed continuously. Figure 1.4(b) compares the jitter performance of EF

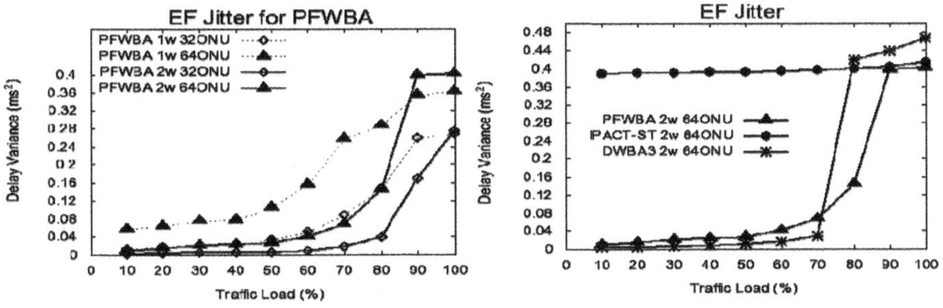

(a) (b)

Fig. 1.4 Delay variance comparison; (**a**) Jitter for PEWBA; (**b**) Jitter Comparison

traffic among the PFWBA, WDM IPACT-ST and DWBA3 with two channels and 64 ONUs for different traffic loads. Simulation results show that the proposed PFWBA outperformed the other two schemes for the EF traffic class.

1.6 Conclusion

The proposed protection architecture, CFT-WDM-EPON, enables feeder fibers to provide a recovery mechanism. If no failures occur in the PON system, then the CFT-WDM-EPON can share the loading of the working feeder fibers. When the failures occur on the OLT or feeder fibers, the backup fibers will recover the failed ones. Additionally, the PFWBA scheme integrates an efficient dynamic wavelength allocation and E-DBA mechanism of the PFEBR to improve the prediction accuracy and system performance. Simulation results show that the PFWBA can reduce the overall end-to-end delay in differentiated traffic.

References

1. IEEE 802.3ah task force home page. Available: http://www.ieee802.org/3/efm.
2. N.J. Frigo, P.P. Iannone, P.D. Magill, T.E. Darcie, M.M. Downs, B.N. Desai, U. Koren, T.L. Koch, C. Dragone, H.M. Presby, and G.E. Bodeep, Wavelength-division multiplexed passive optical network with cost-shared components, *IEEE Photonics Technology Letters*, **6**(11), 1365–1367 (1994).
3. S.J. Park, C.H. Lee, K.T. Jeong, H.J. Park, J.G. Ahn, and K.H. Song, Fiber-to-the-home services based on wavelength division multiplexing passive optical network, *Journal of Lightwave Technology*, **22**(11), 2582–2591 (2004).
4. A. Banerjee, Y. Park, F. Clarke, H. Song, S. Yang, G. Kramer, K. Kim, and B. Mukherjee, Wavelength-division-multiplexed passive optical network (WDM-PON) technologies for broadband access: a review [Invited], *Journal of Optical Networking*, **4**(11), 737–758 (2005).
5. M. McGarry, M. Maier, and M. Reisslein, WDM Ethernet passive optical networks, *IEEE Communications Magazine*, **44**(2), 15–22 (2006).

6. M. McGarry, M. Maier, and M. Reisslein, An evolutionary WDM upgrade for EPONs, Technical Report (Arizona State University), (2005).
7. A.R. Dhaini, C.M. Assi, M. Maier, and A. Shami, Dynamic Wavelength and Bandwidth Allocation in Hybrid TDM/WDM-EPON Networks, *Journal of Lightwave Technology*, **25**(1), 277–286 (2007).
8. I.S. Hwang, Z.D. Shyu, L.Y. Ke, and C.C. Chang, A Novel Early DBA Mechanism with Prediction-based Fair Excessive Bandwidth Reallocation Scheme in EPON, *Computer Communications*, **31**(9), 1814–1823 (2008).
9. F.T. An, K.S. Kim, D. Gutierrez, S. Yam, E. (S.T.) Hu, K. Shrikhande, and L.G. Kazovsky, SUCCESS: A next-generation hybrid WDM/TDM optical access network architecture, *Journal of Lightwave Technology*, **22**(11), 2557–2569 (2004).
10. K.S. Kim, D. Gutierrez, F.T. An, and L.G. Kazovsky, Batch scheduling algorithm for SUCCESS WDM-PON, GLOBECOM – *IEEE Global Telecommunications Conference*, **3**, 1835–1839 (2004).
11. K.S. Kim, D. Gutierrez, F.T. An, and L.G. Kazovsky, Design and performance analysis of scheduling algorithms for WDM-PON under SUCCESS-HPON architecture, *Journal of Lightwave Technology*, **23**(11), 3716–3731 (2005).
12. K.H. Kwong, D. Harle, and I. Andonovic, Dynamic bandwidth allocation algorithm for differentiated services over WDM-EPONS, 9th *IEEE Singapore International Conference on Communication Systems*, 116–120 (2004).
13. J. Zheng, Efficient bandwidth allocation algorithm for Ethernet passive optical networks, IEE Proceedings Communications, **153**(3), 464–468 (2006).
14. E.S. Son, K.H. Han, J.H. Lee, and Y.C. Chung, Survivable network architectures for wavelength-division-multiplexed passive optical networks, *Photonic Network Communications*, **12**(1), 111–115 (2006).
15. X.F. Sun, Z.X. Wang, C.K. Chan, and L.K. Chen, A novel star-ring protection architecture scheme for WDM passive optical access networks, Conference on Optical Fiber Communication, Technical Digest Series, **3**, Article number 1501381, 563–565 (2005).
16. H. Nakamura, H. Suzuki, J.I. Kani, and K. Iwatsuki, Reliable wide-area wavelength division multiplexing passive optical network accommodating gigabit ethernet and 10-Gb ethernet services, *Journal of Lightwave Technology*, **24**(5), 2045–2051 (2006).
17. W. Willinger, M.S. Taqqu, and A. Erramilli, A bibliographical guide to self-similar traffic and performance modeling for modern high-speed networks, Stochastic Networks: Theory and Applications, Oxford University Press, Oxford, 339–366 (1996).
18. S. Blake, D. Black, M. Carlson, E. Davies, Z. Wang, and W. Weiss, An Architecture for Differentiated Services, IETF RFC 2475 (1998).
19. X. Bai and A. Shami, Modeling Self-Similar Traffic for Network Simulation, Technical report, NetRep-2005-01 (2005).
20. ITU-T Recommendation G.114, One-way transmission time, May 2003.

Chapter 2
Evaluation on Data Modeling Languages for Standardization of NETCONF-Based Network Management: Application of an Evaluation Framework for Data Modeling Languages in Network Management Domain

Hui Xu, Debao Xiao, Yanan Chang, Xiaoqiong Wu and Limiao Chen

Abstract With evolution of the Internet, complexity of computer networks has greatly increased, when more and more network resources need to be effectively managed. The aim of this chapter is then to establish an evaluation framework to measure the capabilities of data modeling languages in adapting to the requirements of ever-evolving network management and apply it to examine the possibility of XML Schema and YANG as NETCONF-based data modeling languages for the purpose of standardization.

Keywords Data modeling languages · NETCONF-based · Network management · Evaluation framework · XML Schema

2.1 Introduction

With evolution of the Internet, complexity of computer networks has greatly increased, when more and more network resources need to be effectively managed. Meanwhile, considering the emergence of NETCONF for the sake of next generation network management, its upper three layers (transport protocol, RPC and operations) have been standardized to a degree, but its content layer has not been standardized yet. Given the current proprietary nature of the configuration data being manipulated, the specification of this content depends on the NETCONF implementation. Therefore, it is expected that a separate effort to study NETCONF-based data modeling should be undertaken.

In this case, the definition of Information Model (IM) and Data Model (DM) needs to be seriously taken into account for network management solutions. IMs always model Managed Objects (MOs) at a conceptual level and are

H. Xu (✉)
Institute of Computer Network and Communication, Huazhong Normal University, Wuhan, Hubei, P.R. China, 430079

S.-I. Ao et al. (eds.), *Trends in Communication Technologies and Engineering Science*, Lecture Notes in Electrical Engineering 33, DOI 10.1007/978-1-4020-9532-0_2,

protocol-neutral, while DMs are defined at a concrete level, implementing in different ways and are protocol-specific. As for each network management model, a data modeling language is quite necessary for description of the managed resources.

To the best of our knowledge, few studies have been done on describing the capabilities of data modeling languages in network management domain [1–4]. Obviously, the work on evaluating data modeling languages in the interest of next generation network management is comparatively indispensable. However, the fact is that a reused evaluation framework for management data modeling languages is still greatly lacking.

The aim of this chapter is then to establish an evaluation framework to measure the capabilities of data modeling languages in adapting to the requirements of ever-evolving network management and apply it to examine the possibility of XML Schema and YANG as NETCONF-based data modeling languages for the purpose of standardization.

2.2 Proposed Evaluation Framework

Nowadays, data modeling is under hot research, but it is still a preparatory period for corresponding research on next generation network management, especially NETCONF-based one. It becomes necessary to put forward a universal evaluation framework for data modeling languages to level their capabilities in satisfying the requirements of future network management. And our proposed evaluation framework is based on a set of criteria, which are modeling approaches, interoperability, readability, conformance, data representation, extensibility and security considerations.

2.2.1 Modeling Approaches

Four main modeling approaches should be considered, including data-oriented one, command-oriented one, object-oriented/object-based one and document-oriented one. The data-oriented approach models all management aspects through data objects, and at least two operations ("get" and "set") should be defined. The command-oriented approach defines a large number of management operations, specifying not the details but the commands to get/set selected information. The object-oriented/object-based approach combines the data-oriented approach and the command-oriented approach in view of integration. The document-oriented approach represents state information, statistics information and configuration information of a device as a structured document.

Future management data modeling language should implement an integration of various modeling approaches, a possible scenario of which is a data-oriented view for monitoring, a command-oriented view for operations and a document-oriented view for configuration. Note that, the very language should avoid implementing the same function with simple combination of these approaches.

2.2.2 Interoperability

With the development of next generation networks, management in an environment with a heterogeneous set of devices becomes a trend. Hence, standardization of DMs for network management should work at a higher level, making them more close to IMs. Following this way, data modeling languages should accordingly provide interoperability, which is consistent with the understanding of MOs already learned by network operators.

2.2.2.1 Protocol Independence

Protocol independence means that the language defines management data supporting any protocol instead of belonging to some specific protocol. In other words, the DM defined by this language can be implemented on any platform that installs different protocols.

In order to integrate with existing network management technologies, DMs should be defined by protocol-neutral modeling languages that can be mapped on different underlying protocols.

2.2.2.2 Naming Independence

Naming independence is a desired mechanism provided by the language that specifies how name collisions are handled, and thus uniquely identifies attributes, groups of attributes, and events.

Since a data modeling language for network management is required to be protocol-independent, and protocols typically use different approaches to name instances, it has to support multiple instance naming systems. Being naming independence, the language needs to think about the relationships between DMs. More efforts should then be made to ensure implementation of the language not being interfered by problems of different objects from multiple modules with the same name.

2.2.3 Readability

It is desired that management data modeling languages should be easily understood by both operators and computers. Human readability is necessary for convenient study and use. And machine readability is required to accelerate automatic process of network management, since both DMs and IMs are now being complemented by semantic models, where the meaning of concepts used in network management and relationships existing between them are made explicit [5].

2.2.3.1 Human Readability

Human readability is the capability by which administrators can directly read and understand representations including input and output (requirements, responses, error messages, etc).

Only if administrators conveniently read and understand meanings of the DM, can they efficiently write and use it. This also does favor to the interoperation between DMs and administrators. It is then desirable that all DMs used for a network management solution are well formed according to the data modeling language.

2.2.3.2 Machine Readability

Machine readability refers to the feature that description of the relevant DM can be understood by computers, thus related applications can be quickly developed. Its implementation largely depends on semantic expressiveness, and its speed also has very close relation with the parse-ability of machines. Note that, each data modeling language has a different level of semantic expressiveness, which includes several facets like concepts, relations and behaviors, and it is not easy to measure semantic expressiveness. Nowadays, this problem can be temporally reduced to a problem of integrating different management data modeling languages.

Future network management protocol aims in enabling the system to automate its management process. From this point of view, semantic expressiveness is quite essential for better machine readability. For example, the behavior defined by data modeling languages should be well understood, so that the automation requirements towards network management can be promoted and become much more promising.

2.2.4 Data Representation

Traditional network management solutions are often not strong in configuration management mostly as a result of modeling problems related to data representation, such as configuration objects being not easily identified. Consequently, the level of data representation ability should be taken into consideration.

2.2.4.1 Diversity of Data Types

Diversity of data types implies that data types should be diverse enough so that the modeling language can support various data. Hence, data with a suitable type can be clearly described and understood for users.

More structured data types are needed to make DMs much simpler to design and implement in the field of network management. It is said to be better that data types defined by a management modeling language should be as various as possible and emphasis should be placed on creating application-level ones especially for the configuration.

2.2.4.2 Specification of Configuration Data, State Data and Statistics Data

Configuration data is the set of read-write data, while both state data and statistic data are read-only, only different in the scope of practical use. The DM specified for a network device should identify what is configuration data, what is state data and what is statistic data, without the trouble to separate container elements.

When a device is performing configuration operations, a number of problems would arise if state data and statistic data were included in configuration data, and in order to account for these issues, future network management protocol should recognize the differences among state data, statistic data and configuration data, and provides operations for each. Thus as for the data modeling language, it then becomes necessary to make a clear distinction between (a) configuration data and (b) state data and statistic data.

2.2.5 Conformance

When defining MOs, it is also quite necessary to describe machine-readable conformance for the DM.

2.2.5.1 Backward Compatibility

Backward compatibility illuminates that new versions of the data modeling language can be used to define the relevant DM for the purpose of network management as the older one used to do.

This capability is quite important, for the reason that it eliminates the need to start over when a new data modeling language is used. From the viewpoint of network management, it means that new versions of the data modeling language that define management content can be rolled out in a way that does not break existing supporters.

2.2.5.2 Versioning

Versioning explains that each version of the data modeling language is complete, thus easy to control. This capability promotes the maintenance of backwards compatibility and does not need to change to the new language if it is also backwards compatible.

2.2.5.3 Definition of Event Notification Messages

Definition of event notification messages needs to be ensured by the data modeling language to allow a single definition of notification content to be sent either asynchronously or synchronously.

Network management protocols are desired to support asynchronous notifications, and with regard to a future data modeling language, not only notification messages but also types of the events should be clearly identified.

2.2.5.4 Definition of Error Messages

Definition of error messages indicates that error messages generated by network management applications should be recognized by the data modeling language.

Error messages, which are created by applications as a result of performing network management operations against the related DM, need to be included in the modeling language.

2.2.6 Extensibility

With increase of isomerization and complexity of the Internet, there is a great need for data modeling languages to have the ability to extend data structures, data types, elements and attributes easily for the practice of developing related software or systems to manage future heterogeneous networks.

2.2.6.1 Extensibility of Data Structures

Extensibility of data structures shows that the data modeling language has the capability to add new data structures with no need to affect available ones, and thus expresses the relations among data effectively and operates on data effortlessly.

2.2.6.2 Extensibility of Data Types

Extensibility of data types reveals that data types defined by the modeling language can be extended easily, so that the language can support various kinds of management data both simply and clearly.

Considering application of future protocols to manage next generation networks, especially for the use of configuration management, more and more new data types should be added to satisfy different presentation needs.

2.2.6.3 Extensibility of Elements and Attributes

Extensibility of elements and attributes means that types of element nodes and attributes defined by the data modeling language shouldn't be too fixed to extend. When there is a need to add new types of elements or new attributes for existing elements, the operation of "creation" should be done properly conveniently.

Objects of great variety need to be managed in next generation network management, which means the demand of adding object types. Hence, the data modeling language should have this capability, in order that everyone can manage the objects both simply and effectively.

2.2.7 Security Considerations

Security cannot be ignored by modeling languages to ensure confidentiality and integrity of management data.

2.2.7.1 Granularity of Access Control

Granularity of access control refers to the precision of accessing data from the relevant DM. There are mainly two levels of granularity, which are coarse one and fine one. Using coarse granularity of access control, a bulk of data can be retrieved and edited from the DM, such as getting the whole data from MIB. And fine granularity refers to a detailed operation to a small part of data, such as elements.

Both coarse granularity and fine granularity have their advantages and disadvantages. For example, implementation of coarse granularity is simple, while reusability is very poor. Hence, the tradeoff between coarse granularity and fine granularity becomes quite necessary for data modeling especially when merging and mapping information across multiple systems or data stores, since granularity may not match in the process of mapping.

2.2.7.2 Lock Mechanism

Lock mechanism cannot be ignored by management data modeling languages in order to guarantee security of the configuration.

As to some devices, it is quite hard to determine which parameters are administratively configured and which are obtained via mechanisms such as routing protocols. Taking configuration management into consideration, an implementation should figure out how users lock an entire configuration database, even if users do not have "write" access to the entire database. Furthermore, it is also of great importance to a partial lock of a configuration data store. Although it's not clear how serious this problem is, the solution is now an open issue.

2.3 Language Presentation

In order to validate the proposed evaluation framework, for one thing, three typical data modeling languages in network management domain being measured for comparison, which are Structure of Management Information (SMI) with its different versions for SNMP, Managed Object Format/Common Information Model (MOF/CIM) for WBEM, Structure of Management Information, next generation (SMIng) for both SNMP and COPS-PR, will be presented in chronological order with a brief introduction.

2.3.1 Structure of Management Information

Originally developed from the similar concept in OSI network management, SMI (including two versions, SMIv1 and SMIv2) defines organization, composing and identifier used in the framework of SNMP.

Unfortunately, SMI has some drawbacks, which hinder the application of SNMP to manage future networks. The root lies in the fact that SMI uses a data-oriented approach to model all management aspects. First of all, SMI is insufficient to represent hierarchical network configuration data, which is one of the main reasons for SNMP being used mostly in monitoring for fault management and performance management but hardly used for configuration management. Second, SMI is such a conceptually simple language that it is usually quite difficult to be used for modeling complex management operations.

2.3.2 Managed Object Format/Common Information Model

Firstly proposed by DMTF in 1997, CIM is the core part of WBEM, which is developed to solve the problem of heterogeneous management environment. CIM provides a data modeling environment in the form of object-oriented design diagrams and a language-neutral description of the model known as the MOF, which is also object-oriented.

Some other characteristics need to be further considered, which are (a) MOF only defines information of types and descriptions, hence it is quite difficult to extend; and (b) CIM specification describes the mappings from MOF to other data modeling languages in network management domain, but the syntactic and semantic conformance in the mapping process remains difficult to achieve.

2.3.3 Structure of Management Information, Next Generation

Proposed by IRTF, the SMIng project started in 1999, aiming to address some drawbacks of SMIv2 and create a new kind of management data modeling language to integrate SMI and Structure of Policy Provisioning Information (SPPI), avoiding the use of ASN.1. Applying an object-based approach to model MOs, SMIng has more advantages in expressiveness compared to SMI, such as better capability in defining data types, improved table definition, consideration of some operations, definition of attribute groups and event groups. Additionally, it is also possible to define extensions, which specify new elements by providing the syntax with which they should comply.

However, due to disagreement of both the SMIng syntax and the relationship between SMIng and SMI, SMIng didn't finally become a standard data modeling language for network management.

2.4 Validation

In order to justify the validity of the proposed evaluation framework, we apply it to compare three typical management data modeling languages, which have been introduced in Section 2.3. And the characteristics of these languages will then be summarized through comparison based on given criteria for evaluation, during the process of which, usability of the proposed framework can also be validated by the result.

2.4.1 Comparison

Using the proposed framework, the comparison of data modeling languages available in network management domain is performed. Table 2.1 demonstrates the comparison result in terms of the criteria for evaluation.

Note that, as for modeling approaches, we especially distinguish object-based approach from object-oriented approach, since the former one is an incomplete version of the latter one. And our measurement of criteria except for modeling approaches is classified as the following four levels.

Table 2.1 Comparison result

Criteria	SMI	MOF/CIM	SMIng
Modeling approaches			
Data-oriented	√		
Command-oriented			
Object-based			√
Object-oriented		√	
Document-oriented			
Interoperability			
Protocol independence	–	–	*
Naming independence	–	–	–
Readability			
Human readability	*	*	+
Machine readability	*	+	+
Data representation			
Diversity of data types	*	–	–
Specification of configuration data, state data and statistics data	–	–	–
Conformance			
Backward compatibility	+	*	–
Versioning	–	+	*
Definition of event notification messages	++	–	++
Definition of error messages	–	–	–
Extensibility			
Extensibility of data structures	–	–	–
Extensibility of data types	+	+	+
Extensibility of elements and attributes	–	–	+
Security considerations			
Granularity of access control	+	–	+
Lock mechanism	–	–	–

- A minus sign (–) means that the language does not have such a capability
- An asterisk sign (*) denotes that the language is weak in this capability
- A plus sign (+) is used when the language is good at this capability
- Two plus sign (++) is placed when the language completely possesses this capability

2.4.2 Summary

As is indicated in Table 2.1, some facts in terms of criteria can be gained as follows.

(1) Almost all the current data modeling languages are weak in features related to data representation, extensibility and security considerations, which reveals that these three capabilities are quite important when modeling management data and need to be taken into consideration fairly earlier than others.

(2) Only a few facets of both interoperability and conformance are involved in some particular languages, one example of which is SMIng, only focusing on half of these facets. As for next generation network management, this level is still far from the final aim.

(3) All these languages attach some importance to readability, especially for machine readability, on which they adopt different methods to promote their semantic expressiveness.

It can then be concluded from Table 2.1 that, SMIng does best implementation of most criteria, while SMI and MOF/CIM follow SMIng in the capabilities of data modeling.

2.5 NETCONF-Based Data Modeling

NETCONF-based network management is now under hot research, which has emerged as a promising approach to standardize XML-based network management for the purpose of automation. NETCONF has overcome the weaknesses of SNMP, and it provides a better configuration of IP network devices due to the effective use of XML and related technologies. However, NETCONF-based data modeling is still under study, since the specification of NETCONF content depends on the implementations [6].

2.5.1 XML Schema

Although NETCONF happens to use XML encoding and optional XPath filtering, it is indeed a network management protocol. Consequently, the conceptual management data available on a device should be understood by operators. NETCONF needs its own data modeling language, just like SNMP needs SMIv2. Since NETCONF is designed to meet special requirements of the configuration, its data modeling language needs to support special features not found in other languages. For example, it has to deal with different contexts such as *running*, *candidate* and *startup* configs, which respectively represent three different states of the configuration.

Due to the drawbacks of previous data modeling languages, IETF has taken XML-based data modeling into consideration, especially in IETF NETCONF Working Group (WG) [7]. Thus, NETCONF-based data modeling focuses on XML Schema and other XML-based languages.

Adopting a document-oriented approach, XML Schema [8] seems to be promising as a data modeling language for the purpose of standardizing NETCONF-based network management. S. Chisholm etc propose the use of XML Schema to define NETCONF content [9]. A framework for defining NETCONF content is then provided using a meta-model and XML Schema, which aims to build on existing well-deployed technologies and overlap management specific semantics to ensure high-quality interoperable NETCONF content definition.

2.5.2 Yang

YANG is a data modeling language based on the SMIng syntax, being used to model semantics and organization of configuration and state data manipulated by the NETCONF protocol, NETCONF Remote Procedure Calls (RPCs), and NETCONF notifications [10].

Aiming to provide readability as the highest priority, YANG focuses on semantics and can be directed mapped to XML content. Although YANG is now limited in scope of usage, being applied only to NETCONF, it gains experience from existing implementations, maximizes utility within this scope and can be extended in the future.

NETCONF Data Modeling Language (netmod) WG [11] proposed by IETF aims in supporting the ongoing development of IETF and vendor-defined data models for NETCONF, since NETCONF needs a standard content layer and its specifications do not include a modeling language or accompanying rules that can be used to model the management information that is to be configured using NETCONF. The main purpose of this WG is to provide a unified data modeling language to standardize NETCONF content layer, by defining a "human-friendly" language and emphasizing readability and ease of use. The very language seems to be able to serve as the normative description of NETCONF DMs. Thus from this point of view, the WG plans to use YANG as its starting point for this language.

2.6 Application of Proposed Framework to NETCONF-Based Data Modeling

Using our proposed evaluation framework, we compare the possible NETCONF data modeling languages that are XML Schema and YANG with SMIng, existing top-one data modeling language, the result of which is presented in Table 2.2.

Table 2.2 Evaluation result of XML schema and YANG through comparison using the proposed framework

Criteria	XML Schema	YANG	SMIng
Modeling approaches			
Data-oriented			
Command-oriented			
Object-based		√	√
Object-oriented			
Document-oriented	√	√	
Interoperability			
Protocol independence	++	–	*
Naming independence	++	–	–
Readability			
Human readability	+	++	+
Machine readability	*	+	+
Data representation			
Diversity of data types	++	++	–
Specification of configuration data, state data and statistics data	++	++	–
Conformance			
Backward compatibility	+	+	–
Versioning	++	++	*
Definition of event notification messages	–	++	++
Definition of error messages	–	++	–
Extensibility			
Extensibility of data structures	++	++	–
Extensibility of data types	++	++	+
Extensibility of elements and attributes	++	++	+
Security considerations			
Granularity of access control	++	++	+
Lock mechanism	–	++	–

First of all, as is demonstrated in Table 2.2, it can be summarized that, most properties of XML Schema surpass those of previous data modeling languages, especially in aspects such as interoperability, data representation and extensibility, which have been quite well-known by its numerous users. However, its machine readability is not so satisfying, for the reason that, what it is accomplished in is not semantic expressiveness but content definition. Furthermore, compared to its wide application, XML Schema is both too complicated and excessively general as a data modeling language for special use only in the scope of network management. On the other hand, definition of a NETCONF-based management DM is much more than an XML instance document description, or in other words, XML Schema is still not expressive enough with a view to NETCONF-based data modeling.

All these reasons above lead to the fact that, there are still no DMs defined by XML Schema yet. Currently, IETF Operations & Management (OPS) Area [12] is focusing the solutions on SMI-to-XML Schema conversion and XSD for accessing

SMIv2 DMs, since SNMP-based network management has been supplemented with a lot of proprietary MIB modules defined by different device vendors, and discarding MIB objects and SMI syntax when designing a new DM does reduce this benefit from experience of so many years. But more work has to be done on standardizing a NETCONF-based data modeling language for network management.

Furthermore, it can also be seen from Table 2.2 that, compared to XML Schema, YANG is a NETCONF-specific data modeling language, taking semantics such as notification message definitions, error message definitions and lock mechanism into consideration. All these features make YANG much easier to describe DMs in a way that maps to NETCONF in a very straight-forward manner and has therefore been chosen as the best approach up to now [10].

Note that, YANG is weak in interoperability, as is indicated in Table 2.2, since it is now limited in applications only to NETCONF-based network management. However, YANG is desired to be extended in the future, which will promote its protocol independence and naming independence. Additionally, description by YANG is at a semantic level, and XSD can be generated from the DMs defined in YANG.

2.7 Conclusions and Future Work

This chapter establishes an evaluation framework for data modeling languages in network management domain, validates it by summarizing the characteristics of existing languages through comparison, and finally applies it to the case of NETCONF-based data modeling. This framework is universal and can be reused to study data modeling in the field of next generation network management. Future work includes further study on application of this framework to NETCONF-based data modeling, aiming to promote the standardization progress of data modeling languages for NETCONF-based network management.

Acknowledgments The authors would like to thank Zhixia Zhao, MA, Huifen Liu, MA, and Jun Yue, MA, for their work on preparation of some related material and helpful comments on the initial version of this chapter.

References

1. J. Schoenwaelder, Overview of the 2002 IAB Network Management Workshop, RFC 3535 (2003).
2. J.E. López de Vergara, V.A. Villagrá, J.I. Asensio, and J. Berrocal, Ontologies: Giving semantics to network management models, *IEEE Network*, 17(3), 15–21 (2003).
3. J.E. López de Vergara, V.A. Villagrá, and J. Berrocal, Applying the web ontology language to management information definitions, *IEEE Communications Magazine*, 42(7), 68–74 (2004).
4. J. Schoenwaelder, Protocol Independent Network Management Data Modeling Languages – Lessons Learned from the SMIng Project, draft-schoenw-sming-lessons-01, (2007), work in progress).
5. A. Pras et al., Key research challenges in network management, *IEEE Communications Magazine*, 45(10), 104–110 (2007).

6. R. Enns, NETCONF Configuration Protocol, RFC4741, (2006).
7. IETF, NETCONF Working Group (June 10, 2008), http://www.ops.ietf.org/netconf/
8. W3C, XML Schema (January 18, 2008), http://www.w3.org/XML/Schema
9. S. Chisholm, A. Clemm, and J. Tjong, Using XML Schema to define NETCONF Content, draft-chisholm-netconf-model-08, (2008), work in progress.
10. M. Bjorklund YANG - A data modeling language for NETCONF, draft-ietf-netmod-yang-00, (2008), work in progress).
11. IETF, NETCONF Data Modeling Language Working Group (May 12, 2008), http://www.ietf.org/html.charters/netmod-charter.html.
12. IETF, Operations & Management Area (April 19, 2007), http://www.ops.ietf.org/

Chapter 3
Ad Hoc Multiple Source Routing

Su-Kit Tang and Dongyang Long

Abstract In this paper, we propose a Multiple Source Routing (MSR) protocol to maximize network throughput and minimize the protocol overheads in wireless ad hoc networks. MSR constructs multiple source routing paths between a source node and a sink node that resolves the scalability, privacy and efficiency problems caused by long paths for DSR in large networks. MSR also ensures the efficiency of request packet propagation in route discovery operation. Our simulation results reveal that MSR performs at a satisfactory level in dense networks in terms of connectivity and transmission efficiency. Network resources can be utilized for network throughput significantly at a cost of minimal protocol overheads comparatively.

Keywords Network · Ad Hoc · Multipath · Routing

3.1 Introduction

On-demand routing protocols, such as AODV [1] and DSR [2], have been proposed for efficient communication in wireless ad hoc networks. These protocols discover the most feasible path and maintain routes between two nodes when communication is needed. Single path is constructed by broadcasting request packets from a source node to a sink node. However, in any network, there may be more than one route to sink nodes. To utilize the network resources, many on-demand multipath routing protocols have been proposed [3–8]. They build maximal multiple disjoint paths between two nodes for performance improvement, such that network throughput can be increased. Figure 3.1 shows a transmission of data packets by three disjoint paths from source node s to sink node t. Three data packets can be sent by three different paths. In the meantime, the amount of energy consumed for the delivery of same amount of data by one single path can be shared by more nodes in different paths.

S.-K. Tang (✉)
Department of Computer Science, School of Information Science and Technology, Sun Yat-Sen University, Guangzhou, China

S.-I. Ao et al. (eds.), *Trends in Communication Technologies and Engineering Science*, Lecture Notes in Electrical Engineering 33, DOI 10.1007/978-1-4020-9532-0_3,

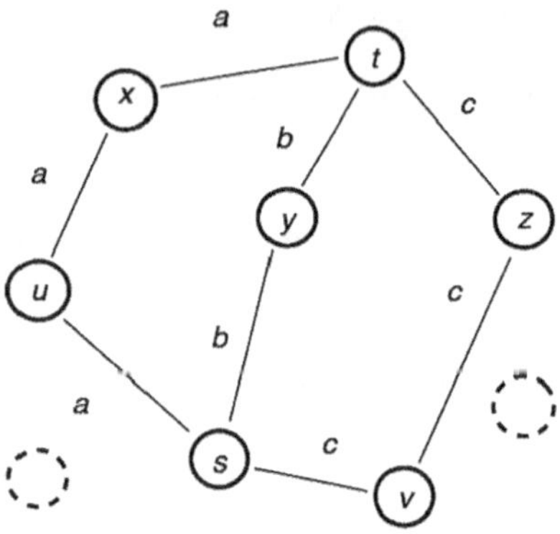

Fig. 3.1 Transmission of data packets by disjoint paths

This optimizes the usage of paths discovered in path discovery and maintenance operations. Therefore, the routing costs imposed to routing protocols for topology changes can be minimized.

In this paper, we propose a Multiple Source Routing (MSR) protocol. MSR constructs maximal disjoint paths by extending Split Multipath Routing (SMR) [3] with implicit source routing [9]. MSR constructs multiple source routing paths that resolves the scalability, privacy and efficiency problems caused by long paths for DSR in large networks. MSR also ensures the efficiency of request packet propagation in route discovery operation. In Section 3.2, we review some routing algorithms for wireless ad hoc networks that are crucial to MSR. In Section 3.3, MSR will be proposed and discussed. We also have a simulation in Section 3.4 to verify the correctness of MSR. From the simulation result, we observe that MSR can dramatically increase the network throughout at a satisfactory level. The amount of overheads it imposes is significantly reduced. Lastly, we conclude this paper.

3.2 Related Works

MSR maximizes network throughput in wireless ad hoc networks, by extending SMR with implicit source routing. SMR, based on DSR, builds maximal disjoint paths. Therefore, DSR is also crucial to MSR. DSR [2] is a single path routing algorithm that setups a path from sender to destination by forwarding the address of each node in a path to the destination. Sender will maintain a route cache that it has learned. If a route is found in the cache, it uses the route to transmit the packet. If a route is not found, it initiates a route request and the route request will be propagated in the ad-hoc network until it reaches the destination. When the destination receives the route request, it will return the route record (The path), in which it contains all intermediate nodes' addresses along the trip, to the sender. Significantly, DSR suffers a scalability problem that the propagation of route record may involve unexpected size of source routing information. This overhead creates impact on limited wireless bandwidth, exposes addresses of nodes involved in routing, and consumes resources of processing addresses. Therefore, Hu proposed the use of implicit source routing for single path routing in wireless ad hoc networks that preserves the advantage of source routing while avoiding the associated per-packet overhead in most cases [9]. It introduced a cache table in each participating nodes. This cache table maintains a list of next hop addresses for each particular routing path, indexed by a flow identifier. A source may then send any packets headed by a flow identifier in lieu of a source route. This avoids of the overhead caused by the source route. In SMR, an extension to DSR, disjoint paths are constructed by spreading request paths in a network. Its objective, which is similar to ours, is to build maximally disjoint paths for load balancing transmissions over the network. Data traffic is split into multiple routes to avoid congestion and to use network resources efficiently. Similarly, a number of solutions that build multiple disjoint paths for performance improvement have been proposed too [3–8]. They use various approaches to construct disjoint paths. However, they are extensions to DSR. They also suffer from unexpected length of route record in large networks.

3.3 The Protocol

In this paper, we propose a Multiple Source Routing (MSR) that constructs maximal disjoint paths from a source node to a sink node. Disjoint paths constructed by MSR are paths that will not intersect with each others. To ensure the efficiency of request packet propagation in a network, each node on one path will be assigned with an address of composite numbers. This address assists in locating a node in a path among all disjoint paths. When a source node initiates communication with a sink node, request packets will be generated and broadcasted by the source node. Request packets are spread across the wireless network until they reach the sink node. As request packets travel, intermediate nodes, receiving these packets, are required to process and rebroadcast the packets. A simple processing job on request packets involved in a node is the addressing operation that extends a path by assigning address of a disjoint path to the node. Since communication in wireless network is done by broadcasting, nodes have to make correct and accurate decision on request packet processing and forwarding. Otherwise, duplicated request packets of a sink node would occur.

Each intermediate node a request packet has visited will be assigned with an address. Each disjoint path is identified by a unique path id, *pid*, which is generated by the source node. A disjoint path is always determined by *pid* so nodes in the same path would always have the same *pid*. Along with a hop counter c of a disjoint path, an address of a node in a path can be determined. As a request packet travels, c is incremented at each node. This maintains the sequential order of addresses in a path. Disjoint paths of same session will denote by b, which groups the paths by source and sink nodes for one communication.

For instance, MSR creates a disjoint path by spreading request packet r. When a source node, s, initiates a transmission session, it broadcasts a request packet r.

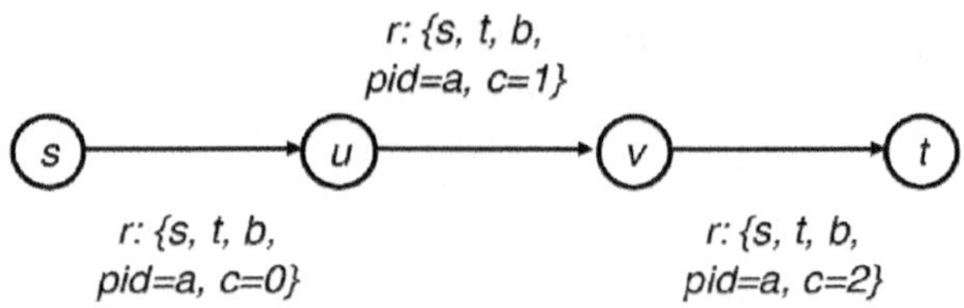

Fig. 3.2 Multiple source routing operation

A source node will randomly generate a large number for *pid*. *pid* is assumed to be unique as it is sufficiently large and collision-free. The hop count, *c*, is set to 0, indicating the position of this node in a disjoint path as well as the length of the path at sink nodes. Neighbor nodes receiving *r* will verify whether *r* is new at current time unit by *pid* and *c*. If it is new, *c* are incremented and *r* will be rebroadcasted. Otherwise, *r* will be ignored. This avoids the broadcast storm problem discussed in [10]. This processing repeats all the way in a disjoint path at nodes receiving *r* until the sink node, *t*, is reached. Thus, a disjoint path *p* is created. Figure 3.2 shows an overview of the multiple source routing operation.

In Fig. 3.2, a disjoint path is constructed and initiated by a source node *s* in session *b*. First, *s* broadcasts a request packet *r* that containing $pid = a$, and $c = 0$. Node *u* receives *r* and rebroadcast *r* after incrementing *c* by one. Subsequent node *v* receiving *r* will increment the path variable *c* by one, and rebroadcast *r*. Gradually, sink node *t* will receive *r*, from node *v*, containing $pid = a$, and $c = 2$, which indicates a 2-hop disjoint path identified by $pid = a$. The disjoint path $\{t \rightarrow u \rightarrow v \rightarrow s\}$ is constructed. A node receiving a request packet *r* will propagate and maintain an entry for *r* in its cache table until *r* expires. Entries in cache table are maintained in cache for a period of time. It records all disjoint paths that the node resides until timeouts. Since *pid* and *c* can uniquely identify a node in a path, quick routing decision can be made.

We define disjoint paths as non-intersecting paths, running from a source node to a sink node. These paths do not consist of any common intermediate nodes in the same session if they are parallel (See Fig. 3.1). To ensure that paths are disjoint with each others for a sink node, a path discovery heuristic is required to verify the originality of request packets. This enables a node to determine whether a request packet should be accepted. Note that disjoint paths for different source or sink nodes are assumed to be of no conflict of interest, so they would not be considered.

1. A request packet will only be accepted if cache table does not show an entry of the same source and sink nodes at the same session. A request packet is used to initiate a communication from a source node to a sink node. When a node receives the request packet, it can determine if the request packet should be accepted by looking up its cache Table 3.1 a disjoint path entry is not found, the request will be accepted, processed and rebroadcasted.
2. A request packet will be accepted if cache table shows an entry of the same source and sink nodes at the same session that is for different disjoint paths but of bigger hop counts. Request packets are running directionless and a node *v* may

Table 3.1 Summary of MSR performance against DSR

Network size (m^2)	500 × 500		1500 × 1500		3000 × 3000	
Routing protocols	DSR	MSR	DSR	MSR	DSR	MSR
Packet delivery ratio (%)	99.99	99.986	99.972	99.971	99.996	99.996
Packet delivery latency (sec/pkt)	68.3	68.32	68.26	68.46	70.13	69.87
Overheads (bytes)	16.7	16.69	118.73	38.36	39.38	22.23
Throughput (kilobytes)	2305.23	7157.01	2229.99	2997.51	273.28	302.84

receive two request packets initiated by the same source node and destined at the same sink node at same session. Since *pid* is collision-free, these two request packets would have their own unique *pid*. Running both request packets through node v would lead to a single delivery path only. Therefore, node v can only take either one of the two request packets. Otherwise, this evaluation of request packets raises an optimization issue in this situation.

3. A request packet will be accepted if cache table shows an entry of the same source and sink nodes at the same session for the different disjoint path that its hop counter is not better than the newly received one. Cache Table will be updated with this request packet. In this paper, the term optimal is defined to be minimal hop count as the amount of energy consumed in a transmission would be less, assuming that the energy consumption for one broadcast operation is same for all nodes. The shorter the path is, the less the energy it consumes, in one transmission. Based on this rationale, a node will accept a request packet of smaller hop counter. This will trigger a rebroadcast of request packet with updated information for the change of the path segment constructed previously if better request comes after. In case of same hop counters, a request packet from a farthest distance is in preference comparatively. This can encourage the construction of shortest path in a dense area. However, determination of the distance of neighbor nodes is out of scope of this research.
4. A request packet will not be accepted if cache table shows an entry of the same source and sink nodes at the same session for the same disjoint path with smaller hop counter. Suppose that node u is somewhere after node v in a path construction from source node s to sink node t at time session b, shown in Fig. 3.3. Request packet r_2 of ($pid_2 = x$, $c_2 = y{+}2$) from node v is running back to node u. Cache Table of node u shows r_0 of ($pid_0 = x$, $c_0 = y$) where $pid_0 = pid_2$ and

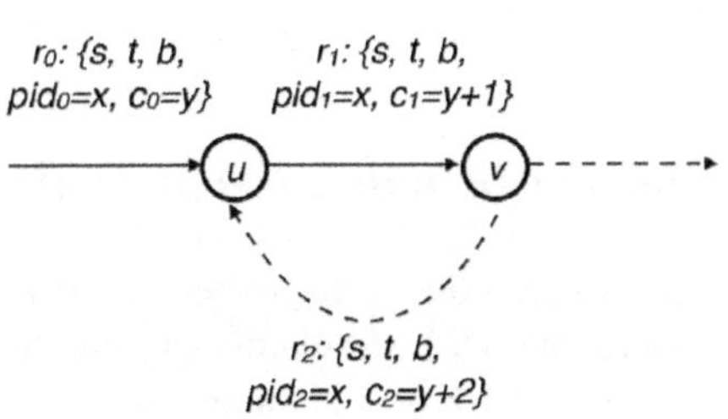

Fig. 3.3 A node u rejects a loop-back request packet r from node v during path construction in MSR algorithm

```
Algorithm MSR(r)
if this is a sink node then
  t ← this
  s ← source node (destination)
  b ← now
  pid ← random_number
  c ← 0
  broadcast r = (t, s, b, pid, c)
else if this is an intermediate node then
  result = Cache_table_lookup(r)
    if result = null then
      Cache_table_add(r)
    else if result = path_found then
      if r and result are from diff paths then
        if r is better than result then
          Cache_table_replace(r)
        end if
      else if r and result are same path then
        if r.c > result.c then
          do nothing...loop
        else if r is better than result then
          Cache_table_replace(r)
        end if
      end if
    end if
    broadcast r = (t, s, b, pid, c + 1)
 else if this is a source node then
   result = Cache_table_lookup(r)
   if r and result are not shadow then
     Cache_table_add(r)
   end if
 end if
```

Fig. 3.4 Algorithm MSR(r)

$c_2 > c_0$. Node u will not accept r. In addition, a condition $pid_0 = pid_2$ holds. We can see that both requests are of same originality and it is a loop.

5. A shadow path is a path that is a segment of another path. Suppose that there are two request packets running the same disjoint path, only one with smaller hop counter will be accepted. This minimizes the length of a disjoint path.

Based on the situations discussed above, we have the pseudo-code of MSR presented in Fig. 3.4. The MSR will take a request packet r as a parameter. Depending on the type of a node, MSR(r) will do corresponding tasks. If it is a sink node, it initiates r and broadcast r. If it is an intermediate node, it follows the core heuristics to process r. If it is a source node, it accepts request packet r if it is not from a shadow path.

3.4 Performance Evaluation

To evaluate the performance of MSR, we implemented MSR in ns2 [11], an open source network simulator, and conducted a set of simulations using the following settings: IEEE 802.11b standard at MAC layer implementation; 100 nodes randomly placed on areas of 500×500 m^2, 1500×1500 m^2, 3000×3000 m^2 for different network density environments. This demonstrates the efficiency of MSR against DSR in terms of overheads; 100 randomly generated scenarios for each area. In our simulation, a run was conducted for each scenario at each area and collected data was averaged over these 100 scenarios.

In our simulation, we consider four metrics to evaluate the performance of MSR.

(a) Packet Delivery Ratio (*PDR*): The rate of delivering data packets successfully to sink nodes, which is defined as:

$$PDR = \frac{R}{S} \times 100\%$$

where R is the number of received packets by the sink nodes and S is the number of sent packets by the source nodes.

(b) Packet Delivery Latency (*PDL*): The amount of time required to deliver a data packet from the time it is generated in a source node to the time it arrives at a sink node, which is defined as:

$$PDL = \frac{D}{R}$$

where D is the total amount of packet delivery time and R is the number of received packets sent by source nodes.

(c) Overheads: The total amount of header data required by the routing protocols for the delivery of a data packet along a routing path.

For the delivery of a data packet, the overheads of DSR is the number of times a routing path that are required to transmit. The overheads of DSR is calculated by the product of average length of routing path, the length of address node and the average hop counts. The average hop count is determined by the number of sent and forwarded packets divided by the number of sent packets. Since the destination node address is also included in a path, the length of a path is determined by adding one to the average hop counts. Therefore, we define the overheads of DSR as:

$$Overheads_{DSR} = \left(\frac{S+F}{S} + 1\right) \times A \times \left(\frac{S+F}{S}\right)$$

where F is the number of forwarded packets by intermediate nodes and A is the length of node address.

The overheads of MSR is the number of times routing information, which includes a time session, a disjoint path id, a hop counter and a packet type, that are required to transmit. The overhead of MSR is calculated by the length of routing information times the average hop counts, which is defined as:

$$Overheads_{MSR} = H \times \left(\frac{S+F}{S}\right)$$

where H is the routing information required by MSR.

Fig. 3.5 Simulation results of MSR against DSR

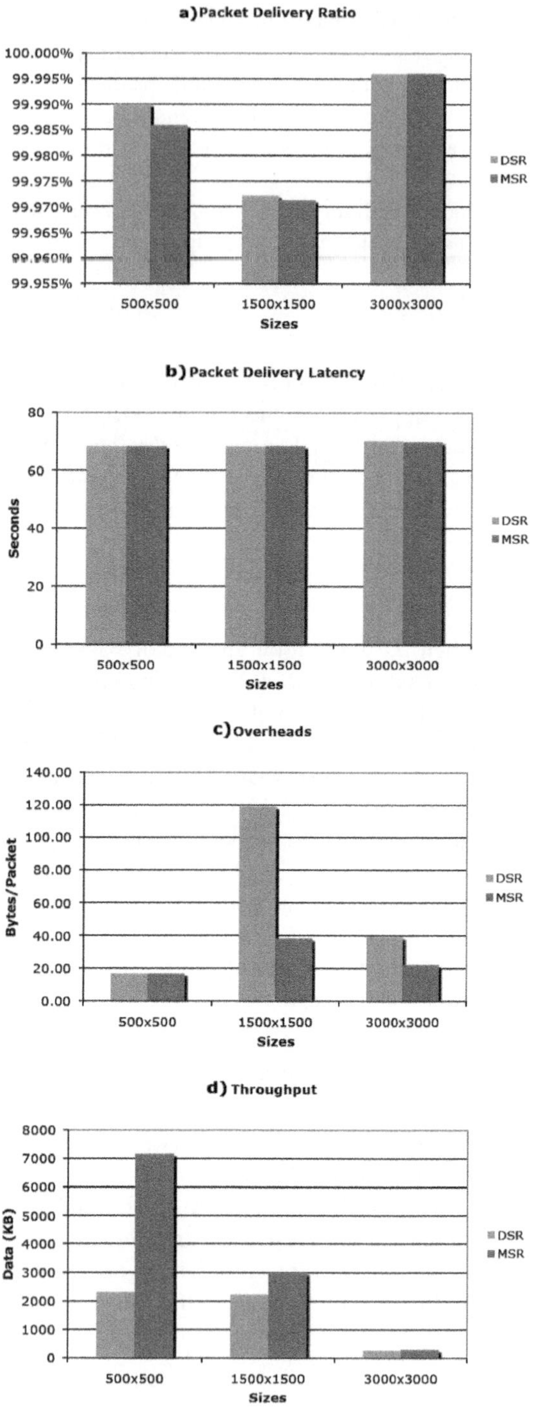

(d) Throughput: The total amount of data delivered to sink nodes successfully in a certain period of time. In our simulation, we use the same transmission rate for MSR and DSR. To obtain a throughput gained by disjoint paths, MSR will generate packets for all available paths in each transmission.

We observe that MSR does work as efficient as DSR in terms of packet delivery ratio and packet delivery latency. Recall that both of them deliver packets by routing paths in on-demand manner. In Fig. 3.5(a), even though PDR of two protocols reaches 99.9%, but DSR could perform better than MSR. It is because the utilization of network capacity in high dense networks by MSR may create traffic congestion as the number of disjoint paths increases. Therefore, packets may be dropped. In sparse networks, this problem is resolved. The number of disjoint paths that can be constructed is limited. Connectivity among nodes is comparatively lower and the amount of traffic generated is limited too. But in Fig. 3.5(b), PDL does not show any difference between two protocols. In Fig. 3.5(c), MSR outperforms in terms of overheads required in routing operation. The amount of overheads MSR generated is significantly less than DSR does. In dense networks, as the length of routing paths is short, the amount of overheads from DSR and MSR is just slightly different. Once the network size increases, the length of routing path gets long and the amount of overheads increases. Overheads of DSR increase. Since the overhead of MSR is constant, its overhead remains at low level. In the meantime, we observe that the transport capacity of a network increases by network connectivity. Disjoint paths maximize the network throughput. In Fig. 3.5(d), in high dense networks, MSR shows a good result that three times more than DSR in throughput is achieved. As the connectivity of network nodes determines the number of disjoint paths, in sparse networks the throughput of MSR gets close to DSR. A summary of our simulation results is shown in Table 3.1.

In this paper, the efficiency of routing protocols is determined by the amount of protocol overheads generated in the network during transmission and the network throughput. Figure 3.6 shows the summary of efficiency gained by MSR over DSR.

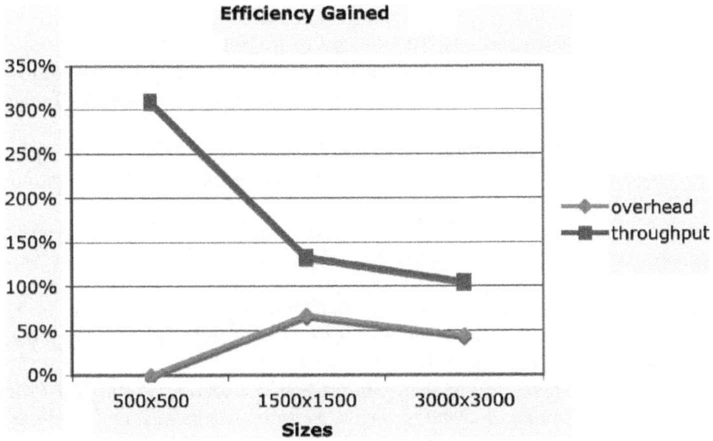

Fig. 3.6 Efficiency gained of MSR over DSR in terms of overheads and throughput

As can be seen that efficiency gained by MSR over DSR is significant in dense networks as connectivity between network nodes is comparatively high. Therefore, the network throughput by MSR shows a downward trend from dense networks to sparse networks. But it is an upward trend from small networks to large networks.

3.5 Conclusion

In this paper, we proposed a multiple disjoint paths construction algorithm, called Multiple Disjoint Path (MDP), to utilize network resources. MDP constructs maximal disjoint paths using implicit source routes. It resolves scalability, privacy and efficiency problems caused by long paths for DSR in large networks. Our simulation reveals that MDP performs at satisfactory level. From the aspects of connectivity and transmission efficiency, efficiency gained by MDP over DSR is significant in dense networks due to high connectivity. Network resources are utilized as network throughput is significantly increased. In the meantime, the amount of overheads in packet header required by the routing protocol is significantly reduced comparatively.

Acknowledgements This work was partially sponsored by the Natural Science Foundation of China (Project No. 60273062, 60573039) and the Guangdong Provincial Natural Science Foundation (Project No. 04205407, 5003350). Mr. Su-Kit Tang is a Doctor candidate of the Sun Yat-Sen University, GuangZhou, China. He is also with the Macao Polytechnic Institute, Macao, China (email: sktang@ipm.edu.mo). Professor Dongyang Long is with the Sun Yat-Sen University, GuangZhou, China (email: issldy@mail.sysu.edu.cn)

References

1. C.E. Perkins, E.M. Belding-Royer, and S.R. Das, "Ad Hoc On-demand Distance Vector (AODV) Routing", *IETF RFC 3561*, 2003.
2. D. Johnson, D. Maltz, and Y. Hu, "Dynamic Source Routing in Ad Hoc Wireless Networks", *Internet Draft*, draft-ietf-manet-dsr-10.txt, work in progress, July 2004.
3. S.J. Lee and M. Gerla, "Split Multipath Routing with Maximally Disjoint Paths in Ad Hoc Networks", *IEEE International Conference on Communications*, Helsinki, Finland, Vol. 10, 2001, pp. 3201–3205.
4. M.K. Marina and S.R. Das, "Performance of Route Caching Strategies in Dynamic Source Routing", *Proceedings of the 2nd Wireless Networking and Mobile Computing (WNMC)*, Mesa, AZ, USA, April 2001, pp. 425–432.
5. W. Wei and A. Zakhor, "Robust multipath source routing protocol (RMPSR) for video communication over wireless ad hoc networks", *IEEE International Conference on Multimedia and Expo*, Taipei, Taiwan, Vol. 2, June 2004, pp. 1379–1382.
6. N. Wisitpongphan and O.K. Tonguz, "Disjoint Multi-path Source Routing in Ad Hoc Networks: Transport Capacity", *Proceeding of IEEE Vehicular Technology Conference*, Florida, USA, Vol. 4, October 2003, pp. 2207–2211.
7. B. Yan and H. Gharavi, "Multi-Path Multi-Channel Routing Protocol", *Proceedings of the Fifth IEEE International Symposium on Network Computing and Applications*, Massachusetts, USA, 2006, pp. 27–31.

8. A. Srinivas and E. Modiano, "Minimum energy disjoint path routing in wireless ad-hoc networks", *Proceedings of the 9th annual international conference on Mobile computing and networking*, California, USA, September 2003, pp.122–133.
9. Y.C. Hu and D. Johnson, "Implicit Source Routes for On-Demand Ad Hoc Network Routing", *Proceeding of the 2nd ACM International Symposium on Mobile Ad Hoc Networking & Computing*, Long Beach, CA, USA, pp. 1–10, October 2001.
10. S.Y. Ni, Y.C. Tseng, Y.S. Chen, and J.P. Sheu, "The Broadcast Storm Problem in a Mobile Ad Hoc Network", *ACM/IEEE Mobicom*, Seattle Washington, USA, August 1999, pp. 151–162.
11. Network Simulator (NS-2), Available at http://www.isi.edu/nsnam/ns/index.ht.

Chapter 4
Efficient BER Improvement Mechanism for Wireless E1/E1 ATM Links

G.V.K. Sasirekha and K.M. Gangaraju

Abstract In the wireless scenarios where the links are non-engineered, the typical BERs are of the order 1 in 100 to 1 in 1000. It is essential to incorporate a Forward Error Correction (FEC) with a robust synchronization scheme that achieves a BER improvement. An efficient BER enhancement mechanism is proposed including block synchronization scheme for E1/E1 (ATM) links with 1 in 10 day probability of false alarm and 0.9999 probability of detection with minimal or no bandwidth overhead. A study of the various forward error control schemes was done and RS encoding/decoding was the selected FEC code for the implementation.

Keywords E1/E1 ATM links · Forward Error Correction · BER improvement · RS codes · Block synchronization

4.1 Introduction

In the wireless scenarios where the links are non-engineered, the typical BERs are of the order 1 in 100 to 1 in 1000. Usage of commercial E1/E1 ATM switches, meant for wired media, in wireless environment results in poor performance. It is essential to incorporate a Forward Error Correction (FEC) with a robust synchronization scheme that achieves a BER improvement from 10^{-3} to 10^{-8}. The proposed BER improvement mechanism works transparently in an existing communication link and therefore can be seamlessly integrated to a communication network. The system, "Link Performance Enhancer" (LPE) is designed as a System On Programmable Chip (SOPC) wherein the complete logic for cell level processing is implemented in a single Field Programmable Gate Array (FPGA). Figure 4.1 shows the typical usage scenario of the proposed BER improvement mechanism in wireless E1 (ATM) environment.

Some of the prior work in this regard can be found in [1–4]. In [1] authors describe error protection for ATM-based wireless networking systems using Rate

G.V.K. Sasirekha (✉)
Center for Artificial Intelligence and Robotics Bangalore, Bangalore, India

S.-I. Ao et al. (eds.), *Trends in Communication Technologies and Engineering Science*, Lecture Notes in Electrical Engineering 33, DOI 10.1007/978-1-4020-9532-0_4,

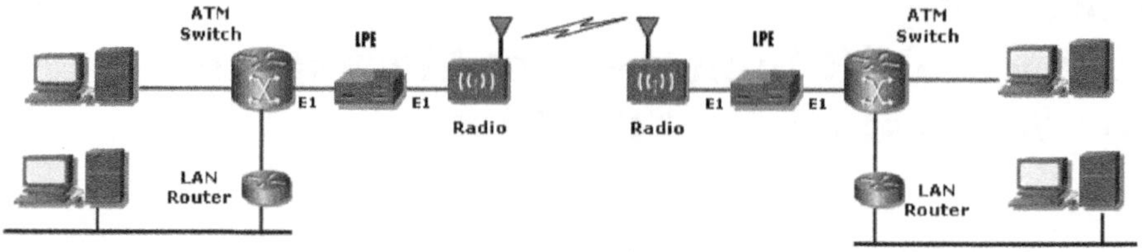

Fig. 4.1 Usage scenario of LPE

Compatible Punctured Code (RCPC). Reference [2] discusses protocol aided concatenated forward error control for wireless ATM using RS encoding/decoding at ATM Adaptation Layer (AAL). Reference [3] describes a Reed-Solomon encoder/decoder ASIC for wireless links. Reference [4] describes an error control scheme for tactical ATM which is applied at ATM layer. Though [4] describes implementation at AAL level, it involves complicated processing requiring segmentation & reassembly, common part convergence sub layer, service specific convergence sub layers for AAL1, AAL2, AAL5, etc. Implementation at ATM layer would allow a choice of FEC for payload only, header only, header + payload options. However, for multimedia applications the preferred choice would be to error protect both header and payload. Our objective is to propose a physical layer implementation in which the ATM cells are concatenated into data blocks for the encoding. This results in a simplified scheme.

Towards this objective, an efficient BER enhancement mechanism is proposed including block synchronization scheme for E1/E1 (ATM) links with 1 in 10 day probability of false alarm and 0.9999 probability of detection with minimal or no bandwidth overhead.

A study of the various forward error control schemes was done and RS encoding/decoding was the selected FEC code for the implementation. RS codes have high code efficiency, and can correct both burst and random errors [5]. Availability of Intellectual Property (IP) eases the hardware implementation tremendously.

4.2 Theoretical Background on RS Coding

Reed-Solomon (RS) codes are *non-binary cyclic* codes with symbols made up of m-bit sequences, where m is any positive integer having a value greater than 2. RS (n, k) codes on m-bit symbols exist for all n and k for which

$$0 < k < n < 2^m + 2$$

where k is the number of data symbols being encoded, and n is the total number of code symbols in the encoded block (refer [5, 6] for details). For the most conventional RS (n, k) code

$$(n, k) = (2^m - 1,\ 2^m - 1 - 2t)$$

where t is the symbol-error correcting capability of the code, and $n\text{–}k = 2t$ is the number of parity symbols.

The input error rate to output error rate improvement for various RS codes has been discussed in [5, 6]. The selected Error Correction Code (255, 249) with $t = 3$ and RS (255, 235) with $t = 10$ satisfies this requirement of BER improvement from 10^{-3} to 10^{-8} or better.

A mechanism is needed for identification of block boundaries. For this we propose a unique synchronization pattern to be prefixed to each block before transmitting. The receiver logic can hunt for this pattern and find the block boundaries. This is described in detail in Section 4.3.

4.3 Description of the Architecture and Design

The sub-blocks of the BER improvement mechanism are as indicated in Fig. 4.2. It contains the switch interface, media interface and the logic for RS encoder/decoder, idle cell insertion/detection logic, block synchronization pattern insertion and detection logic. In the forward path, the E1 frames from an E1/E1 (ATM) switch are received by the switch interfaces. The E1 frame boundaries are identified by detecting the frame sync pattern. The data from the 30 user slots of E1 frame is collected for further processing. The processed i.e. encoded and block sync added data is then sent to the media interface for further transmission to the radios. In the receive path, the data received from the media interface block boundaries are identified by the block synchronization pattern detection logic named "Sync Detect" in Fig. 4.2. Sync Detect logic gives the required block alignment marker. The aligned block of data is then decoded. This data is transmitted to the switch via switch interface card.

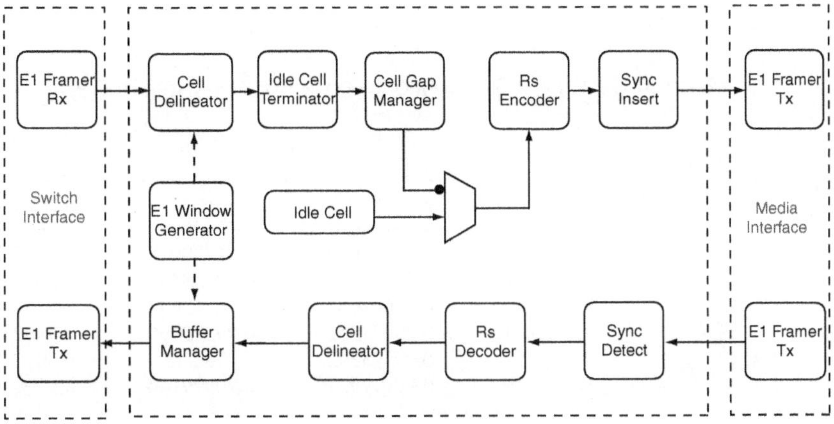

Fig. 4.2 Block diagram of LPE

The design details of the block synchronization mechanism are presented following subsections.

4.3.1 Block Synchronization Design Issues

The design of the synchronization involves the addressing twin issues. They are:

(a) Selection of the length of unique word i.e. pattern.
(b) Selection of frequency of transmission of unique word

Table 4.1 indicates the probability of detection P_d for different unique word lengths n in bits. The entries in the table are generated using the formula given below

$$P_d = \sum\nolimits_{i=0}^{t} {}^{48}C_i P_e^i (1 - P_e)^{48-i}$$

where t is the error tolerance threshold and P_e is the probability of error of the channel.

P_d for BER of 1 in 10^2 and for $t = 0$

$$P_d = (1 - P_e)^{48} = (1 - .01)^{48} = 0.61729014$$

while for a $t = 1$:

$$P_d = 48\, P_e(1 - P_e)^{47} + 0.61729014$$

It may be observed from Table 4.1 that the probability of missing sync is for $n = 48$ and a threshold $t = 5$ is

$$= (1 - 0.9999914) = 0.0000086 = 8.6 \times 10^{-6} \text{ or 1 in 116279.}$$

Table 4.1 Probability of detection

t	n = 32	n = 48	n = 64
0	0.72498033	0.61729014	0.52559648
1	0.95931741	0.91658233	0.86537603
2	0.99600655	0.98762643	0.97348770
3	0.99971252	0.99862990	0.99605647
4	0.99998392	0.99988029	0.99953297
5	0.99999927	**0.99999144**	0.99995437

4.3.2 Calculation of Probability of Occurrence of False Alarm

The Probability of False alarm P_f can be calculated using the formula

$$P_f = \frac{1}{2^n} \sum\nolimits_{i=0}^{t} {}^{i}C_n$$

Table 4.2 shows the probability of false alarm.

Table 4.2 Probability of false alarm

t	n = 32	n = 48	n = 64
0	2.3283e-010	3.5527e-015	5.4210e-020
1	7.6834e-009	1.74082970e-013	3.5236e-018
2	1.2316e-007	4.18154399e-012	1.1281e-016
3	1.2780e-006	6.56292797e-011	2.3714e-015
4	9.6505e-006	7.56916307e-010	3.6815e-014
5	5.6537e-005	**6.84024215e-009**	4.5014e-013

Further, Let $T = 488.28 \times 10^{-9}$ i.e. bit time at E1 data rate.

$N = 255{\times}8{\times}6$; i.e. one sync word sent for every 6 encoded blocks where each encoded block is 255 bytes i.e. 255 × 8 bits.

$$D = ((T \times N)/P_f)/(24 \times 60 \times 60)$$

where D is the no. of days within which 1 false alarm can occur. Therefore, for the chosen unique word length of 48 bits and the frequency of transmission of once for every 6 blocks of 255 bytes, D is 10 days. Table 4.3 shows the number of days per false alarm The FEC mechanism discussed in the processing block are discussed for two scenarios.

Scenario 1: E1 wireless links. The E1 frame synchronization pattern is detected by the switch interface. The data from the 31 slots (excluding frame sync, slot 0) is encoded by a RS-encoder (255, 249) to form 255 byte blocks. For every 6 consecutive blocks, 6 bytes of block synchronization information is inserted. The overhead by encoding and block sync insertion is 42 bytes (6 × 6 + 6). The number of bytes gained for every six blocks by dropping E1 frame sync slot is 48 bytes (there are about 48 frames in 6 blocks). Therefore, the mechanism can do the BER

Table 4.3 Number of days per false alarm

t	n = 32	n = 48	n = 64
1	9.0029	3.9736e+005	1.9631e+010
2	0.5616	1.6542e+004	6.1317e+008
3	0.0541	1.0540e+003	2.9169e+007
4	0.0072	91.3879	1.8789e+006
5	0.0013	**10.1127**	1.5367e+005

improvement with no overhead. For the choice of any other FEC scheme overhead can be similarly computed. The E1 frame synchronization pattern can be re-inserted at the receiver as the block sync marker is a multiple of E1 frame sync marker i.e. $(255\times6+6)/32$ is an integer.

Scenario 2: E1 ATM wireless links. The E1 frame synchronization pattern is detected by the switch interface. The data from the E1 frame excluding frame sync (slot 0) and signaling (slot 16) is encoded by a RS-encoder (255, 235) to form 255 byte blocks. For every 6 consecutive blocks, 6 bytes of block synchronization pattern is inserted. The overhead by encoding and block sync insertion is 126 bytes $(6 \times 20 + 6)$. The no. of bytes gained for every six blocks by dropping slot 0 and slot 16 i.e. E1 frame sync slot is 96 bytes (there are about 48 frames in 6 blocks). Therefore, the mechanism can do the BER improvement with overhead of 30 bytes per 6 blocks. To combat this overhead the idle cells are removed before transmission. Idle cells can be re-inserted on the receive side and fresh E1 frames can be formed to be sent to the E1 ATM switch interface. The procedure for idle cell removal is described in the Section 4.3.3.

4.3.3 Idle Cell Removal

The removal of idle cells involves finding the cell boundaries and then checking the header information. "Cell Delineation" is the process that allows identification of cell boundaries from data received at an interface to a channel. Traditionally, framing of packets in digital transmission technologies employ a "framing word" which is a unique synchronization sequence located at a specific point within a packet that can be used to indicate the frame boundary. Since ATM uses relatively small packet size, 53 bytes, compared with other transmission technologies, employing a "framing word" would prove to be costly. Instead, ATM makes use of the HEC byte, which has a dual purpose by also providing single-bit error correction and multi-bit error detection over the header bits. This eliminates the requirement for a specific "frame word".

As indicated in ITU-T recommendation 1.432.1, the HEC byte in the cell header is an 8-bit Cyclic Redundancy Check (CRC) based on a generator polynomial $G(x) = x^8+x^2+x+1$. Before transmitting a packet, the transmitter must calculate an appropriate code word for insertion into the HEC field. The code word is the remainder of the product of $x^8 \times A(x)$ (the polynomial representing the first four octets in the cell header), divided by (modulo-2) the generator polynomial, $G(x)$. $^8A(x) = D(x)G(x)+H(x)$. The ATM Cell Header is illustrated in the Fig. 4.3.

The ATM Cell Delineation process is described in ITU-T Recommendation Document 1.432. Idle cells have their VPI –VCI fields as zero.

4.3.4 Implementation

The proposed LPE was implemented using a System On Programmable Chip approach on an ALTERA Cyclone II device. The RS encoder/decoder

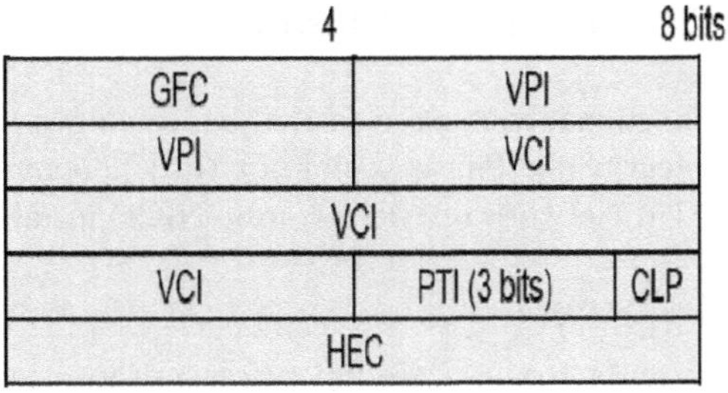

Fig. 4.3 Format of HEC octet

"megafunction" is an intellectual property of ALTERA. This has an advantage of quick prototyping, flexibility to make it adaptive, and also facilitates easy translation to a structured ASIC.

4.4 Test Results

This section presents the theoretical calculations, experimental setups and the observed measurements.

4.4.1 Theoretical Calculations

Cell Error Rate (CER) corresponding to the required Bit Error Ratio (BER) was calculated. This is required since the standard tests (O.191) give only the CER.

A cell error is declared when one or more bits in the ATM payload gets corrupted. Cell errors are detected using CRC in O.191 tests. ITU-T O.191 tests were carried out to find various error parameters and the latency. Channel Simulator was used to mimic the radio link characteristics in the laboratory settings.

$$CER = \sum\nolimits_{i=1}^{t} \left\{ {}^{384}C_i \, P_e^i \, (1 - P_e)^{384-i} \right\}$$

$$n = 384 \text{ (ATM Payload size)}$$
$$P_e = \text{Probability of bit-error } (\approx \text{BER})$$

Calculated CERs corresponding to the required BERs are given in Table 4.4.

Table 4.4 Computed BER vs. CER

BER	CER
10^{-3}	0.31899942
10^{-8}	3.8399926×10^{-6}

4.4.2 Functional Testing

Functional tests were carried out as shown in Figs. 4.4 and 4.5 to validate the data integrity and Phase Lock Loop (PLL) operations. These tests were carried out using ATM Network Interface Cards (NICs) installed on general purpose machines.

Test Setup 1

In setup 1 one of the links (transmit to receive) is transported through the LPE as shown in Fig. 4.4. The other link is connected directly. Permanent Virtual Circuit (PVC) is configured between the end-hosts in a point-to-point manner. The NIC is configured in such a way that any IP packet destined for the other ends host machine is directed to the PVC. This setting provides an IP over ATM connection between the hosts, thus enabling the use of any IP based services to test the LPE.

Ping application was used to test the functionality of the LPE. This program sends an ICMP echo request to a host and expects back an ICMP echo reply. This gives an opportunity to find dropped packets. A packet drop can occur because of faulty buffer management design. This test also verifies the data integrity issues of the device. Any modifications to the data transported through the LPE will result in a corrupted cell, which will be discarded at the receiving end and notified.

Ping test was performed with various payload sizes and no packet drops were observed, thus proving the functionality of the device in terms of logic and data integrity.

Test Setup 2

In test setup 2, both transmit and receive of the NICs are via. the LPE as shown in Fig. 4.5. Permanent Virtual Circuit (PVC) is configured between the two PCs in a point-to-point manner. The NIC is configured in such a way that any IP packet

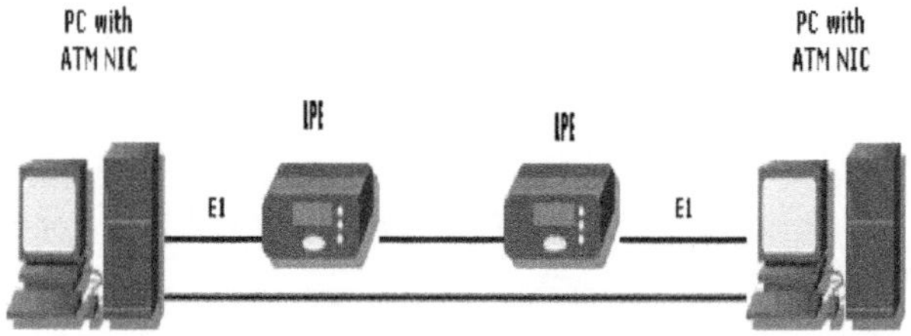

Fig. 4.4 Functional test setup 1

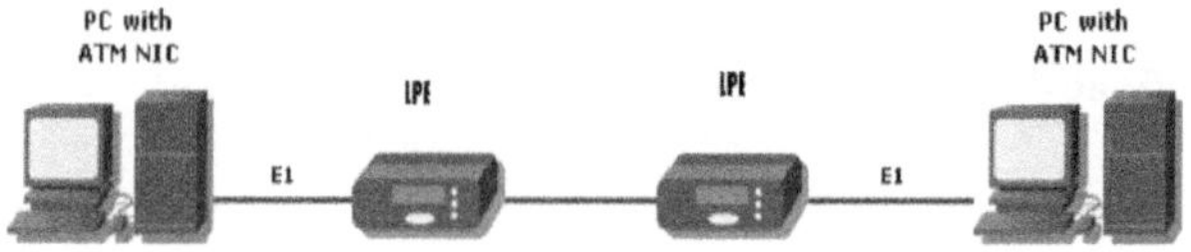

Fig. 4.5 Functional test setup 2

destined for the other PC is directed to the PVC. The same setting that was used for setup 1 was used, providing an IP over ATM connection between the hosts.

This setup is characterized by multiple PLLs in the round trip path. Each of the interface cards uses a PLL to extract clock from the received data. This test emulates the normal usage scenario.

4.4.3 Performance Tests

Performance tests were carried out to determine the BER improvement and latency under various traffic load conditions. The test setup employed for this purpose is shown in Fig. 4.6.

Referring to Fig. 4.6, a Network Analyzer was used to generate and monitor ATM E1 traffic. ITU-T O.191 tests were carried out to find various error parameters and the latency. Channel Simulator was set to a BER of 1 in 1000. The observed and calculated CER corresponding to this BER are shown in row 1 of Table 4.5.

In other setups (with LPE), the switch interface was connected to the network analyzer and the media interface connected to the channel simulator (shown in Figs. 4.7 and 4.8). The setting of channel simulator was still 1 in 1000 BER.

- Load testing: Measures the system's ability to handle various types of traffic loads.
- Latency testing: Measures the round trip latency of the transmission path.
- Volume testing: Subjects the system to larger amounts of data to determine its point of failure

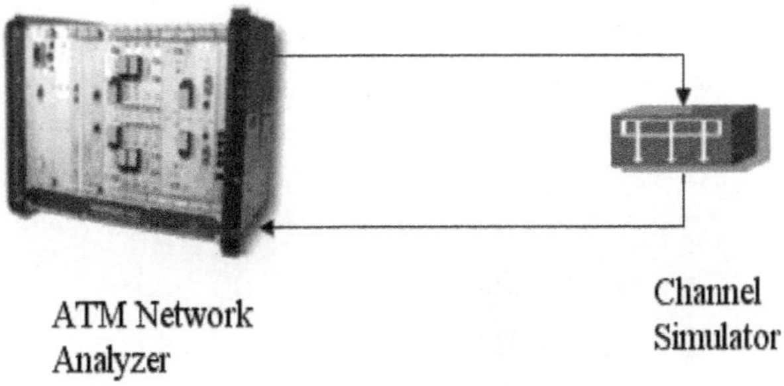

Fig. 4.6 Performance test setup

Table 4.5 Observed CER

BER	Observed CER			
2 min	5 min	10 min	20 min	
10^{-3}	0.3166	0.3161	03163	0.3156
10^{-8}	4.08×10^{-6}	9.17×10^{-6}	7.449×10^{-6}	5.94×10^{-6}

Fig. 4.7 Performance test: single LPE

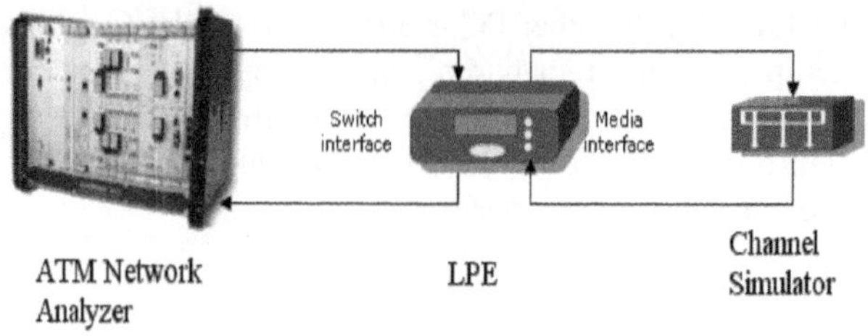

The settings of the parameters were as follows

Traffic Profile

- Type: CBR
- Start Idle: 0
- Period: 0
- Utilization: 50
- Bandwidth: 0.959999

Block Size

- 16384

ATM QoS Test was run for various time intervals in setup with and without cell hardener. The following error-related network performance parameters (defined in Recommendations I.356) were measured:

- Cell Error Ratio
- Cell Loss Ratio
- Severely Errored Cell Block Ratio
- Cell Mis-insertion Rate

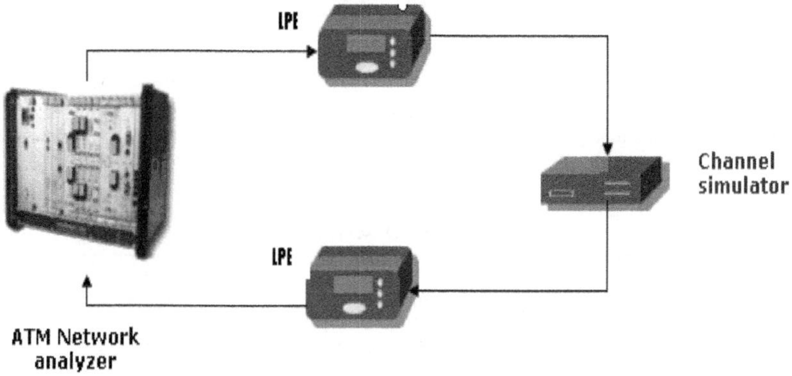

Fig. 4.8 Performance test: two LPEs

The CERs observed in these setups are shown in row2 of Table 4.5 which correspond to the BER of 1 in 10^8. This shows that the expected performance improvement was achieved. The measured values of CER matched with the theoretical calculations.

4.5 Conclusions

A scenario which necessitates usage of Link Performance Enhancer (LPE) has been presented. A design of the LPE was implemented, tested and evaluated for its performance. The results obtained are in agreement with the theoretical analysis thus validating our design.

References

1. J.I. Park and J. Caffery, "A Protocol Aided Concatenated Forward Error Control for wireless ATM". ieeexplore.ieee.org/iel5/7793/21424/00993338.pdf Wireless Communications and Networking Conference, 2002. Volume 2, Issue, Mar 2002 Page(s): 613–617.
2. E.W. Biersack, "A Simulation Study of Forward Error Correction in ATM Networks", Computer Communications Kaufman K, Cohen D, Yagel R. Volume graphics. IEEE Computer 1993; 26(7):51–64. Review, 22(1), January 1992.
3. F. Borgonovo and A. Capone, "Comparison of Different Error Control Schemes for Wireless ATM". ieeexplore.ieee.org/iel5/6459/17299/00797869.pdf Wireless Communications and Networking Conference, 1999. Volume 1, Issue, 1999 Page(s):466–470.
4. S.V. Pizzi et al. "Error Control for Tactical ATM" ieeexplore.ieee.org/iel4/5850/15596/00722160.pdf Military Communications Conference, 1998. MILCOM 98. Proceedings, Volume 2, Issue, 18–21 Oct 1998 Page(s):397–408.
5. B. Sklar Reed Solomon Codes, http://www. http://www.informit.com/content/images/art_sklar7_reed-solomon/elementLinks/art_sklar7_reed-solomon.pdf
6. B. Sklar, *Digital communications: Fundamentals and Applications*. II Edition, Prentice Hall, Englewood Cliffs (2001), ISBN 0-13-084788-7.

Chapter 5
Performance of Ad Hoc Routing Protocols in Mobile WiMAX Environment

Md. Saiful Azad, Farhat Anwar and Md. Arafatur Rahman

Abstract Worldwide Interoperability for Microwave Access (WiMAX) is a technology that bridges the gap between fixed and mobile access and offer the same subscriber experience for fixed and mobile user. Demand for such type of mobile broadband services and applications are growing rapidly as it provides freedom to the subscribers to be online wherever they are at a competitive price with other significant facilities such as increasing amounts of bandwidth, using a variety of mobile and nomadic devices etc. The earliest version of WiMAX is based on IEEE 802.16 and is optimized for fixed and nomadic access, which is further extended to support portability and mobility based on IEEE 802.16e, also known as Mobile WiMAX. However, frequent topology changes caused by node mobility make routing in Mobile WiMAX networks a challenging problem. In this study, we focus upon those routing protocols especially designed for wireless networks. Here, we compare the performance of four ad hoc routing protocols (AODV, DSR, OLSR and ZRP) for Mobile WiMAX environment under the assumption that each of the subscriber station has routing capabilities within its own network. From our simulation, we found that ZRP and AODV protocols outperform DSR and OLSR.

Keywords AODV · DSR · Mobile WiMAX · OLSR and ZRP

5.1 Introduction

Today's broadband Internet accesses are restricted to wireline infrastructure using DSL, T1 or cable-modem based connections. However, these wireline infrastructures are considerably more expensive and time consuming to deploy than a wireless one. Moreover, in rural areas and developing countries, providers are unwilling to install the necessary equipment (optical fiber or copper-wire or other infrastructures) for broadband services with little profit. Broadband Wireless Access (BWA)

M.S. Azad (✉)
Department of ECE, Faculty of Engineering, IIUM, Kuala Lumpur, Malaysia

S.-I. Ao et al. (eds.), *Trends in Communication Technologies and Engineering Science*, Lecture Notes in Electrical Engineering 33, DOI 10.1007/978-1-4020-9532-0_5,

has emerged as a promising solution for "last mile" access technology to provide high speed connections. IEEE 802.16 standard for BWA and its associated industry consortium, Worldwide Interoperability for Microwave Access (WiMAX) forum promise to offer high data rate over large areas to a large number of users where broadband is unavailable. This is the first industry wide standard that can be used for fixed wireless access with substantially higher bandwidth than most cellular networks [1, 2]. In the recent years, IEEE 802.16 working group has developed a number of standards for WiMAX. The first standard IEEE 802.16 was published in 2001 focused on the frequency range between 10 and 66 GHz and required line-of-sight (LOS) propagation between the sender and the receiver [3]. IEEE 802.16-2004, has been developed to expand the scope to licensed and license-exempt bands from 2 to 11 GHz. IEEE 802.16-2004 specifies the air interface, including the Media Access Control (MAC) of wireless access for fixed operation in metropolitan area networks. Support for portable/mobile devices is considered in IEEE 802.16e standard, which is published in December 2005. We are considering IEEE 802.16e standard in our study, which is also known as mobile WiMAX.

A number of ad hoc routing protocols are already designed to provide communication in wireless environment, e.g. AODV, OLSR, DSDV, ZRP, LAR, LANMAR, STAR, and DYMO. Performance comparison among some set of routing protocols are already performed by the researchers such as among PAODV, AODV, CBRP, DSR, and DSDV [4], among DSDV, DSR, AODV, and TORA [5], among SPF, EXBF, DSDV, TORA, DSR, and AODV [6], among DSR and AODV [7], among STAR, AODV and DSR [8], among AMRoute, ODMRP, AMRIS and CAMP [9], among DSR, CBT and AODV [10], among DSDV, OLSR and AODV [11], and many more. For our study and performance comparison, we choose AODV, DSR, OLSR and ZRP routing protocols. AODV and DSR are selected, because, both of them are on-demand routing protocol with different mechanisms. OLSR is a purest example of proactive routing protocol and ZRP is a hybrid protocol which combines the characteristics of both proactive and reactive scheme.

For performing the simulation, we assume that each of the subscriber station maintains routing table for its own network, so that it can send data directly to the destination without the help of base station. However, if one subscriber station has to send data to a station located in another network, it must send data through the base station and vice versa.

5.2 Ad Hoc Routing Protocols

5.2.1 Ad Hoc On-Demand Distance Vector (AODV)

Ad-hoc On-demand distance vector (AODV) [12] is a variant of classical distance vector routing algorithm, based on DSDV [13] and DSR [14]. It shares DSR's on-demand characteristics in that it also discovers routes on an as needed basis via a similar route discovery process. However, AODV adopts traditional routing tables;

one entry per destination which is in contrast to DSR that maintains multiple route cache entries for each destination. The initial design of AODV is undertaken after the experience with DSDV routing algorithm. Like DSDV, AODV provides loop free routes in case of link breakage but unlike DSDV, it doesn't require global periodic routing advertisement.

Apart from reducing the number of broadcast resulting from a link break, AODV also has other significant features. Whenever a route is available from source to destination, it does not add any overhead to the packets. However, route discovery process is only initiated when routes are not used and/or they expired and consequently discarded. This strategy reduces the effects of stale routes as well as the need for route maintenance for unused routes. Another distinguishing feature of AODV is the ability to provide unicast, multicast and broadcast communication.

AODV uses a broadcast route discovery algorithm and then the unicast route reply massage. The following sections explain these mechanisms in more detail.

5.2.1.1 Route Discovery

When a node wants to send a packet to some destination node and does not locate a valid route in its routing table for that destination, it initiates a route discovery process. Source node broadcasts a route request (RREQ) packet to its neighbors, which then forwards the request to their neighbors and so on. Figure 5.1 demonstrates the broadcast of RREQ across the network. To control network-wide broadcasts of RREQ packets, the source node use an expanding ring search technique. In this technique, source node starts searching the destination using some initial time to live (TTL) value. If no reply is received within the discovery period, the TTL value is incremented by an fixed amount. This process will continue until the threshold value is reached.

When an intermediate node forwards the RREQ, it records the address of the neighbor from which first packet of the broadcast is received, thereby establishing a reverse path. When the RREQ reaches a node that is either the destination node or an intermediate node with a fresh enough route to the destination, it replies by unicasting the route reply (RREP) towards the source node. As the RREP is routed back along the reverse path, intermediate nodes along this path set up forward path entries to the destination in its route table and when the RREP reaches the source

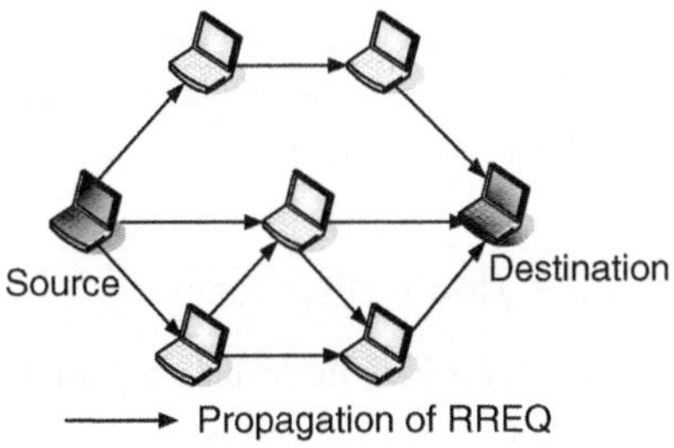

Fig. 5.1 Propagation of RREQ throughout the network

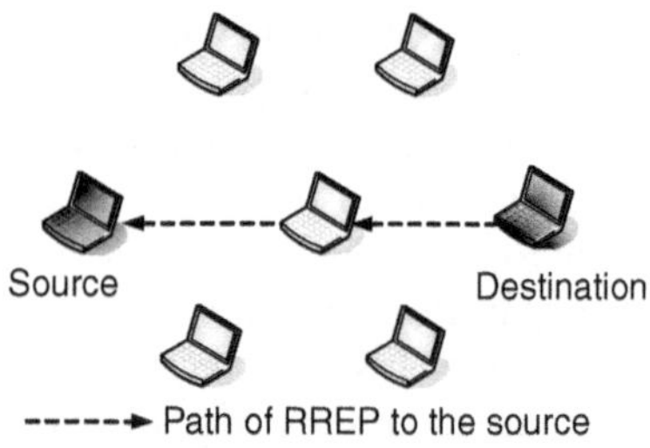

Fig. 5.2 Reply of RREP towards the network

node, a route from source to the destination is established. Figure 5.2 indicates the path of the RREP from the destination node to the source node.

5.2.1.2 Route Maintenance

A route established between source and destination pair is maintained as long as needed by the source. If the source node moves during an active session, it can reinitiate route discovery to establish a new route to destination. However, if the destination or some intermediate node moves, the node upstream of the break remove the routing entry and send route error (RERR) message to the affected active upstream neighbors. These nodes in turn propagate the RERR to their precursor nodes, and so on until the source node is reached. The affected source node may then choose to either stop sending data or reinitiate route discovery for that destination by sending out a new RREQ message.

5.2.2 Dynamic Source Routing (DSR)

The Dynamic Source Routing (DSR) [15] is one of the purest examples of an on-demand routing protocol that is based on the concept of source routing. It is designed specially for use in multihop ad hoc networks of mobile nodes. It allows the network to be completely self-organizing and self-configuring and does not need any existing network infrastructure or administration. DSR uses no periodic routing messages like AODV, thereby reduces network bandwidth overhead, conserves battery power and avoids large routing updates. Instead DSR needs support from the MAC layer to identify link failure.

DSR is composed of the two mechanisms of Route Discovery and Route Maintenance, which work together to allow nodes to discover and maintain source routes to arbitrary destinations in the network. The following sections explain these mechanisms in more details.

5.2.2.1 Route Discovery

When a mobile node has a packet to send to some destination, it first checks its route cache to determine whether it already has a route to the destination. If it has

Fig. 5.3 Propagation of route request message across the network

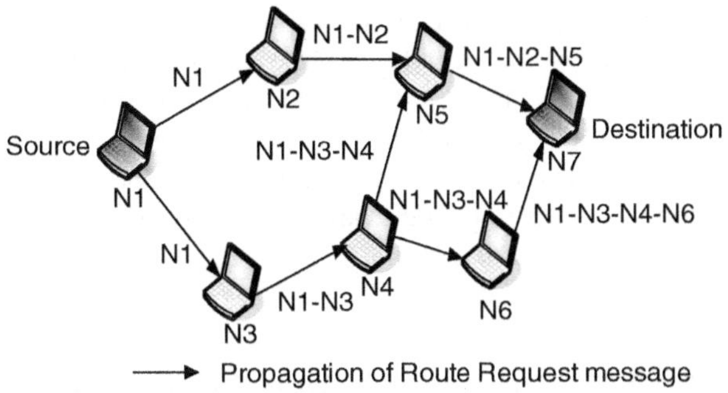

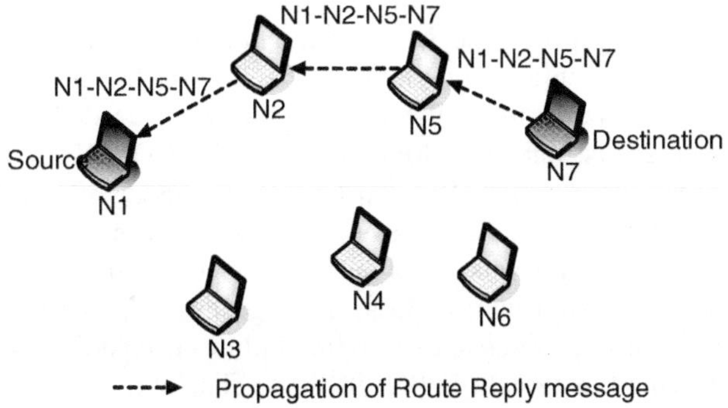

Fig. 5.4 Propagation of route reply message towards the source

an unexpired route, it will use this route to send the packet to the destination. On the other hand, if the cache does not have such a route, it initiates route discovery by broadcasting a route request packet.

Each node receiving the route request packet searches throughout its route cache for a route to the intended destination. If no route is found in the cache, it adds its own address to the route record of the packet and then forwards the packet to its neighbors. This request propagates through the network until either the destination or an intermediate node with a route to destination is reached. Figure 5.3 demonstrates the formation of the route record as the route request propagates through the network.

Whenever route request reaches either to the destination itself or to an intermediate node which has a route to the destination, a route reply is unicasted back to its originator. Figure 5.4 illustrates the path of the RREP from the destination node to the source node

5.2.2.2 Route Maintenance

In DSR, route is maintained through the use of route error packets and acknowledgments. When a packet with source route is originated or forwarded, each node transmitting the packet is responsible for confirming that the packet has been received by the next hop. The packet is retransmitted until the conformation of receipt is received. If the packet is transmitted by a node the maximum number of times and yet no receipt information is received, this node returns a route error message to the source of the packet. When this route error packet is received, the hop in error is removed from the host's route cache and all routes containing the hop are truncated at that point.

5.2.3 Optimized Link State Routing (OLSR)

The Optimized Link State Routing (OLSR) [16] protocol inherits the stability of the pure link state algorithm and is an optimization over the classical link state protocol, adopted for mobile ad hoc networks. It is proactive in nature and has the advantage of having routes immediately available when needed. The key concept used in this protocol is that of multipoint relays (MPRs). MPRs are selected set of nodes in its neighbor, which forward broadcast messages during the flooding process. OLSR reduces the size of control packet by declaring only a subset of links with its neighbors who are its multipoint relay selectors and only the multipoint relays of a node retransmit its broadcast messages. Hence, the protocol does not generate extra control traffic in response to link failures and additions. The following section describes the functionality of OLSR in details.

5.2.3.1 Neighbor Sensing

For detecting the neighbor, each node periodically broadcasts its HELLO messages, which contains the information of the neighbors and their link status. The protocol only selects direct and bidirectional links, so that the problem of packet transfer over unidirectional links is avoided. HELLO messages are received by all one-hop neighbors, but they are not relayed further. These messages permit each node to learn the knowledge of its neighbors up to two hopes and help performing the selection of its multipoint relays.

5.2.3.2 Multipoint Relay Station

Each node of the network selects its own set of multipoint relays from periodically broadcasted hello messages. The MPR set is selected in a manner that a subset of one hops neighbors, which covers its entire two hop neighbors. For example, in Fig. 5.5, node N2 selects nodes N1 and N6 to be the MPR nodes. Since these nodes cover all the nodes (N7, N8, N9 and N4), which are two hops away from it. Multipoint relays of a node are stated in its subsequent HELLO messages, so that the information reaches to the multipoint relays themselves. Multipoint relay set is recalculated when either a change in the neighborhood is detected or a change in the two hop neighbor set with bi-direction link is detected.

Fig. 5.5 An example of multi point relay (MPR) selection

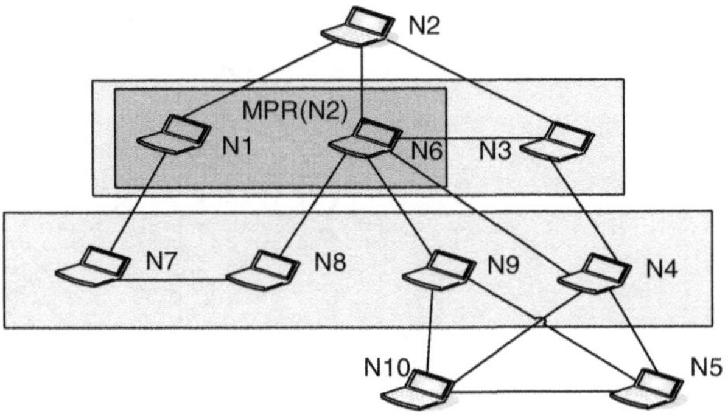

5.2.3.3 MPR Information Declaration

Each node in the network periodically broadcasts specific type of control messages called Topology Control (TC) message to build the intra-forwarding database needed for routing packets. A TC message is comprised of MPR selection set and with a sequence number, incremented when the MPR selector set changes. Information gained from TC messages is used to build the topology table in which it records the information about the topology of the network. A node records information about the multipoint relays of other nodes in this table and then based on this information, the routing Table 5.1 calculated.

Table 5.1 Comparison of ad hoc routing protocol

Performance parameters	AODV	DSR	OLSR	ZRP
Time complexity	O(2d)	O(2d)	O(d)	$O(I)^{II}$ $O(2d)^{III}$
Communication complexity	O(2d)	O(2d) or O^{I}	O(N)	$O(Z_N)^{II}$ $O(N+V)^{III}$
Routing structure	Flat	Flat	Flat	Flat
Routing acquisition	OD	OD	Pro	Hybrid
Loop free	Yes	Yes	Yes	Yes
Multicast capability	Yes	No	No	No
Routing metric	SP	SP	SP	SP
Flood for route discovery	Yes	Yes	No	Partially^{IV}
Multipath possibilities	No	Yes	No	No
Periodic broadcast	Yes	No	Yes	Yes

I: periodic update interval; N: number of nodes in the network; OD: On-demand; Pro: Proactive; SP: Shortest Path; V: number of nodes on the route reply path; ZN: number of nodes in a zone; d: diameter of the network.

5.2.4 Zone Routing Protocol (ZRP)

Zone Routing Protocol (ZRP) [17] is a hybrid protocol which combines the advantages of both proactive and reactive schemes. It was designed to mitigate the problems of those two schemes. Proactive routing protocol uses excess bandwidth

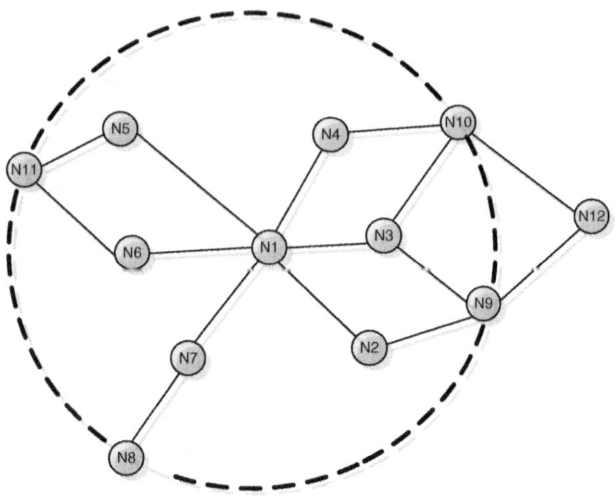

Fig. 5.6 A routing zone of radius 2

to maintain routing information, while reactive protocols suffers from long route request delays and inefficient flooding the entire network for route determination. ZRP addresses these problems by combining the best properties of both approaches.

Each node in ZRP, proactively maintains routes to destinations within a local neighborhood, which is referred as a routing zone. However, size of a routing zone depends on a parameter known as zone radius. Figure 5.6 illustrates an example of routing zone (for node N1) of radius 2 hops. Nodes N1 through N11 are members of node N1's routing zone, whereas node N12 lies outside. Here, N8 through N11 are border nodes, whereas nodes N2 through N7 are interior nodes.

5.2.4.1 Intrazone Routing Protocol (IARP)

In ZRP, each node maintains the routing information of all nodes within its routing zone. Nodes learn the topology of its routing zone through a localized proactive scheme, referred as an Intrazone Routing Protocol (IARP). No specific protocol is defined to serve as an IARP rather it can include any proactive routing protocol, such as distance vector or link state routing. Different zone may operate with different proactive routing protocols as long as the protocols are restricted within the zone. A change in topology only affects the nodes inside the zone, even though the network is quite large.

5.2.4.2 Interzone Routing Protocol (IERP)

The Interzone Routing Protocol (IERP) is responsible for reactively discovering routes to the destination beyond a node's routing zone. This is used if the destination is not found within the routing zone. The route request packets are transmitted to all border nodes, which in turn forward the request if the destination node is not found within their routing zone. IERP distinguish itself from standard flood search by

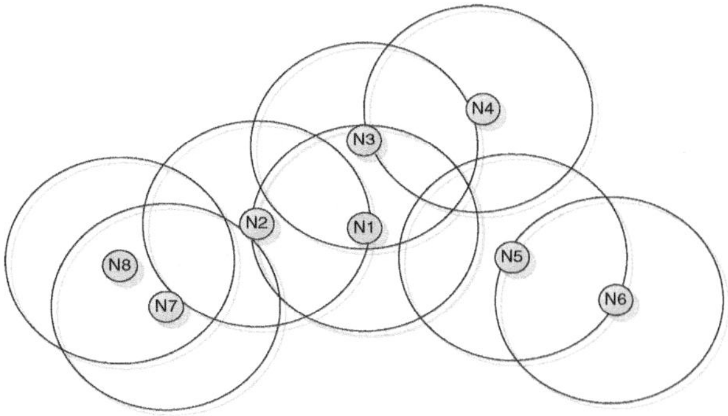

Fig. 5.7 An example of IERP operation

implementing this concept, called bordercasting. The bordercasting packet delivery service is provided by the Bordercast Resolution Protocol (BRP). An example of route discovery procedure is shown in Fig. 5.7. Assume, source node N1 wants to communicate with node N8. To find a route, N1 first checks whether N8 is within its routing zone. Since N8 lies outside N1's routing zone, N1 bordercasts a route request to all its border nodes (i.e. N2 and N3). Nodes N2 and N3 then determine that N8 is not in their routing zones and therefore bordercast the request to their border node. Node N7, which is a border node of N2, recognizes N8 as one of its interior node and responds to route request, indicating the forwarding path N1 → N2 → N7 → N8.

For detecting link failure and new neighbor nodes, ZRP relies on a protocol provided by the Media Access Control (MAC) layer, known as Neighbor Discovery Protocol (NDP). If MAC level NDP is not supported, the functionality must be provided by IARP. NDP transmits HELLO beacons at regular intervals to advertise their presence. After receiving a beacon, neighbor Table 5.1 updated. If no beacon was received from a neighbor within a specified time, the neighbor is considered as lost.

A performance comparison among the protocols depicted above is given below in Table 5.1.

I. Cache hit (postfailure)
II. For Intrazone Routing
III. For Interzone Routing
IV. Only if the destination is outside the source's routing zone

5.3 Simulation Environment

The overall goal of this simulation study is to analyze the performance of different existing ad hoc routing protocols in Mobile WiMAX environment. The simulations have been performed using QualNet version 4 [18], a software that provides scalable simulations of Ad hoc Networks. In our simulation, we consider a network of 50 nodes (one source and one destination) that are placed randomly within a 1000 m × 1000 m area and operating over 500 s. Multiple runs with different seed numbers are conducted for each scenario and collected data is averaged over those runs.

A two-ray propagation path loss model is used in our experiments with log-normal shadowing model. The MAC802.16 is chosen as the medium access control protocol. The specific access scheme is CSMA/CA with acknowledgements. In order to fully guarantee the service types, we configure 8 queues at the network layer. Unsolicited Grant Service (UGS) service type is considered to support real-time data streams consisting of fixed-size data packets issued at periodic intervals.

The node movements (except base station) in these experiments are modeled using the random waypoint mobility model [14] with mobility speed ranging from 10 to 100 km/h. We choose this range because WiMAX support medium mobility unlike cellular system [19].

5.4 Simulation Results

Figure 5.8 shows the packet delivery ratio of AODV, DSR, OLSR and ZRP as a function of mobility speed. All these four protocols have packet delivery ratio of 100% when the nodes are stationary. However, packet delivery ratio decline when nodes begin to move. Routes between nodes break, because of this frequent movement and packet delivery ratio drops. When looking at the packet delivery ratio (Fig. 5.8) it can easily be observed that ZRP and AODV perform much better than DSR and OLSR. AODV demonstrate better performance when node mobility is between 20 and 50 km/h because of its on-demand routing nature. ZRP shows better performance in higher mobility than other three protocols. DSR and OLSR show nearly the same behavior. However, in highly mobile situation, DSR demonstrate poor performance than other three protocols, since DSR relies on cache routes which are prone to be stale with increase in speed.

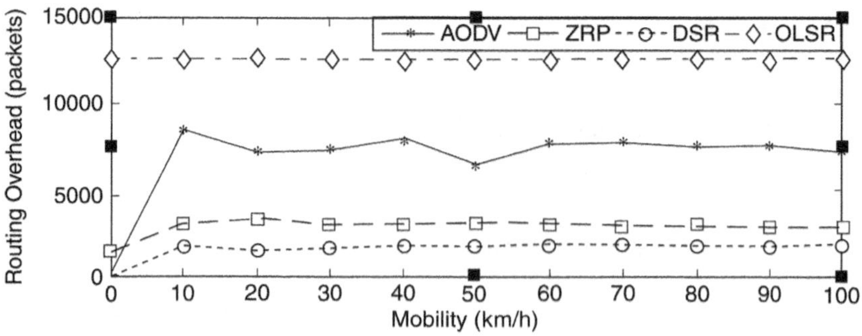

Fig. 5.8 Packet delivery ratio

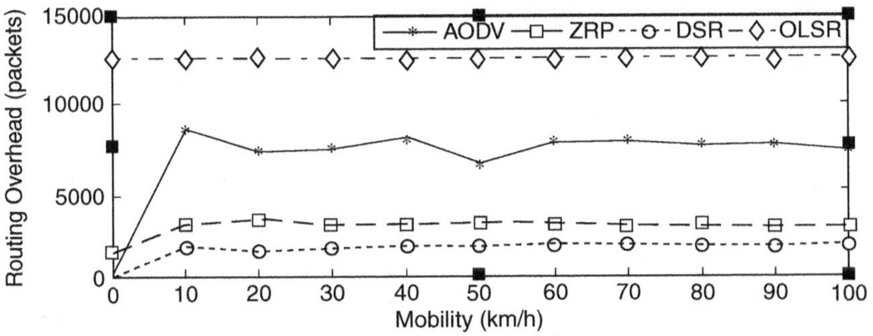

Fig. 5.9 Routing overhead

Figure 5.9 shows the number of routing protocol packets sent by each protocol obtaining the packet delivery ratios shown in Fig. 5.8. AODV, ZRP and DSR have less routing overhead when the nodes are stationary. However routing overhead increases when the nodes begin to move. DSR has considerably less overhead because of its on-demand routing nature. ZRP requires sending more routing packets due to its proactive scheme, namely the frequent hello packets to update the routing table within the local zone than DSR. Though AODV uses on-demand routing scheme, it always has higher routing overhead than DSR. Due to aggressive caching, DSR will most often find a route in its cache and therefore rarely initiate a route discovery process unlike AODV. OLSR demonstrates almost constant routing overhead in different mobility scenarios (0–100 km/h), which is higher than other three protocols.

Figure 5.10 shows the average end-to-end delay from the source to the destination's application layer. OLSR and ZRP demonstrate less delay than other two protocols due to their proactive nature. They regularly update their routing table. In case of AODV and DSR, which are reactive in nature, they have higher delay. Among these two reactive routing protocols, AODV demonstrate better performance. In

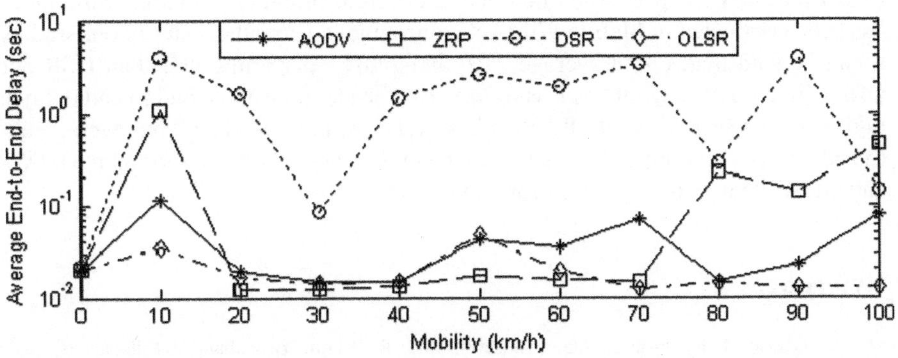

Fig. 5.10 Average end-to-end delay

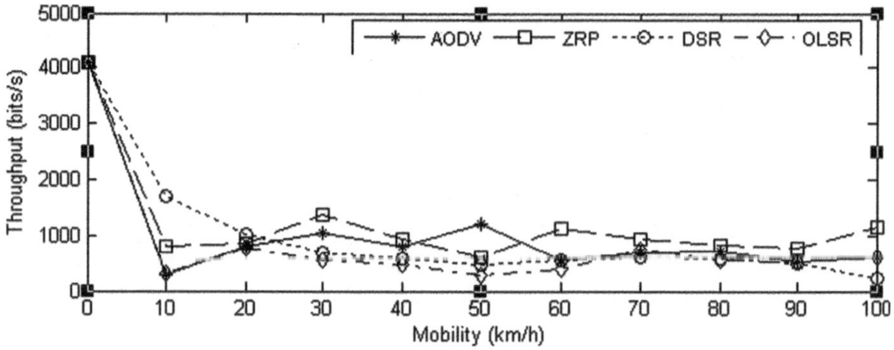

Fig. 5.11 Throughput

higher mobility scenarios (80–100 km/h), AODV has lower delay than ZRP. DSR performs worst, because DSR often uses stale routes due to the large route cache, which leads to frequent packet retransmission and extremely high delay times.

Figure 5.11 shows the throughput comparison of AODV, DSR, OLSR and ZRP. We measure the "throughput" at the receiver. When the nodes are stationary, all four protocols provide almost same throughput which is around 4000 bps. Throughput decline when nodes begin to move. From the figure it can easily be observed that ZRP and AODV perform better than DSR and OLSR. Although in higher mobility scenario (60–100 km/h) AODV, DSR and OLSR demonstrate nearly same performance. AODV demonstrate better performance when node mobility is between 20 and 50 km/h. ZRP shows better performance in higher mobility than other three protocols. DSR performs better than OLSR in low mobility. However, OLSR demonstrate better performance in higher mobility.

5.5 Conclusion

A performance comparison of four different ad hoc routing protocols (AODV, DSR, OLSR and ZRP) is performed here using different mobility scenarios. Simulation has been conducted in Mobile WiMAX environment. From the result of our studies, it can be said that, on an average ZRP and AODV perform better than DSR and OLSR. In case of DSR, it has less routing overhead, but average end to end delay is higher. However in case of OLSR, it has higher routing overhead, but average end to end delay is less. For other metrics (packet delivery ration and throughput), DSR and OLSR demonstrate poor performance.

References

1. A. Ghosh, D.R. Wolter, J.G. Andrews, and R. Chen, Broadband ad hocaccess with WiMax/802.16: Current performance benchmarks and future potential. *Communications Magazine, IEEE*, vol. 43, 2, February 2005, pp. 129–136.

2. M.A. Hasan, "Performance evaluation of WiMAX/IEEE 802.16 OFDM physical layer," *Masters Thesis*, Dept. of Electrical and Communications Engineering, Communications Laboratory, Helsinki University of Technology, Finland, June 2007.
3. K. Lu and Y. Qian, "A secure and service-oriented network control framework for WiMAX networks," *IEEE Communications Magazine*, May 2007, pp. 124–130.
4. A. Boukerche, Performance evaluation of routing protocols for ad hoc networks, *Mobile Networks and Applications*. Kluwer Academic Publishers, Dordrecht, vol. 9, pp. 333–342 (2004).
5. D. Broch, A. Maltz, D.B. Johnson, Y.C. Hu, and J. Jetcheva, A performance comparison of multihop ad hoc network routing protocols. In: *Proceedings of MOBICOMM*, October 1998.
6. S.R. Das, R. Castaneda, and J. Yan, Simulation based performance evaluation of mobile ad hoc network routing protocols. In: *Proceedings of Seventh International Conference on Computer Communications and Networks*, 1998.
7. S.R. Das, C.E. Perkins, and E.M. Royer, Performance comparison of two on-demand routing protocols for ad hoc networks. In: *Proceedings of INFOCOM 2000, Tel-Aviv, Israel*, March 2000.
8. H. Jiang, "Performance comparison of three routing protocols for ad hoc networks," *Communications of the ACM*, vol 37, August 1994.
9. J. Broch, D.A. Maltz, D.B. Johnson, Y. Hu, and J. Jetcheva, A performance comparison of multi-hop ad hoc network routing protocols. In: *Proceedings of the Fourth Annual ACM/IEEE International Conference on Mobile Computing and Networking*, MobiCom'98, October 1998, pp. 25–30.
10. A. Boukerche, Performance comparison and analysis of ad hoc routing algorithms. In: *Proceedings of IPCCC 2001*, USA, April 2001, pp. 171–178.
11. S. Azad, A. Rahman, and F. Anwar, A performance comparison of proactive and reactive routing protocols of Mobile Ad-hoc NETwork (MANET), *Journal of Engineering and Applied Sciences* 2(5), 891–896 (2007).
12. C.E. Perkins and E.M. Royer, Ad hoc on-demand distance vector routing. In: *Proceedings of the 2nd IEEE Workshop on Mobile Computing Systems and Applications*, New Orleans, LA, February 1999, pp. 90–100.
13. C.E. Perkins and P. Bhagwat, "Highly dynamic destination-sequenced distance-vector routing (DSDV) for mobile computers," SIGCOMM, London, UK, August 1994, pp. 234–244.
14. S.J. Lee, M. Gerla, and C.C. Chiang, On-demand multicast routing protocol," Ad hoc Adaptive Mobility Laboratory, Computer Science Department, University of California, Los Angeles, IEEE, 1999, http://www.cs.ucla.edu/NRL/wireless
15. D.B. Johnson, and D.A. Maltz, Dynamic source routing in ad hoc networks. In: T. Imielinski and H. Korth (eds.), *In Mobile Computing*, Kluwer Publishing Company, Dordrecht, Ch. 5, pp. 153–181, 1996.
16. T. Clausen, P. Jacquet, A. Laouiti, P. Mulethaler, A. Qayyum, and L. Viennot, Optimized link state routing protocol. In: *Proceedings of IEEE INMIC*, Pakistan, 2001.
17. S. Giannoulis, C. Antonopoulos, E. Topalis, and S. Koubias, ZRP versus DSR and TORA: A comprehensive survey on ZRP performance, *Emerging Technologies and Factory Automation, 2005. ETFA 2005*, vol 1, September 2005.
18. QualNet Network Simulator, http://www.scalable-networks.com
19. "Mobile WiMAX: Personal broadband services for enhancing lifestyles and productivity." *White Paper, WiMAX Forum*, 2006.

Chapter 6
Predictive Mobility Management with Delay Optimizations in 802.11 Infrastructure Networks

B. Issac, K.A. Hamid and C.E. Tan

Abstract In 802.11 wireless infrastructure networks, as the mobile node moves from the current access point to another, the active connections will not be badly dropped if the handoff is smooth and if there are sufficient resources reserved in the target access point. The predictive mobility management scheme we propose has primarily – mobility prediction block, delay management block and resource management block that aids the handoff. In a 5×5 symmetric grid of access points, within a 6×6 grid of regions, by location tracking and data mining, we predict the mobility pattern of mobile node with good accuracy. Active pre-scanning of mobile nodes, pre-authenticating neighbouring access points and pre-reassociation using mobility prediction are used to reduce the probe delay and authentication delay and reassociation delay respectively. The model implements reservation in two stages by using mobility prediction results and traffic type, so that sufficient resources can be reserved when the mobile node does the handoff. The overall mobility management scheme thus improves the quality of service and enables smooth handoff. Elaborate performance simulation is done in Java to verify the proposed model.

Keywords Delay management · Mobility prediction · Mobility management · Resource reservation management

6.1 Introduction

The mobility management in wireless networks is a challenging area of research as the moving node should enjoy seamless handoff in its traversal. It covers the options for storing and updating the location information of mobile node or users who are connected to the system and allocating the right amount of resources. Once an accurate mobile path prediction can be done, it can help increase the efficiency

---

B. Issac (✉)
Faculty of Engineering, University Malaysia Sarawak, 94300 Kota Samarahan, Sarawak, Malaysia
e-mail: bissac@swinburne.edu.my

S.-I. Ao et al. (eds.), *Trends in Communication Technologies and Engineering Science*, Lecture Notes in Electrical Engineering 33, DOI 10.1007/978-1-4020-9532-0_6,

of wireless network, by effectively allocating resources to the most probable access point(s) (that can be the next point of attachment) instead of blindly allocating excessive resources in the neighborhood of mobile nodes [1, 2]. In this chapter we want to come up with a predictive mobility management system which could eventually make mobility on an 802.11 network more proactive with minimum packet loss and delay, when compared to the existing technologies.

6.2 Related Works and Technologies

Park and Dadej had proposed an Adaptive Handover Control Architecture (AHCA) in [3]. The architecture has been inspired by the related research in the field of Mobile IP handoff control, such as Programmable Handoffs [4], Policy-Enabled Handoffs [5], and other adaptive or feedback-based control approaches [6].

There are two different approaches available for reducing the probe delay in 802.11 networks. They are – limiting the number of channels to be probed and reducing the waiting time for each channel to be probed. The channel mask scheme and the (neighbour graph) NG-pruning scheme limit the number of channels to be probed by introducing the channel mask/cache and the NG, respectively. SyncScan and MultiScan also reduce the number of channels to be probed [7].

In another work, Pack has proposed a predictive authentication scheme by using a modified version of IEEE 802.1 × key distribution model. Here a Frequent Handoff Region (FHR) is defined which is the set of adjacent APs around a mobile node. The FHR and Proactive Neighbour Caching (PNC) schemes are based on a similar concept where there is proactive propagation of the mobile node's context to the selected APs [7–8].

Liu and Maguire have proposed a predictive mobility management algorithm for supporting mobile computing and communications [9]. Sabeur et al. proposed a Mobile Party protocol, which is a new scheme for mobility management in the context of mesh networks, where it integrates routing and mobility management functionalities together [10]. Mihovska et al. investigated the policy-based mobility management scheme proposed for the WINNER system (a resource management system) [11]. Uskela gives a framework that shows how the different definitions of the mobility management relate to each other [12]. Sun and Sauvola discusses the effects of mobility on both the architectures and protocols for network communications [13].

6.3 Mobility Management Architecture

The Predictive Mobility Management Scheme (PMMS) architecture as proposed in Fig. 6.1 has the following blocks – Mobility Prediction Block, Delay Management Block with sub-blocks like Authentication Delay Management Block, Scan Delay Management Block and Reassociation Delay Management Block, Resource Reservation Block and Handoff block.

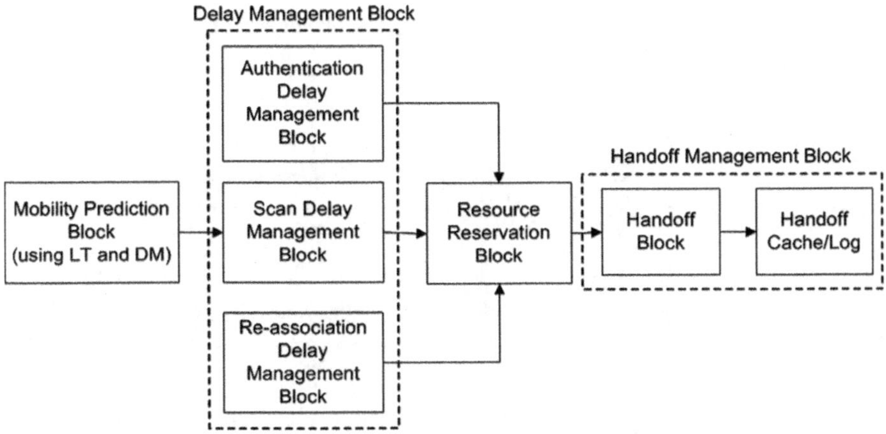

Fig. 6.1 The PMMS architecture

The mobility prediction block uses a centralized mobility prediction technique that combines location tracking without GPS and data mining technique using a Mobile Path Prediction Server (MPPS). We call it *Location Tracking with Data mining Prediction Scheme* (LTDMPS) as in [14]. Location tracking is done through a central server that receives the data regarding neighbouring access points (APs) from the mobile node which does regular active scanning. Data mining is done by the same server by using the mobility path history of mobile node movements and by extracting mobility patterns from that history. We have taken some ideas from [15], but implemented it differently.

The delay management block contains three sub-blocks namely – Authentication Delay Management Block, Scan Delay Management Block and Re-association Delay Management Block. An empirical measurement of these delays can be found in the paper by Mishra et al. [16].

In resource management, two stages of reservations are done on the next AP, that the mobile node is planning to connect. In handoff management block, the actual handoff takes place. The handoff scheme hands a MN over to a new AP only if the current signal level drops below a threshold (RSSI handoff lower threshold) and the next AP RSSI is stronger ((RSSI handoff upper threshold) than the current one by a given hysteresis margin. If the resources reserved are sufficient for the accessing MN, then the MN enjoys a smooth handoff. A Handoff cache/log can be constructed to store the previous handoff decisions, to speed up future decisions. This cache or log would be referred to as past history of the mobile node movements.

Consider the access points numbered 0–24, arranged in a 5×5 symmetric matrix format. Surrounding each access point (AP) there are four regions. These regions are numbered from R0 to R35 and is positioned around the 25 access points, as a 6×6 matrix as shown in Fig. 6.2. The mobile node can move around any AP in the surrounding four regions or can move to other adjacent regions of neighbouring AP.

A Mobile Path Prediction Server (MPPS) is connected to the local area network where the access points are connected, to make the mobility prediction in

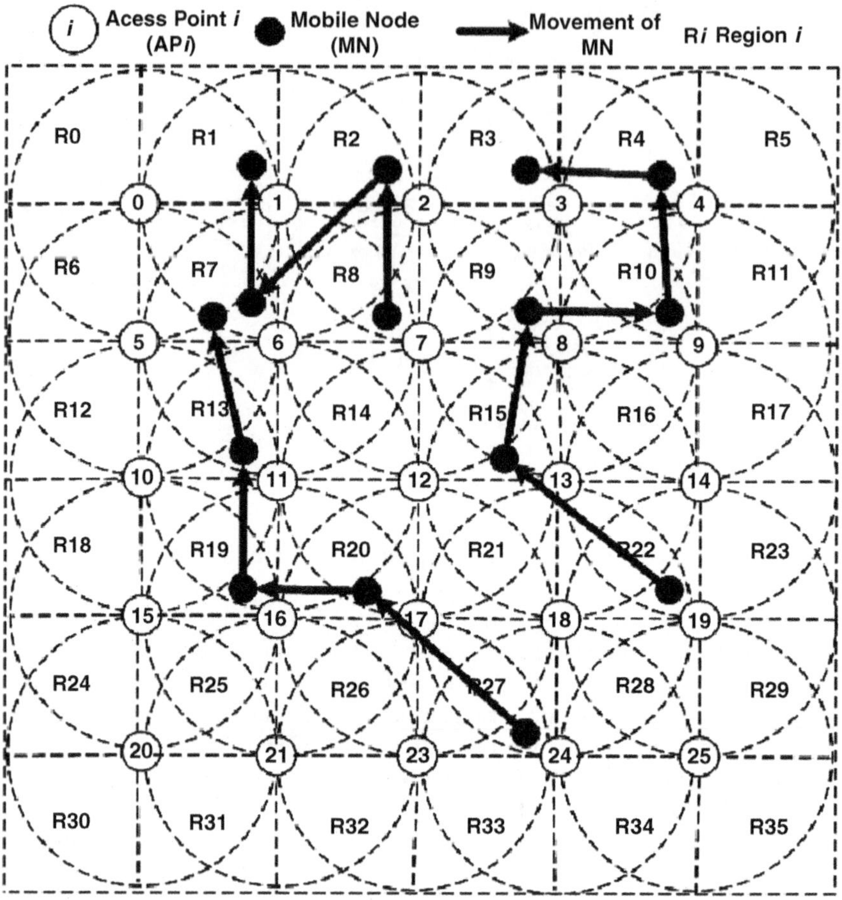

Fig. 6.2 Access points and mobile regions arranged in a matrix format showing the three mobile node paths

a centralized way. At frequent intervals of time, the mobile node transmits the Received Signal Strength Indication (RSSI) of the surrounding APs to the MPPS.

6.4 Analysis of Mobility Prediction

Java simulation is run to create 10,000 mobile paths and this is stored in a central repository. A random waypoint mobility model is used so that the mobile node moves randomly in all directions. This data set is used for data mining and to create the mobility patterns. Later, a test set of 30 random mobile paths were created (with a maximum of six access points and a minimum of three access points) and tested

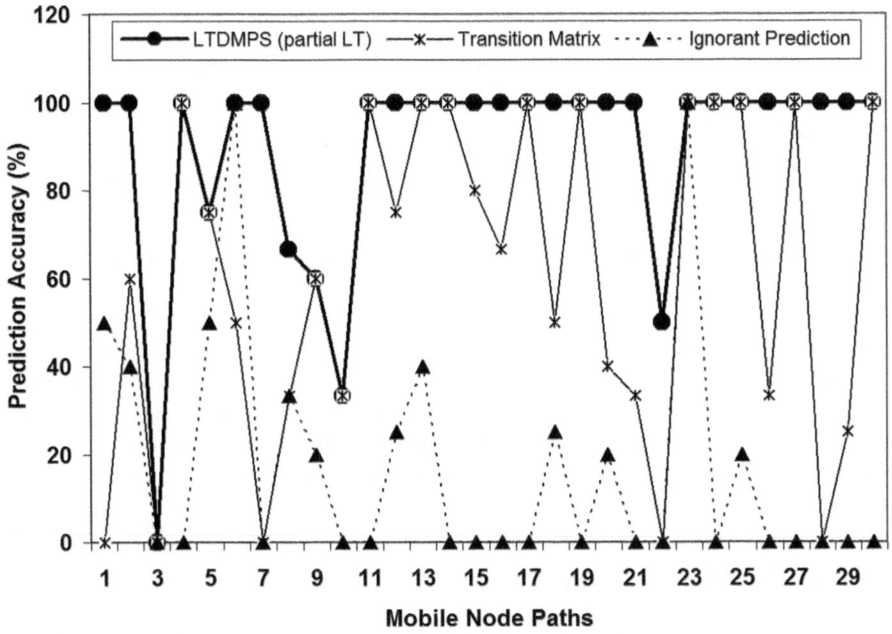

Fig. 6.3 Prediction accuracy graph for 30 random sample paths for LTDMPS (Partial location tracking and data Mining), TM and IP schemes

for prediction accuracy as shown in the comparison graphs. The average prediction accuracy was later tested for the 10,000 paths also.

For 30 paths, the average accuracy of our mobility prediction scheme LTDMPS (Partial Location Tracking and Data Mining) was 89.5% as in Fig. 6.3. We compared our prediction method with two other methods, as baseline schemes. The first prediction method is called Mobility Prediction based on Transition Matrix (TM). In this method, a cell-to-cell transition matrix is formed by considering the previous inter-cell movements of the mobile nodes or users [17]. The second prediction method is the Ignorant Prediction (IP) scheme. This approach disregards the information available from the movement history [18]. A comparison line graph in Fig. 6.3 shows the prediction accuracy for the three different schemes, in relation to the 30 mobile paths. It is very clear that LTDMPS (our proposal) which averages 89.5% prediction accuracy for 30 paths is much better compared to TM (60.5% accuracy) and IP (17.4% accuracy) schemes.

6.5 Analysis of Delay Management and Resource Management

Handoff delay measured for different client wireless adapter on MNs and access points (APs) from different vendors as per [16] and [7] are as follows in Table 6.1. This shows the delay scenario we face with our current 802.11 technology.

Table 6.1 Delays in different wireless products

AP	Cisco			Soekris			Lucent		
MN	P	A	R	P	A	R	P	A	R
Lucent	37.2	3.08	5.07	196.9	1.77	1.77	81.2	0.00	1.73
Cisco	399.8	3.56	4.13	349.9	4.48	1.79	347.3	1.49	1.09
ZoomAir	195.6	2.40	8.84	191.3	2.37	1.77	347.5	0.00	3.09

Legend (all in msec): P – probe delay; A – authentication delay; R – reassociation delay.

In Table 6.1, we can find the average delay values to be as follows: Average Probe Delay = 238.52 msec; Average Authentication Delay = 2.12 msec; Average Reassociation Delay = 3.25 msec. It clearly shows that probe delay is the biggest of all delays. It is known that for running multimedia applications in a wireless network effectively, the average handoff delay should be around 50 msec [7, 16].

6.5.1 Delay Management

There are three delays that a mobile node faces as it moves around from one AP to another – Probe delay, Authentication and Reassociation delay. These delays play an important role during the process of MN handoff to a new AP. Reducing these delays would hasten and smoothen the handoff process [19].

1. The access points would be sending information about itself (identity, transmission channel, etc.) to the MPPS server and updates the server if there is any change. Once a MN attaches itself to the first AP, the server updates the MN about its neighbours, through the AP of attachment. MN that is located in a region does active multi-channel scanning (based on information from server) where it sends probe requests for each channel scanned to locate the neighbour APs and notes their AP's RSSI values along with the channel of transmission. It sends this information to the MPPS server. Once the MN moves, it already knows the channel where it located the next AP and avoids channel scans. The central MPPS server can also update the MN (in case of any change or lack of information), as the server is constantly updated of an AP's channel of transmission. This would reduce the probe delay significantly.
2. When the MN gets connected to any AP, it would initiate authentication with all the neighbouring APs, horizontally, vertically and diagonally. Later when it moves to another AP there would be no authentication delay, as the MN is already authenticated. Once connected to the new AP, the MN parallely initiates authentication with all the neighbouring APs of current AP. This would reduce the authentication delay greatly as MN moves further.
3. Once the next AP prediction is known (based on the prediction algorithm discussed as in [14]), the process of reassociation is initiated by sending the reas-

sociation request, with the *confirm flag signal* as FALSE. This will initiate the process but will not confirm it. Once the MN or MPPS server knows that the prediction is true, only then the reassociation is confirmed. The confirm flag signal, if not communicated as true (within a time duration) will ensure that the prediction is wrong and that would cancel the reassociation done. Some timer can be set for the cancellation purposes. This would reduce the re-association delay.

6.5.2 Resource Reservation Management

Using the mobility prediction scheme, once the next AP is predicted, steps for resource reservation management can be done. There can be two stages in the reservation process – the first stage reservation and the second stage reservation.

First Stage Reservation: In the first stage reservation, once the next AP prediction is done, the buffers are reserved (say, 5% of free buffer space) explicitly through the MPPS server. The prediction can happen through RSSI sampling (when RSSI values are checked to be greater than the RSSI threshold value) or through data mining (especially when there are two possible paths). The reservation can be done without confirmation, where only a blocking is done. Later as the MN moves toward a particular AP and when RSSI value is greater than the RSSI handoff threshold value, the reservation confirmation can be given through *a confirm flag signal*, depending on prediction result.

Second Stage Reservation: Based on the Type of Service or ToS field (in IPv4 packets) or Traffic Class field in combination with Flow Label field (in IPv6 packets) the packets in a network can be classified. The ToS field helps the host to specify a preference for how the datagram would be handled as it goes through the network. Some specific application protocols for audio (or video) can be assigned higher priority than the other protocols/applications. In the second stage reservation, when the MN sees the packets that is going through it to be of a particular type, that information is communicated by MN to the MPPS server which in turn notifies the next AP (or all the AP's that MN could move to – i.e. all current AP neighbours $\cap$ APs in the next/moved region, which is a maximum of 3 APs) and some buffer (say, 5% of free buffer space) is marked for reservation in them. In a sense, only blocking of buffer space is done, which can be released for other emergencies, if any. The actual reservation will only be done when a confirmation is given later, through *a confirm flag signal* [20, 21].

6.6 Performance Simulation Results

The performance graphs shown are based on Java simulation (the parameters of which are taken from empirical measurements or from related simulations) and shows the anticipated results. Friis free space equation [22] is used for RSSI calculation as MN moves around, as in equation (6.1):

$$P_r(d) = \frac{P_t * G(r) * \lambda^2}{(4 * \pi)^2 * d^2 * L} \tag{6.1}$$

where, P_r(d) is the received power, P_t is the transmitted power, G(t) and G(r) are transmitter and receiver antenna gain, λ is the wavelength, d is the T-R separation in meters, and L is the system loss factor not related to propagation ($>= 1$).

Queuing theory formula is used to calculate mean delay experienced by packets [22], as in equation (6.2):

$$T = \frac{1}{a} + \frac{1}{1 - \frac{b}{a}} \tag{6.2}$$

where, *a* is the mean processing capacity and *b* is the mean packet arrival rate.

Performance graphs are done by considering 10 random mobile paths as in Table 6.2. Some of the simulation parameters used are shown in Table 6.3. We used Java package edu.cornell.lassp.houle.RngPack with uniform pseudo-random number generators to generate random numbers. Ranmar Class which is a lagged Fibonacci generator proposed by Marsaglia and Zaman is used [23].

If the beacon interval is 100 msec, the average probe delay of IEEE 802.11b with 11 channels is 1100 ms and 802.11a with 32 channels is 3200 msec. That indicates that to probe one channel it is taking around 100 msec. Figure 6.4 shows the stacked column graph that shows the probe/scan delay, authentication delay and reassociation delay respectively for 10 mobile node paths as listed in Table 6.2. The first set of 10 values are the probe delay, the second set of 10 values are the authentication delay and the third set of 10 values are the reassociation delay. For a given mobile path, the values are stacked on top of the other. The average probe delay for 100 paths under simulation was found to be 9.9 msec.

As an example, for 802.11b and 802.11a, all the channels would be probed to look for next AP. This can be lesser when only few channels are probed (like channel

Table 6.2 Predicted path accuracy table for a 10 path sample using LTDMPS (i.e. full location tracking and data mining)

Regular path (Curr AP→ Next AP)	Predicted path (Predicted AP [Original AP])	Accuracy (%)
6 → 0 → 1	6[6] → 0[0] → 1[1]	100
14 → 8 → 2 → 1 → 0 → 5	14[14] → 8[8] → 2[2] → 1[1] → 0[0] → 5[5]	100
16 → 21 → 15	16[16] → 15[15] → 20[20]	100
10 → 5	10[10] → 5[5]	100
3 → 2 → 1 → 0 → 6	3[3] → 2[2] → 1[1] → 0[0] → 5[5]	100
21 → 15 → 10	21[21] → 15[15] → 10[10]	100
12 → 6 → 0 → 1	12[12] → 6[6] → 0[0] → 1[1]	100
7 → 1 → 0 → 6	7[7] → 1[1] → 0[0] → 5[5]	100
4 → 3 → 2 → 1 → 6 → 0	4[4] → 3[3] → 2[2] → 1[1] → 0[0] → 5[5]	100
7 → 8 → 2 → 1	7[7] → 2[2] → 3[2] → 1[1]	66.6

Table 6.3 Simulation parameters

Simulation variables	Values
buffer_size	100 MB
audio_file_size	20 MB
text_file_size	5 MB
minChannelTime	2 msec
maxChannelTime	6 msec
pingTime	3 msec (low load)
onewayTime	2 msec (low load)
RSSImax	100 mW
RSSIthreshold	3 mW to 4mW
RSSIhandoff	1 mW to 2mW
antenna_height	1.5 m
wireless_frequency	914e+6 Hz
bandwidth	2* 1e6 Hz
receivepower_threshold	1.427e-08 W (100m)
trans_power	100 mW
proc_cap	1,000,000 packets
arrivalrate_max	950,000 packets/sec
arrivalrate_avg	650,000 packets/sec
arrivalrate_min	150,000 packets/sec
loadMax	>= 10 nodes
loadAvg	>= 5&& < 10 nodes
loadMin	>= 1&& < 5 nodes

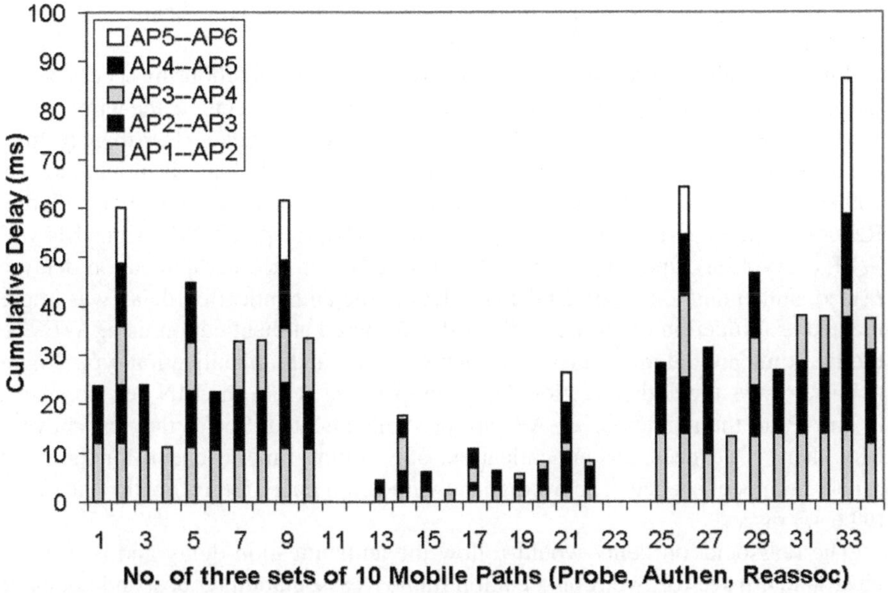

Fig. 6.4 The stacked column graph showing three sets of 10 mobile paths showing scan/probe delay, authentication delay and reassociation delay

1, 6 and 11) or if configured that way. In our new proposal, the MN gets to know all its neighbour AP (in a given region) and the channel that they are transmitting, which enables it to cut down the probe delay completely as the AP identity and channel number is already known before the next scan. A normal ping or probe request from mobile node would confirm the AP's presence and that would take on the average 3 msec to 5 msec, depending on the load. Except for the initial scan when a mobile node enters the wireless network, the probe delay later can be minimized to less than 10 msec.

The probe delay (T_{pd}) of mobile node's active scan can be expressed as, $N \times \text{MinChannelTime} \leq T_{pd} \leq N \times \text{MaxChannelTime}$, where N is the number of channels scanned. The optimal value of MinChannelTime (if no response by this time, MN scan next channel) can be around 2–4 msec and that of MaxChannel-Time (MN stops probing by this time, if responses received) can be 6–8 msec for the multi-channel active scanning. This can speed things up. The load distribution is done randomly during the mobile node movements in the ten mobile node paths considered in Table 6.2. The classification of the network load on a Basic Service Set (BSS) is done as follows. If there are 5 or less than 5 nodes at an access point, the network traffic load is classified as low. If the number of attached nodes are between 5 and 10, then the network load is classifies as medium. But if the number of attached nodes is equal to or greater than 10, then it is classified as heavily loaded BSS.

In subsequent scans the scan delay is lessened, as the mobile node already has the knowledge of access points, along with its channel. Based on the network load other parameters like collision, load delay, transmission rate and other traffic variables would also be affected. In fact, scan delay $= f$(load, MinChannelTime, Max-ChannelTime, AP identity, Channel identity).

Since the mobile node authenticates itself with all the neighbouring access points, the authentication delay is only significant when the MN first enters the wireless network. After that there would be negligible authentication delay (less than 10 msec or so), as it is already authenticated to the possible next AP that it is moving to. The authentication delay includes delay for four message exchanges, namely Challenge-Request, Challenge-Response (AP send nonce, RN), Response (MN signs RN with WEP encryption), and Approval (AP verifies RN). In fact, authentication delay $=$ f(load, authentication type). In the simulation, the authentication delay was implemented as a function of the network load and shared authentication using WEP for the first time and later as function of network load and authentication type. Figure 6.4 also shows the authentication delay comparisons. Once the MN gets connected to an AP, all the neighbouring APs are pre-authenticated. So for the next move of MN, there is no need for authentication, only a ping time to check for the target access point's existence. The average authentication delay for 100 paths was found to be 4.3 msec.

The reassociation delay would follow the authentication delay and that would be around 4 msec for two re-association frame (reassociation request and response) exchanges under light load conditions. It would be around 10–12 msec when IAPP (Inter Access Point Protocol) is implemented, where there would be around six frames (reassociation request and response, IAPP frames like – send security block, ACK security block, Move request, Move response) exchanged. Figure 6.4 also shows the reassociation delay comparisons. In fact, reassociation delay $= f$ (load, IAPP protocol). In the simulation, the reassociation delay was implemented as a function of the network load and IAPP. The average reassociation delay for 100 paths was found to be 12.2 msec.

Figure 6.5 shows the resource allocation for the proposed mobility management scheme. The default buffer size on access point is taken as 100 MB. Voice and text traffic are identified from packets and treated differently. The resource reservation is done in two stages. The first stage reservation (5% of free buffer) is done based on mobility prediction and second stage reservation (5% of free buffer) is done based on traffic type. Voice traffic gets higher buffer reservation. In the simulation, the resource reservation was implemented as a function of prediction result and traffic type (i.e. f (prediction result, traffic type)).

As shown in Fig. 6.6, the handoff is smoother with the new proposal, as the overall delay is around 50 ms or lesser and it is good for VoIP communications. The total handoff delay is calculated in simulation using the formula, total_delay = scan_time + authentication_time + reassociation_time + load_delay + packet_delay + prediction_delay, where scan_time is the scan delay, authentication time is the authentication delay, reassociation time is the reassociation delay, load delay is due to network load (and possible collisions/re-transmissions), packet delay is the mean

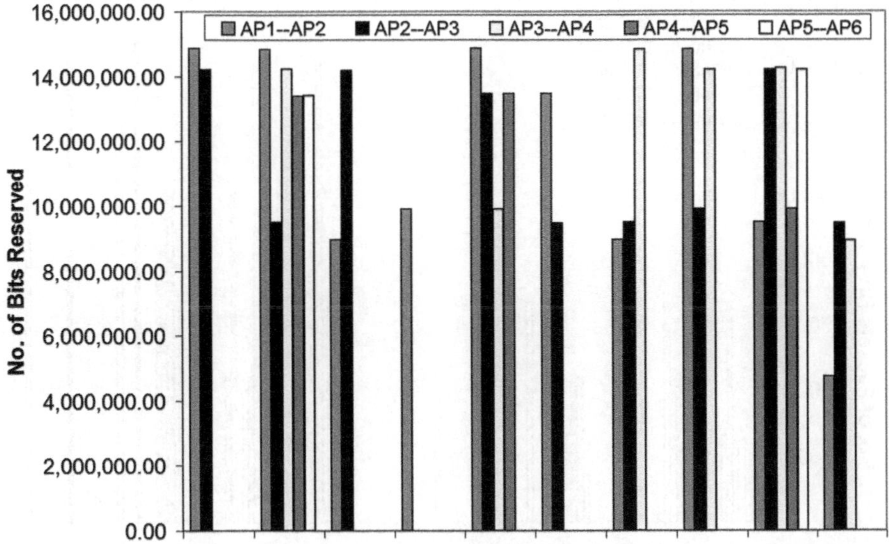

Fig. 6.5 Resource allocation in the proposed mobility management architecture for 10 mobile node paths

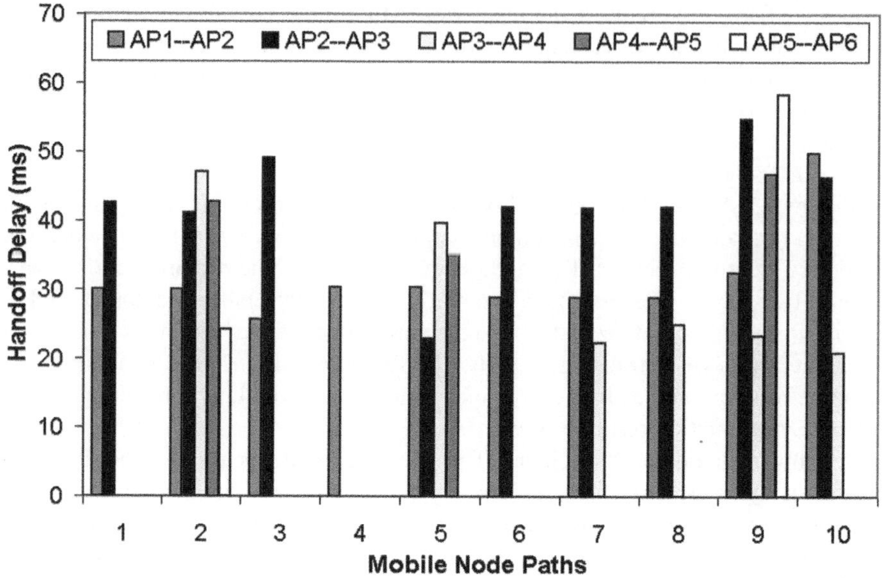

Fig. 6.6 Handoff delay graph with the proposed mobility management architecture for 10 mobile node paths

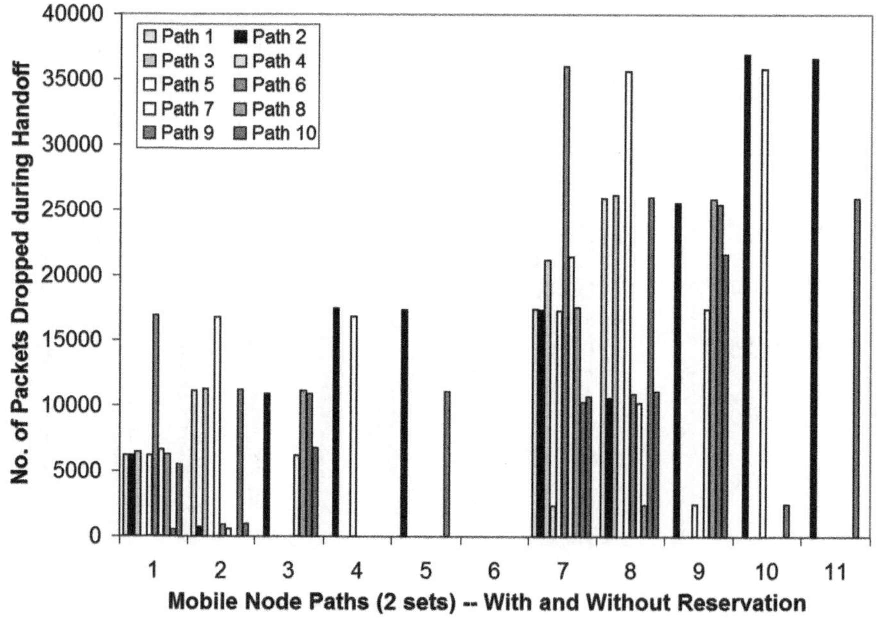

Fig. 6.7 Bits dropped during handoff with the proposed mobility management architecture for 10 mobile node paths – with reservation (*first half*) and without reservation (*second half*)

delay experienced by a packet (which is a function of mean processing capacity and mean packet arrival) and prediction delay is caused by wrong next access point prediction. Overall, the graph shows a mean handoff delay of 35.4 msec for 10 paths as shown in Fig. 6.6 and also 34.5 msec for 100 mobile node paths.

Wireless packets dropped in handoff are a function of handoff delay and the amount of resources available. Thus the packets dropped during handoff can be reduced if the handoff delay is reduced and the available buffer size in next AP is more. As the processing capacity is 1000 packets in around 1 msec, on account of delay, significant number of packets can be dropped. As per simulation, 5% of free AP buffer is normally allotted for any traffic just before connection establishment. The data traffic dropped is shown as in Fig. 6.7, with and without reservation, where there is a great reduction in bits dropped with prior reservations done in the new approach.

6.7 Conclusion

We have proposed a predictive mobility management scheme (PMMS) which uses mobility prediction, delay management and resource management. Mobility prediction is done through location tracking and data mining. The delay management consists of probe delay management (where pre-scanning is done), authentication delay management (where pre-authentication is done) and reassociation delay management (where pre-reassociation is done). The simulation performance graphs indicate that the approach can significantly lessen the handoff delay to around 50 msec, which is tolerable for voice applications.

References

1. Saygin, Y. and Ulusoy, O. (2002) "Exploiting data mining techniques for broadcasting data in mobile computing environments", IEEE Transactions on Knowledge Data Engineering, 2002, pp. 1387–1399.
2. Gok, G. and Ulusoy, O. (2000). "Transmission of continuous query results in mobile computing systems", Information Science, pp. 37–63.
3. Park, T. and Dadej, A. (2002). "Adaptive handover control in IP-based mobility networks", Proceedings of the 1st Workshop on the Internet, Telecommunications and Signal Processing, pp. 34–39.
4. Park, T. and Dadej, A. (2002). "Adaptive handover between terrestrial and satellite wireless networks", Proceedings of the Cooperative Research Centre for Satellite Systems Conference, pp. 46–52.
5. Wang, H. J., Katz, R. H. and Giese, J. (1999). "Policy-enabled handoffs across heterogeneous wireless networks", Proceedings of the Workshop on Mobile Computing Systems and Applications, pp. 51–60.
6. Stemm, M. and Katz, R. H. (1998). "Vertical handoffs in wireless overlay networks", *Mobile Networks and Applications*, 3(4): 335–350.
7. Pack, S. Choi, J., Kwon, T. and Choi, Y. (2007). "Fast handoff support in IEEE 802.11 Wireless networks", IEEE Communications Surveys and Tutorials, pp. 2–12.

8. Pack, S. and Choi, Y. (2002). "Fast Inter-AP Handoff using predictive authentication scheme in public wireless LAN", World Scientific.
9. Liu, G. Y and Maguire, G. Q. (1995). A predictive mobility management algorithm for wireless mobile computing and communications. Proceedings of the Fourth IEEE International Conference on Universal Personal Communications, pp. 268–272.
10. Sabeur, M., Al Sukkar, G., Jouaber, B., Zeghlache, D. and Afifi, H. (2007). Mobile Party: A mobility management solution for wireless mesh network. In Proceedings of the Third IEEE International Conference on Wireless and Mobile Computing, Networking and Communications, pp. 1–9.
11. Mihovska, A., Luo, J., Mino, E., Tragos, E., Mensing, C., Vivier, G. and Fracchia, R. (2007). Policy-Based mobility management for heterogeneous networks. Proceedings of the 16th IST Mobile and Wireless Communications Summit, pp. 1–6.
12. Uskela, S. (2002). Mobility management in mobile Internet. In Proceedings of the Third International Conference on 3G Mobile Communication Technologies, pp. 91–95.
13. Sun, J. and Sauvola, J. (2002). Mobility and mobility management: A conceptual framework. Proceedings of the 10th IEEE International Conference on Networks, pp. 205–210.
14. Issac, B., Hamid, K. A. and Tan, C. E. (2007). "A novel mobility prediction in 802.11 infrastructure networks by location tracking and data mining", Proceedings of the 5th International Conference of Information Technology in Asia, pp. 43–50.
15. Yavas, G., Katsaros, D., Ulusoy, O and Manolopoulos, Y. (2005). "A data mining approach for location prediction in mobile environments", *Data & Knowledge Engineering*, pp. 121–146.
16. Mishra, A., Shin, M. and Arbaugh, W. (2003). "An empirical analysis of the IEEE 802.11 MAC layer hand off process", ACM Sigcomm Computer Communication Review, pp. 93–102.
17. Rajagopal, S., Srinivasan, R. B., Narayan, R. B. and Petit, X. B. C (2002). "GPS-based predictive resource allocation in cellural networks", Proceedings of the IEEE International Conference on Networks, pp. 229–234.
18. Bhattacharya, A. and Das, S. K. (2002). "LeZi–Update: An information-theoretic approach to track mobile users in PCS networks", ACM Wireless Networks, pp. 121–135.
19. Ping-Jung, H., Yu-Chee, T. and Kun-Cheng, T. (2006). A fast handoff mechanism for IEEE 802.11 and IAPP Networks, 63rd IEEE Vehicular Technology Conference, pp. 966–970.
20. Chang, R.-S. and Chu, A. (2000). "Supporting quality of service communications for mobile hosts with advanced resource reservations", *Journal of Internet Technology*, 1(1), pp. 1–10.
21. Chang, R. -S. and Lu, H. -Y. (2000). "Predictive resource reservation in wireless cellular networks", Proceedings of the ICS Workshop on Computer Networks, Internet and Multimedia, pp. 130–137.
22. Rapport, T. S. (2002). *Wireless Communications – Principles and Practice*, Pearson Education International, Prentice Hall: Upper Saddle River, NJ.
23. Marsaglia, G. and Zaman, A. (1990). "Class Ranmar – A laggged fibonacci generator", Retrieved 5 January 2007. [Online]: http://www.honeylocust.com/RngPack/doc/edu/cornell/lassp/houle/RngPack/ Ranmar.html

Chapter 7
Decomposition of SQuaRE – Based Requirements for the Needs of SOA Applications

Witold Abramowicz, Konstanty Haniewicz, Radosław Hofman, Monika Kaczmarek and Dominik Zyskowski

Abstract The quality is perceived as a very important aspect of software applications. There exists widely accepted by both IT and business professionals SQuaRE model used to express the quality requirements towards applications. Nowadays, many applications are developed in accordance with the SOA paradigm and taking advantage of reusable components i.e. Web services. Therefore, a need appears to decompose the quality requirements expressed on the level of application to the level of components used to create the application. The problem with such decomposition is twofold. First of all, a quality model for Web services and SOA based applications is still a work in progress and neither of suggested approaches has been accepted as a standard. Second problem is connected with the user-oriented perception of the SQuaRE model and technical perspective of WS. Within this article we show how requirements defined within the SQuaRE model may be mapped to Web services QoS attributes. We advocate that ontological representation of quality models may provide common understanding of the attributes as well as it facilitates decomposition of business requirements to strictly technical ones. Nevertheless, due to the different perception the full automation of this process seems to be impossible.

Keywords Web service · Security quality requirements engineering · SQuaRE · QoS · SOA · Applications · Decomposition of requirements

7.1 Introduction

The quality is perceived as a very important aspect of software applications. There exists widely accepted by both IT and business professionals SQuaRE model (Software product Quality Requirements and Evaluation) [14] used to express the quality requirements towards applications. Nowadays, many applications are developed in accordance with the SOA paradigm and taking advantage of reusable

W. Abramowicz (✉)
Department of Information Systems, Pozna University of Economics, Al. Niepodległo ci 10, 60-967 Poznan, Poland
e-mail: W.Abramowicz@kie.ae.poznan.pl

S.-I. Ao et al. (eds.), *Trends in Communication Technologies and Engineering Science*, Lecture Notes in Electrical Engineering 33, DOI 10.1007/978-1-4020-9532-0_7,

components i.e. Web services. Therefore, an attempt needs to be undertaken to create a SQuaRE based Web services model that would allow to express similar requirements towards the SOA-based applications [1]. The next step should be the decomposition of the requirements expressed towards the SOA-based application into the requirements applied to single components i.e. Web services.

The main goal of this chapter is therefore, to present the SQuaRE based Web services model that could be used to address a situation in which business users are presented with an application built taking advantage of the SOA paradigm and then show how expressed requirements may be decomposed. Quality of an SOA-based application greatly depends on the components that are used as a foundation of implementation. Therefore, to facilitate the overall quality measurement and application design the proposed model performs an attempt to define expectations toward quality of application's components. This allows for common understanding between operational business side of an enterprise and IT personnel which is eligible for business applications development and maintenance.

To achieve the mentioned goals, the structure of this chapter is as follows. First, a motivating scenario explaining the process of business need arrival and mechanism for its translation into a software application is presented. Then, a common definition of quality is presented along with the discussion of its impact on the discussed model and references to quality standards recognized by the software industry. The following chapter is devoted to decomposition and explanation of steps to be taken to prepare the model, technology used and mappings necessary for transition between SQuaRE model and the model of Web service quality ending in an example of real world application with appropriate mappings enlisted with the use of the featured model. Finally, a summary follows.

7.2 Motivating Scenario

To provide a motivation for the work discussed in this work, let us consider the following scenario depicted in the Fig. 7.1.

Business users discover that they need additional applications to perform their tasks. They are not IT experts so they do not know how these applications should be implemented. However, as potential users of the applications in questions, they do know what they expect from these applications regarding the functionality offered as well as the expected quality in use.

Therefore, business users specify a request for application and send it to the IT department. Company rules enforce that each new application request needs to have the following elements:

- description of expected functionalities from a user point of view,
- description of requirements of a user towards the entire application as well as towards each of the functionalities, if appropriate. These requirements should be specified according to some software quality model that the entire company decided to follow. In our example it may be the SQuaRE model. Therefore, the placement of a request equals filling in the appropriate form on the web site.

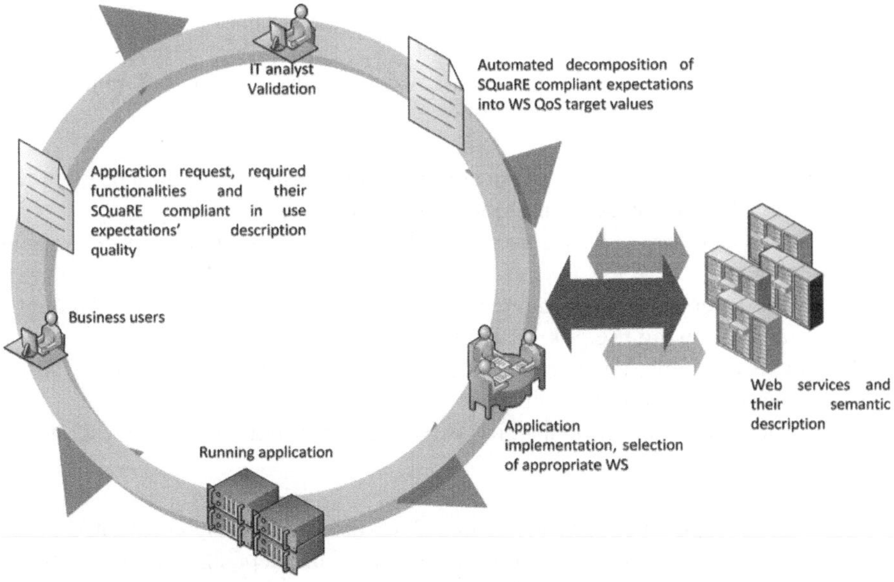

Fig. 7.1 Motivating scenario

The form is sent to the IT department where the IT expert validates the request and checks whether the fulfilment of the required functionalities requires any additional tools. At this point, when all requirements are clear, the configuration phase starts.

According to the SOA paradigm Web services provided by business partners may be utilized to build applications. It may be even possible to build the entire application automatically taking advantage of the automatic composition of Semantic Web services. If the IT expert decides to build the SOA application and utilize Semantic Web services (SWS) to carry out some tasks, a need appears to select the appropriate Web services i.e. fulfilling the required functionalities and meeting the user's requirements. Therefore, the SQuaRE requirements need to be taken into account. However, as they were defined on the application or functionality level and not the component level, the decomposition of quality requirements should take place. If the decomposition is straightforward, the specific constraints on the values of QoS of SWS are derived. If the decomposition is not straightforward, it needs to be investigated which SQuaRE requirements written in a natural language affect values of which QoS attributes. In addition, as we operate in the automated world of SWS, it would be useful to perform the decomposition task automatically or automate it at least to some extent.

In order to realize the above mentioned scenario and capture the Web services specific features the SQuaRE model needs to be appropriately enhanced. In addition, as far as the automation and machine readability is concerned, the use of semantic technologies to represent the SQuaRE as well as SWS QoS model is considered.

7.3 Related Work

Quality related issues of software often consist of a mixture of different quality meanings. Based on [2] they are as follows:

1. Quality as an abstract, meta-physical term – unreachable ideal, which shows the direction where to products are heading but will never get there,
2. Quality as a perspective of a user considering attributes of software in special context of use,
3. Quality as a perspective of manufacturer, seen as compliance with stated requirements and following ISO 9001:1994 [3] view,
4. Quality as a product perspective understood as internal characteristic, resulting from measures of product attributes,
5. Quality as a value based perspective, differing depending on a stakeholder it was defined for.

History and experience gathered within Software Engineering constituted a significant input to the proposed quality model. Therefore, depending on the authors' background and role in software related processes, the understanding on the quality may be different [4]. The first software quality model considered as a standard was developed and published by the International Standardization Organization in 1991 as ISO/IEC 9126 [5]. Ten years later new edition of this standard reviewing quality model and introducing three perspectives: quality in use, external quality and internal quality appeared. At the same time new international initiative (SQuaRE) was set up aiming to develop set of norms ISO/IEC 25000 [6] . This new approach is perceived as a new generation of software quality models [7].

As nowadays, many applications are developed in accordance with the SOA paradigm and taking advantage of reusable components i.e. Web services, therefore, an attempt was undertaken to define Web services quality model .

A Web service may be defined as a computational entity accessible over the Internet (using particular standards and protocols) [8]. A Web service is a technical implementation, an interface to a real-world service defined as a certain activity undertaken on behalf of a certain entity. However, the technical description of a Web service (the interface) is crucial, but not sufficient. What is also indispensable is the description of a real world service and non-functional properties of both a service and a Web service [4, 9].

The non-functional properties (Quality of Service) play a crucial role in almost all service interactions (to mention only selection, discovery and filtering). Non-functional properties of a service may be defined as anything that exhibits a constraint over the functionality [10]. Majority of authors define quality of Web services as Quality of Service using QoS as abbreviation. In many cases proposed quality attributes are analogical to those defined for low level telecommunication services or generic services definitions [11–12]. A survey of quality characteristics and approaches to define Web service quality that occur most often in literature is presented in [4].

In order to decompose the requirements on the level of application to the level of Web service the appropriate relations between the application and Web services model need to be defined. However, to our best knowledge up till know, no attempt in this direction has been undertaken, mainly due to the fact that no commonly accepted standard of the Web services quality has been developed.

7.4 Decomposition

As indicated in the scenario, we assume that business users are provided with an appropriate form that facilitates expressing their requirements.

In order to allow for machine-readability (what is indispensable in case of SWS) and automation as well as make the mechanism flexible, throughout the scenario we use ontologies as a knowledge representation technique. Therefore, the SQuaRE standard has been ontologized and is used as a basis to create the mentioned form. Filling in the form, a user specifies which SQuaRE characteristics are important and, if appropriate, specifies the desired value.

While a user specifies requirements, the system creates in the background appropriate machine-readable representation of the requirements. These requirements refer to the SQuaRE-based Semantic Web services quality model concepts (presented within following sections) as well as the structure of the application (e.g. to certain functionalities, goals that should be fulfilled, or the entire application) and are also stored in the form of ontology. Then, the requirements on the application level are mapped into the level of components used and new requirements on the values of SWS' non-functional properties are created. It is partially possible as we have extended the SQuaRE standard-based ontology first with some concepts specific to SOA-based applications and secondly, added the additional information like e.g. the relations between SQuaRE concepts themselves as well as SQuaRE concepts and SWS QoS concepts.

The derived requirements, also stored in the form of instances of ontology, refer to the SWS QoS ontology (and once again to the functionality and structure ontology) that defines the non-functional properties of SWS. These requirements may be used during discovery and selection of SWS to be used to create the application.

As mentioned, we do not assume that the mappings (decomposition of requirements from the application level to the Semantic Web service level) are done entirely automatically. Our aim is to provide the IT department with a tool that would help in breaking down high-level quality preferences into more detailed preferences towards specific quality properties.

7.4.1 SQuaRE – Based SOA Quality Ontology

Quality considered as product attribute is difficult to trace and unambiguously define. Quality is commonly considered as complex measure. In ISO/IEC 25000 series SQuaRE model (following previous ISO/IEC 9126) authors made distinction

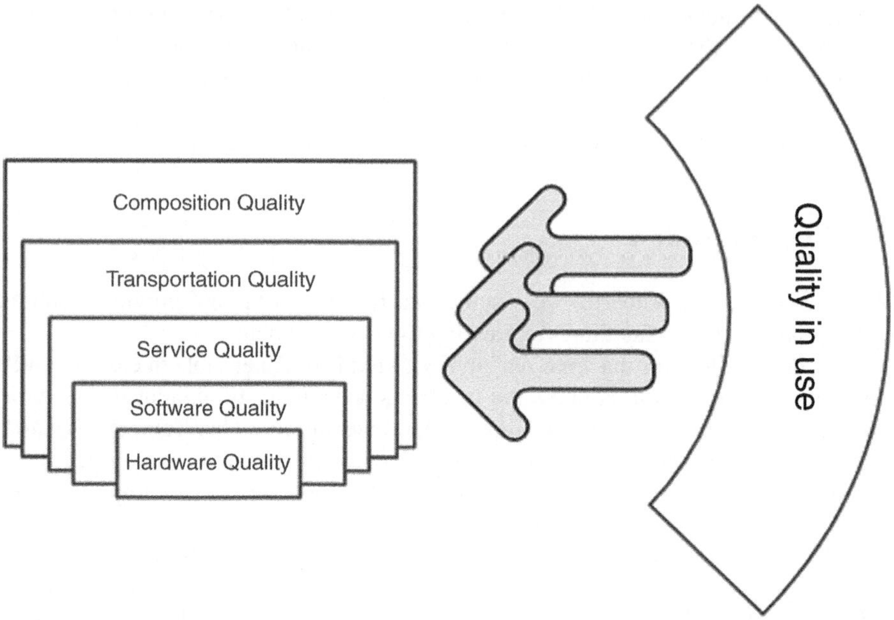

Fig. 7.2 Layers of quality

between software quality and overall quality for user (Software Quality in Use and Quality In Use). Such distinction along with internal and external software quality distinction emphasizes different layers and different responsibilities in providing demanded quality for users. Web services can also be perceived as set of value (and quality) adding layers as shown on Fig. 7.2.

Such an approach allows defining user quality requirements independently from the selected implementation. It is also compatible with IEEE1061 standard defining two fundamental requirements for quality models:

- top-down decomposition – after quality requirements are defined model needs to support their decomposition to measurable attributes
- bottom-up measurement – basing on primitive quality indicators measures quality can be predicted

We assume, that Web services are mainly designed for composition and usage by other software systems, applications or service providers (underpinning service), by compositing software etc.

In terms and in compliance to SQuaRE model, Web service quality can be defined as the extent to which a service used by specific users (brokers/compositors) meets their needs to achieve specific goals including quality in use delivered to software used by end-user.

Due to limited space we do not present SQuaRE-based Web services quality model (full description is contained in [1]).

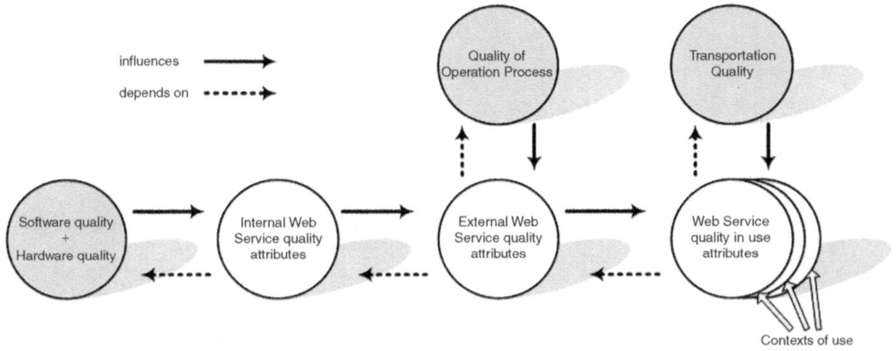

Fig. 7.3 Web service quality views dependencies in lifecycle

SQuaRE based SOA quality ontology reflects the relations between all elements and attributes of SQuaRE-based Web services quality model as presented in the Fig. 7.3. The model consists of three layers: internal Web service quality, external Web service quality and Web service quality in use.

Internal and external Web service quality is characterized by the same eight attributes, but is measured in different contexts (static and dynamic respectively for internal and external quality). Web service quality in use is characterized by six major characteristics. SQuaRE-based SOA Quality ontology was created using OWL language. Its schema is presented in the Fig. 7.4. Our understanding of respective sub-characteristics is given in the Table 7.1.

7.4.2 SWS QoS Ontology

The work already done in the area of quality of Web services has been carefully examined to come up with the most complete list of quality attributes that are essential minima of the featured model. Additionally, the mentioned list is composed of highly abstract attributes in lieu of providing only domain agnostic entities. More, the selected attributes are characterized by the fact that their semantics can not be confused and is easily understood by its user. As a user of the model is an IT analyst, any parallels between the attributes in the model in question and SQuaRE attributes are intended to avoid unwanted ambiguities.

In our approach we take advantage of the ontology presented in [13]. This ontology modelled the relations between different levels of Semantic Web services quality. As presented in the Fig. 7.5, the ontology organizes the non-functional properties of SWS in several groups. First of all, the non-functional properties are divided into qualitative and quantitative. The former are not easily computable, as they are often characterized with textual values that simply describe an opinion of a user about particular parameter. On contrary, the latter parameters are easy to measure as their values are usually numerical and quantifiable.

Fig. 7.4 Ontologization of SQuaRE-based SOA quality model

SWS QoS ontology makes a distinction between quality of execution and quality of result parameter. The difference lays in objectivity of the properties belonging to each group. Quality of result attributes strictly depend on the user past experience with other services and his requirements towards the SWS used. Moreover, it is impossible to reflect all QoR attributes in the ontology, as they are domain-dependent. Different situation is in case of QoE attributes which are mostly generic and may be applied to every Semantic Web service, due to the nature their of usage. The detailed description of concrete parameters was already presented in [13]. In our work we show the mappings between SQuaRE–based SOA quality ontology concepts and concrete parameters that describe the quality of Semantic Web services in a more accurate way.

Table 7.1 Sub-characteristics of SQuaRE-based SOA quality model

Concept	Definition
Security	Characteristics that account for the safe and secure usage of a Web service
Efficiency	Characteristics that bear on the relationship between the level of performance of the service and the amount of resources used, under stated conditions
Functionality	Functions and the parameters needed for proper usage of and obtaining the expected Web service results in a specified conditions.
Reliability	Likeliness of a service to provide expected results within agreed conditions. Reliability depends on the technical infrastructure of a Web service, and on the quality of results produced.
Maintainability	Characteristics of the effort that is required to make specified modifications of in the service. In a particular context of Web services it should be mainly understood as a provider's willingness to change/augment his service in reaction to a user feedback.
Usability	Aspects of user-friendliness and ease of use of a Web service.
Portability	Characteristics that bear on the ease of use of a Web service in different software and hardware contexts.
Interoperability of a Web service	How easy it is for a particular Web service to work with other pieces of software, or how easy it integrates in a particular system.
Safety in use	How risk-proof is the service in a given context of use
Adaptability in use	Aspects of Web service that make it work in changing environment and tolerate deviations in the environment caused by user (robustness).
Security in use	Characteristics that bear on the data confidentiality, reliability of produced results or authentication/authorization issues
Support in use	Aspects of the assistance that may be provided for a user in order to help him complete his task. As the Web service should be rather perceived as a black box, the support of Web service usage may be realized by external means.
Usability in use defines	User satisfaction, productiveness of a Web service or effectiveness in terms of performance.
Context in use	Possibility of Web service usage in different scenarios, and by different types of user. For example, a Web service should provide different levels of security for different types of users.

7.4.3 Decomposition Model/Mappings

The decomposition of SQuaRE driven requirements may not be done entirely automatically, as the SQuaRE model is not detailed when it comes to concrete sub-characteristics of quality perspectives. Another point is that the requirements defined based on SQuaRE model are always domain dependent. Therefore, every user may have different understanding of, for example, usability, portability or efficiency.

However, we are able to give some hints which QoS parameters should be associated with particular sub-characteristics of SQuaRE model. These mappings are presented in the Table 7.2. However, it is important to state that not all

Fig. 7.5 SWS QoS ontology

sub-characteristics may be mapped. It is due to the fact that especially quality of result of a Semantic Web service is not expressible with SWS-QoS ontology. This ontology covers only execution and technical aspects of Semantic Web services. The functionality is again domain dependent which makes expressing its quality a complex task.

The axioms defined in the SWS-QoS ontology may be used to model relationships between QoS properties and SQuaRE sub-characteristics. When the domain is well defined even the formulas denoting translation between these layers may be provided.

The conclusion from this table is that some parameters from SQuaRE-based model can not be translated into simple SWS QoS parameters. This is caused by different paradigms that constitute both models. SQuaRE-based is a high-level model, rather easy understood by typical IT users. SWS QoS describes more detailed, technical aspects of Web services quality and is more proper for engineers, whose task is to know which parameters influence overall quality of a software product.

7.4.4 Example

To present the reader with the actual application of the featured work we consider two real world examples where the work is applicable. First considered application

Table 7.2 Mapping between SQuaRE and SWS-QoS models

SQuaRE sub-characteristic	Mapping	SWS-QoS parameter
Security	Requires that the following properties are implemented within a Web service	Concept security with all subconcepts i.e. authentication, authorization, data confidentiality, data encryption
Efficiency	Impose some restrictions on the values of performance-related attributes – it may be some specific constraint e.g. the duration should be less than 20 minutes or preferences e.g. the duration is important and should be optimized	Performance and all subconcepts – total duration, execution time, resource usage, response time, throughput
Functionality	It affects mainly the functionality offered by a Web service	IOPE and some domain specific characteristic
Reliability	May impose some specific constraints or preferences	Reliability
Maintainability		None relevant concept can be found
Usability		None relevant concept can be found
Portability		None relevant concept can be found
Interoperability of a Web service		None relevant concept can be found
Safety in use		None relevant concept can be found
Adaptability in use	Requires	Robustness (being implemented or high value)
Security in use	Requires	Concept security and all subconcepts
Support in use		None relevant concept can be found
Usability in use defines		None relevant concept can be found
Context in use		None relevant concept can be found

is a car parts catalogue with service information. This application is required for planning service procedures applying to every car. One medium car model uses thousands of parts, each with its technical specification, price, technical drawing etc. Second application is a world natural resources prices calculator. This application is used by large resource consumers and/or resource brokers.

These two applications have different contexts of use and their users have different quality requirements. From the other hand, both applications can be built as standard (standalone) applications with appropriate updating processes in place or as Web service based software. We assume that the user is interested in quality thus both technical and architectural issues are not taken into consideration.

User quality requirements for software quality in use, as defined in SQuaRE model are presented in the Table 7.3. It presents the requirements for both example applications.

Table 7.3 SQuaRE-based requirements defined for two exemplary applications

Quality requirement class	Car-parts catalogue	Natural resource prices calculator
Usability in use	No special requirements	No special requirements
Effectiveness in use	Searching through catalogue requires several queries to find part number. Each service requires looking for 10 to –50 parts.	Resource purchase is being planned as strategic decisions – there is no need to rush.
Productivity in use	User should be able to find 1 part number per minute (including average of 3 requests)	User should be able to generate actual and historical process for selected resources in couple of minutes.
Satisfaction in use	No special requirements	No special requirements
Usability in use compliance	No requirements	No requirements
User types in use	One type of user	One type of user
Context in use	One type of context	One type of context
Environments in use	Standard PC	Standard PC
Contexts in use compliance	No requirements	No requirements
Risks to the operator	Not applicable	Not applicable
Risks to the public	Data provided has to be reliable	Not applicable
Commercial risks in use	If incorrect parts are ordered then garage suffers financial loss	If incorrect decision is taken basing on incorrect data then company suffers huge financial loss. Application should be correcting user inputs.
Risks of software corruption in use	No requirements	Software should be corruption prone
Safety in use compliance	No requirements	No requirements
Confidentiality in use	No requirements	No requirements
Integrity in use	Integrity required	Integrity required
Availability in use	Availability required at level of 98%	Availability required at level of 80%
Identification and authentication in use	No requirements	No requirements
Reliability in use	Reliable data are required	Reliable data are required
Security in use compliance	No requirements	No requirements
Learnability in use	Staff turnover is at medium level – learnability required	Staff turnover very low
Flexibility in use	Required	Not required
Accessibility in use	Application is used in garage – low resolution monitor, not precise mouse or touch-screen etc.	No requirements
Adaptability in use compliance	No requirements	No requirements

When we apply the mappings proposed in previous section we are able to acquire several SWS QoS requirements, for our two applications: throughput 50 queries per minute, execution time 20 s, availability 98%. It is not possible or eligible to suit any additional requirements that can be quantified.

7.5 Summary

Designing new software solution ordered by customer analysts and developers gather quality requirements and describe them using quality model. User quality requirements need to be decomposed as shown in this chapter into characteristics, sub-characteristics and measures and such description can further contribute to establish of external and internal software quality requirements. If one of components is to be replaced by Web service then designer of the solution has to decompose quality requirements for this Web service call as for traditional software module. This means that definition of quality requirements starts from the same set of requirements both for Web service and software module. These requirements define profile, which in case of Web services will be used for discovery and selection of Web service delivering desired quality to end user.

That is one of the reasons quality model for Web services should be compatible with software quality model. Further research should evaluate approaches to publishing quality characteristics for proposed model, evaluate accuracy of sub-characteristics and measures proposed for software quality model.

References

1. Abramowicz W., Hofman R., Suryn W., Zyskowski D.: SQuaRE based quality model, Proceedings of International MultiConference of Engineers and Computer Scientists (IMECS 2008), pp. 827–835, Hong Kong (2008)
2. Kitchenham S., Pfleeger, Software quality: The elisive target, IEEE Software 13(1), 1996
3. ISO 9001:1994, International Standardization Organization, 1994
4. Abramowicz W., Godlewska A., Gwizdała J., Jakubowski T., Kaczmarek M., Kowalkiewicz M., Zyskowski D., A Survey of QoS Computation for Web Service Profiling, 18th International Conference on Computer Applications in Industry and Engineering, Honolulu, Hawai, USA, 2005
5. ISO/IEC 9126:1991 – JTC1/SC7, Information Technology – Software Product Quality, International Standardization Organization, 1991, International Standardization Organization, 1991
6. ISO/IEC 25000:2005 – JTC1/SC7, Software Engineering – Software product Quality Requirements and Evaluation (SQuaRE), International Standardization Organization, 2005
7. Côté M-A, Suryn W., Georgiadou E., Software Quality Model Requirements for Software Quality Engineering, Software Quality Management & INSPIRE Conference (BSI), 2006
8. Preist C., A conceptual architecture for Semantic Web Services, Paper presented at the International Semantic Web Conference, 2004
9. Abramowicz W., Kaczmarek M., Zyskowski D., Duality in Web Services Reliability, Guadeloupe, French Caribbean, 2006

10. J. O'Sullivan, D. Edmond, and A. ter Hofstede, What's in a Service?, Distributed and Parallel Databases, vol 12, 2002
11. Kalepu S., Krishnaswamy S., Loke S-W., Reputation = f(User Ranking, Compliance, Verity), Proceedings of the IEEE International Conference on Web Services, San Diego, California, 2004
12. Evans J., Filsfils C., Deploying IP and MPLS QoS for Multiservice Networks: Theory and Practice, Morgan Kaufmann, San Francisco, 2007
13. Abramowicz W., Haniewicz K., Kaczmarek M., Palma R., Zyskowski D.: NFP Ontology for discovery and sharing web services in distributed registries, Proceedings of 22nd International Conference on Advanced Information Networking and Applications : Workshops, pp. 1416–1421, Okinawa (2008)
14. Suryn W., Abran A., April A., ISO/IEC SQuaRE : The Second Generation of Standards for Software Product Quality, In: The 7th IASTED International Conference on Software Engineering and Applications, California, USA, 2003.

Chapter 8
Utilizing Web Directories for Translation Disambiguation in Cross-Language Information Retrieval

Fuminori Kimura, Akira Maeda, Kenji Hatano, Jun Miyazaki and Shunsuke Uemura

Abstract In this paper, we analyzed the appropriate category level for a Web directory for cross-language information retrieval (CLIR). Our proposed method for CLIR is based on an estimation of the domains of a query using the hierarchical structures of Web directories. Therefore, correct domain estimation is necessary for detecting the appropriate category level of the Web directory. We conducted retrieval experiments using four-category levels to find the appropriate category levels of a Web directory. We found that the 2nd or 3rd levels the most appropriate for CLIR.

Keywords Cross-language information retrieval · Web directory · Appropriate category level

8.1 Introduction

With the worldwide popularity of the Internet, more and more languages are being used for Web documents. And thus, it is now much easier to access documents written in foreign languages. However, existing Web search engines only support the retrieval of documents that are written in the same language as the query, so there is no efficient way for monolingual users to retrieve documents written in non-native languages. There might also be cases, depending on the user's needs, where valuable information is written in a language other than the user's native language. To satisfy these needs using a typical monolingual retrieval system, users have to manually translate the queries themselves using a dictionary. This method can be troublesome for the user. There has been intensive research in recent years on cross-language information retrieval (CLIR), a technique for retrieving documents written in one language using a query written in another language, to eliminate this need.

F. Kimura (✉)
Department of Media Technology, College of Information Science and Engineering, Ritsumeikan University, Kyoto, Japan

S.-I. Ao et al. (eds.), *Trends in Communication Technologies and Engineering Science*, Lecture Notes in Electrical Engineering 33, DOI 10.1007/978-1-4020-9532-0_8,

A variety of methods, including the use of corpus statistics to translate terms and the disambiguation of translated terms, have been investigated and some useful results have been obtained. However, corpus-based disambiguation methods are significantly affected by the domain of the training corpus and may be much less effective for retrieval in other domains. In addition, since the Web consists of documents in various domains or genres, methods used for the CLIR of Web documents should be independent of specific domains. In particular, this paper aims at looking for the most appropriate web directory level to be used as an effective linguistic resource.

8.2 Related Work

The current approaches to CLIR can be classified into three types: document translation, query translation, and the use of inter-lingual representation. The translation of target documents based approach has the advantage of using existing machine translation systems, in which more content information is available for disambiguation. In general, this is a more effective retrieval method than those based on query translation [1, 2]. However, since it is impractical to translate a huge document collection beforehand, and it is difficult to extend this method to new languages, this approach is unsuitable for multilingual, large-scale, frequently-updated document collections found on the Web.

The second approach is a query translation approach [3–5]. In this approach, the system translates only the query terms. Then, the translation cost is lower than the document translation approach. The details of this approach are mentioned in the following paragraph.

The third approach [6] transfers both documents and queries into an inter-lingual representation, such as bilingual thesaurus classes or language-independent vector spaces. This approach requires a training phase that uses a bilingual (parallel or comparable) corpus as training data. It also uses a semantic class of thesaurus[7] or a Latent Semantic Index [8–10]. It does not need to recognize the difference between languages. However, it requires a huge computational cost for learning when the corpus being used is large. Thus, the third approach is also unsuitable for document collections on the Web.

In CLIR, the major approach is a query translation approach. In this approach, the system translates only the query terms. The major problem in using an approach based on the translation and disambiguation of queries is that the queries submitted by ordinary users of Web search engines tend to be very short. They consist of approximately two words on average [11], and are usually just an enumeration of keywords (i.e. there is no context). However, one advantage of this approach is that the translated queries can simply be fed into existing monolingual search engines. In this approach, a source language query is first translated into the target language using a bilingual dictionary. The translated query needs to be disambiguated. One of the major disambiguate approaches is a method that uses a corpus [12]. Our method falls into this category.

We should point out that corpus-based disambiguation methods are significantly affected by the differences between the domain of the query and the corpus. Hull suggests that these differences may adversely affect the retrieval efficiency of methods that use parallel or comparable corpora [13]. Lin et al. conducted comparative experiments between three monolingual corpora that had different domains and sizes, and concluded that a large-scale, domain-consistent corpus is needed to obtain useful co-occurrence data [14].

In relation to Web retrieval, which is the target of our research, the system has to cope with queries on many different topics. However, it is impractical to prepare corpora that cover every possible domain. In our previous paper [15], we proposed a CLIR method that uses the documents in Web directories that have several language versions (such as Yahoo!), instead of using existing corpora, to improve retrieval effectiveness.

8.3 Cross-Language Information Retrieval Using Web Directories

Figure 8.1 outlines of the proposed system. This system consists of a query and the target language versions of a Web directory, each language version of a feature term database, a bilingual dictionary, and the retrieval target document set. The part surrounded by a dotted line on the right side of Figure 8.1 illustrates the components necessary for the translation processing of a query.

The processing in our system can be divided into two phases. One is the preprocessing phase, which extracts the feature terms from each category of a Web directory and stores them in the feature term database in advance. The second one

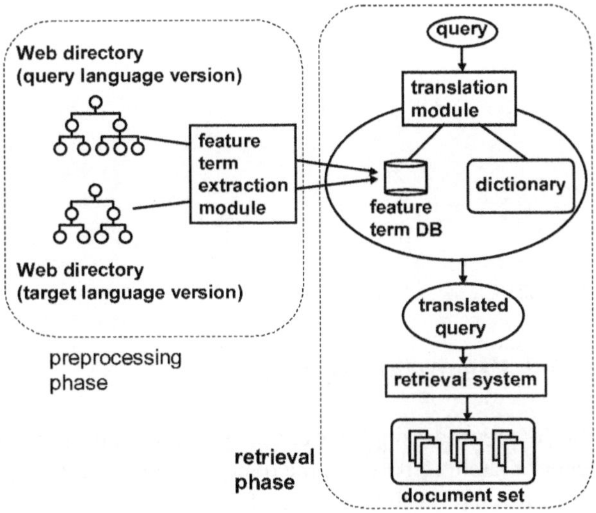

Fig. 8.1 Outline of proposed system

is the retrieval phase, which translates the given query into the target language and retrieves the appropriate documents.

8.3.1 Method of Category Merging

Each category in a Web directory is useful for specifying the fields of the query. However, some categories have an insufficient number of web documents. The system cannot acquire sufficient statistical information to resolve the translation disambiguation. This problem might be caused by one or more of the following reasons; one possible reason is that the topics of some categories are too closely related, and this might be the reason for poor accuracy. Another possible reason is that some categories have an insufficient amount of text in order to obtain statistically significant values for the feature term extraction.

Considering the above observations, we might expect that the accuracy would improve by merging the child categories at some level in the category hierarchy in order to merge some of the similar in topic categories and to increase the amount of text in a category. Figure 8.2 illustrates the outcome of category merging. Each category in a given level in the category hierarchy includes all the sub categories under that category.

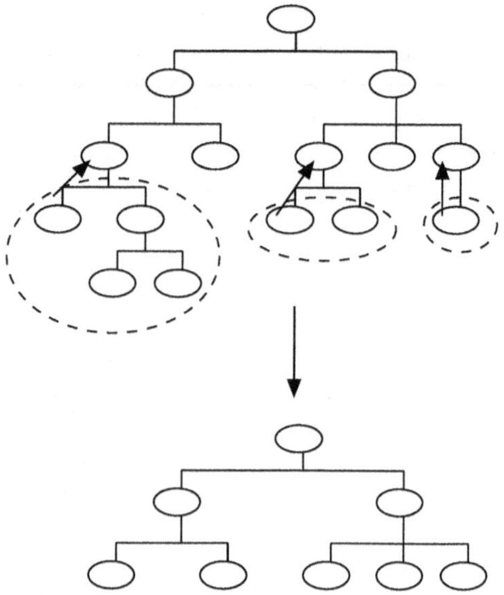

Fig. 8.2 Category merging

8.3.2 Preprocessing Phase

In the preprocessing phase, the system conducts a feature term extraction and category matching between the query and the target languages in advance. This phase is illustrated on the left side of Fig. 8.1. The following procedure is used for this phase:

1. Feature-term extraction
   For each category in all the language versions of a Web directory:

   (a) extract the terms from Web documents in the required category and calculate the weight of the terms.
   (b) extract the top n ranked terms as the feature terms of the category.
   (c) store the feature terms in the feature-term database.

2. Category matching between languages
   For each category in a single language version, estimate the corresponding category in the other language versions.

Note that the category matching method is not the focus of this paper. An arbitrary method can be used for the category matching. For example, we could calculate the similarity between categories based on the extracted feature terms [16], or we could manually match each category. The category pairs acquired by this process are used in the retrieval.

8.3.3 Retrieval Phase

The right side of Fig. 8.1 illustrates the retrieval phase. The details of the processing flow for a retrieval (right side of Fig. 8.1) are illustrated in Fig. 8.3. First, the system estimates the relevant category of the query from the query language version. Second, the system selects a category corresponding to the relevant category. Third, the system translates the query terms into the target language using the feature-term set for the corresponding category. Finally, the system retrieves the appropriate documents using the translated query. The procedure for the retrieval phase is as follows:

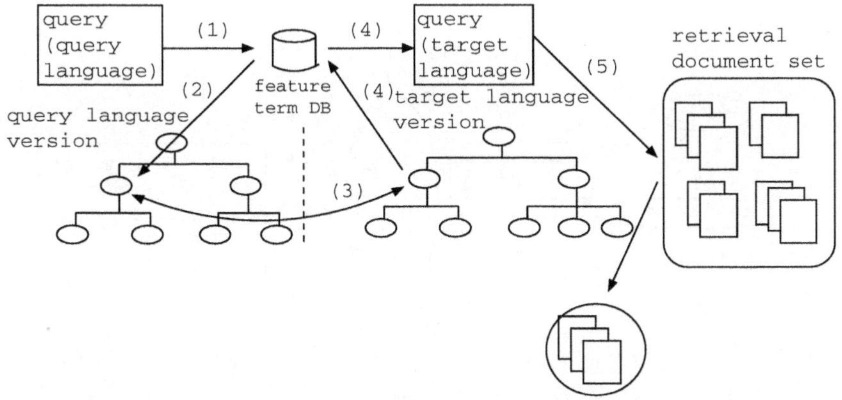

Fig. 8.3 Retrieval flow

(1) For each category in the query language version, calculate the relevance between the query and the feature-term set for the category.
(2) Determine the category with the highest relevance as the relevant category for the query.
(3) Select the category corresponding to the most relevant category from the target language version.
(4) Translate the query terms into the target language using the feature-term set of the corresponding category.
(5) Retrieve appropriate documents using the translated query.

8.3.3.1 Relevant Category Selection

The system calculates the relevance between the query and each category in the query language version (Fig. 8.3 (1)), and determines the most relevant category to the query in the query language version (Fig. 8.3 (2)). The relevance between the query and each category is calculated by multiplying the inner product between the query terms and the feature-term set of the target category by the angle of these two vectors. When there is more than one category whose relevance to the query exceeds a certain threshold, all are selected as relevant categories for the query.

In addition, the corresponding category of the relevant category is selected from the target language version (Fig. 8.3 (3)). The feature term set of the selected corresponding category is used in order to disambiguate the translation candidates.

8.3.3.2 Query Translation

Figure 8.4 illustrates the processing flow for a query translation. This figure is a more detailed example of the information shown in Fig. 8.3 (4). First, for each query term q, the system looks up the term in a bilingual dictionary and extracts all the translation candidates for the featured term. Next, the system checks whether each translation candidate is included in the feature-term set of the corresponding

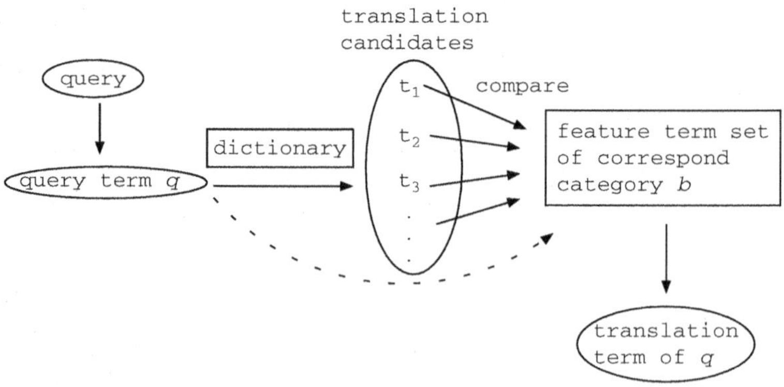

Fig. 8.4 Query translation

category. If it is, the system checks the weight of the candidates in the feature-term set. Lastly, the highest-weighted translation candidate in the feature-term set of the corresponding category is selected as the translation of the featured term.

8.3.3.3 Retrieval of Documents

The system retrieves documents using the queries translated by the method described in this section (Fig. 8.3 (5)). The documents to be retrieved do not have to be those registered in the Web directory. The system may instead use an existing retrieval system.

8.4 Experiments

We conducted experiments on the proposed method using English and Japanese versions of Yahoo! category. In these experiments, we used Japanese queries and retrieved English documents. The purpose of the experiments was to investigate what level of category merging for a Web directory is most effective at improving the precision of CLIR. "We conducted experiments for four different cases, where the category of the Web directory is merged into the top level from the top of the Web directory (hereafter called the "1-lv" for short) or the second level of it ("2-lv") or the third level ("3-lv") or the fourth level ("4-lv")".

We also conducted experiments where no disambiguation of the translations happened for a comparison (hereafter called the "baseline" for short). In the baseline, we used all the translation candidates in a dictionary, except for multi-word terms, as the query terms. We used our proposed method for the processing after the translation of a query. In addition, we conducted experiments using a web-based machine translation service in order to translate the query terms (hereafter called the "web translation" for short). The web translation cannot be directly compared with other cases because the web translation dictionary is different from that in the other cases. Then, we experimented with the web translation for reference. In this experiment, we used "excite translation".[1]

8.4.1 Method of Experiments

In these experiments, we used document sets and queries presented in the CLIR task at The 3rd NTCIR Workshop[2] (hereafter called the "NTCIR3 test collection" for short). NTCIR is an evaluation workshop for retrieval. The CLIR task is one of the tasks in NTCIR. We used two document sets from the NTCIR3 test collection: EIRB010, which consists of several English newspapers published in Taiwan from

---

[1] http://www.excite.co.jp/world/

[2] http://research.nii.ac.jp/ntcir/ntcir-ws3/work-en.html

1998 to 1999, and the Mainichi Daily 1998–1999, which consists of English newspaper articles published in Japan from 1998 to 1999. The Japanese query set for the NTCIR3 test collection, which consists of 50 queries, was used in these experiments.

To resolve any ambiguities, we used the English and Japanese versions of Yahoo! category as the Web directory. We excluded all the sub-categories in the "Regional" category in each version. We eliminated this category because it is unsuitable for translation since it consists of documents written about regions all over the world.

Table 8.1 shows the number of categories in each level. In English, 1-lv had 13 categories, 2-lv had 397 categories, 3-lv had 4066 categories, and 4-lv had 8672 categories. In Japanese, 1-lv had 13 categories, 2-lv had 391 categories, 3-lv had 2953 categories, and 4-lv had 3259 categories. We merged the categories in order to resolve a statistical information shortage. However, some of the merged categories cannot acquire a sufficient amount of statistical information. We eliminated categories that had less than 10,000 feature terms. Table 8.1 also shows the number of categories in each level after eliminating the categories that had less than 10,000 feature terms. After that, in English, 1-lv had 13 categories, 2-lv had 255 categories, 3-lv had 644 categories, and 4-lv had 292 categories. In Japanese, 1-lv had 13 categories, 2-lv had 154 categories, 3-lv had 153 categories, and 4-lv had 42 categories. In addition, the category matching between languages was done manually.

In extracting terms from English Web documents, the terms were transformed into their original form, and stop words were eliminated. We used the stop word list published by Frakes and Baeza-Yates [17], and we used the Japanese morphological analyzer, "Chasen".[3] To extract the terms from Japanese Web documents in these experiments, Chasen was used to separate the sentences, and the system extracted nouns, verbs, adjectives, and unknown terms.

For translation, we used the "EDR Electronic Dictionary: Jpn.-Eng. Bilingual Dictionary".[4] The average number of translation candidates for translating the Japanese queries in the NTCIR3 test collection was 5.17.

We used the queries that were extracted from the "TITLE" fields of the Japanese query set in the NTCIR3 test collection. We used these fields, which contain comparatively fewer terms, because ordinary users generally use only about two terms for a single query [11]. Chasen was used to subject each query to morphological analysis, and we used nouns, verbs, adjectives, and unknown terms as the query terms.

Table 8.1 Number of categories in each level

		1-lv	2-lv	3-lv	4-lv
English	All	13	397	4066	8672
	eliminated	13	255	644	292
Japanese	All	13	391	2953	3259
	eliminated	13	154	153	42

[3]http://chasen-legacy.sourceforge.jp/

[4]http://www2.nict.go.jp/r/r312/EDR/index.html

8.4.2 Lower Bound of Feature Term in the Category

In this experiment, the categories that had less than 10,000 feature terms were eliminated. Essentially, it is better not to eliminate these categories because the information in these categories is lost. However, these categories could possibly be a bad influence, because these categories have insufficient statistical information. Then, it would be necessary to consider which influence was more serious.

Table 8.2 shows the number of 4-lv categories that have more feature terms than each underbound of feature terms. We conducted the retrieval experiment mentioned in 4.0 when the underbound of feature terms are 3,000, 5,000 and 10,000 terms. Non-interpolated average precision was used to present the results shown in Table 8.3. When there were 10,000 terms the average precision was 0.0361, which is the best average overall. This result indicates that it is more important to decide the underbound of feature terms in order to acquire a sufficient amount of statistical information from the categories than to keep the number of categories set. Therefore, the underbound of feature terms was set for 10,000 terms in the experiments thereafter.

Table 8.2 Number of categories in 4 levels for feature terms

Feature term	3,000	5,000	10,000	All
English	1185	674	292	8672
Japanese	233	115	42	3259

Table 8.3 Average precision for underbound of feature terms in each 4-level category.

Feature term	3,000	5,000	10,000
Average precision	0.0301	0.0278	0.0361

8.4.3 Result of Experiments

Table 8.4 shows the results of the experiments mentioned in this section. These results show the non-interpolated average precision of each query for 1-lv, 2-lv, 3-lv, 4-lv, the baseline, and the web translation.

In the case of 1-lv, the system used 13 categories linked from the top page of Yahoo! category. In the case of 2-lv, the system used the child categories of 1-lv. In the case of 3-lv, the system used the child categories of 2-lv. The 4-lv is the child categories of 3-lv. In each case, each category in the top three levels contains all the sub categories.

Table 8.5 shows the results of a T-test between the proposed method and the baseline. We tested to see if there are significant differences between each of the

Table 8.4 Average precision for each query

Query number	1-lv	2-lv	3-lv	4-lv	Base line	Web trans
2	0.0971	0.0870	0.1042	0.1048	0.0270	0.1195
13	0.0222	0.0222	0.0102	0.0222	0.0067	0.0070
20	0.1627	0.2048	0.2390	0.2390	0.1313	0.2321
21	0.0245	0.0245	0.0281	0.0002	0.0189	0.0495
23	0.1962	0.2059	0.2059	0.0001	0.0000	0.0003
39	0.0121	0.0141	0.0040	0.0040	0.0035	0.0000
50	0.0151	0.0212	0.0151	0.0151	0.0137	0.0011
Ave	0.0400	0.0429	0.0462	0.0361	0.0203	0.0377

Table 8.5 Probability value of T-test between proposed method and baseline

Merged level	1-lv	2-lv	3-lv	4-lv
Probability	0.0480	0.0462	0.0296	0.0703

three levels and the baseline. We assumed there was no difference between each of the three levels and the baseline, and tested by using a two-tailed paired T-test.

In addition, Table 8.6 shows the translation list for each query that had a difference in average precision among the three levels.

Table 8.6 Translation list for each query

Query number	lv	Translations
2	1-lv	WTO subscription
		affiliation entry admission
		joint business cooperation
	2-lv	WTO joining subscription
		affiliation entry admission
		joint business cooperation
	3-lv	WTO subscription
20		entry adherence
	1-lv	Nissan Renault funds
		capital fund investment money
		joint business cooperation
	2-lv	Nissan Renault capital
		fund investment money cooperation
	3-lv	Nissan Renault capital
50		fund investment money cooperation
	1-lv	fashion mode style
	2-lv	fashionable clothes vogue
		fashion mode style
	3-lv	fashion mode style

8.4.4 Discussion

8.4.4.1 Effectiveness of Proposed Method

From the average of all queries, the average precision of our proposed method exceeds the average precision of the baseline. This result verified that our proposed method is effective for cross-language information retrieval. Table 8.5 shows the probabilities of all three levels are below 0.05. This means that the assumption of the non-difference between each of the three levels and the baseline is rejected, and there are significant differences. The probability of the 4-lv is not below 0.05, but below 0.10. If this probability tested with a level of significance of 0.10, the assumption of a non-difference between 4-lv and the baseline is also rejected, and there is a significant difference. These results also verified the effectiveness of our proposed method.

8.4.4.2 Translation for Using Category Level

In the 2-lv, there were 16 queries that changed their average precision compared with the 1-lv queries, as Table 8.4 indicates. Eleven queries improved and five queries got worse. In these queries, there were some increases in the number of translations, and decreases in others. Query no. 50 is one of those that increased. In this query, the translation of the Japanese term "fasshon (fashion)" increased to two translations (fashion → fashion, fashinable closes, vogue). In the cases of increase, the queries tend to acquire derivations and synonyms. On the other hand, query no. 20 is one of decreasing ones. In this query, the translation of the Japanese term "teikei (cooperation)" decreased by one translation (joint business, cooperation → cooperation). This tendency indicates that restricting the search to the target fields of the query is effective for acquiring a suitable translation of the query.

In 3-lv, there were 16 queries that changed their average precision compared with the 2-lv queries, as Table 8.4 indicates. 14 queries improved and 11 queries got worse. In query no. 2, the 2-lv had six translations for the Japanese term "kanyu (adherence)", which are "joining", "subscription", "affiliation", "entry", "admission" and "joint business cooperation". On the other hand, 3-lv had only two translations ("subscription", "adherence"). These two translations are proper terms in the diplomacy field. This result shows that restricting to narrower fields is more effective for acquiring suitable translations of the query.

However, excessive restriction also has a risk of being a bad influence. In query no. 50, the 3-lv translations decreased by two terms from the 2-lv translations. These terms improved the average precision compared with 1-lv. This result indicates that if the specified fields of the query are too narrow, there is the possibility of omitting important translations from the field.

8.4.4.3 Appropriate Level of Using Category

Table 8.4 shows that the average precision increased when using the lower-level categories in 3-lv or the levels above it. The significant differences between each level

Table 8.7 Probability value of T-test among proposed method

Merged level	1, 2-lv	2, 3-lv	3, 1-lv	3, 4-lv
Probability	0.2987	0.4103	0.1069	0.2156

were tested by using the T-test and the results are shown in Table 8.7. There are no differences between the neighboring level categories. However, the probability values are significantly different between 1-lv and 3-lv. In the case of 4-lv, the average precision becomes worse than the upper levels.

These results indicate that the precision increases by using a lower-level category. However, using a too lower-level category causes a decrease in precision. The lower-level categories can restrict the target fields. So, these categories are effective in narrowing down the proper translations. There are two factors concerned with this bad influence.

First, excessive restriction increases the probability of failing to acquire translations. Excessive restriction also increases the probability that the restricted field is suitable for some query terms, but does not suit other query terms. This increases the probability that some query terms cannot acquire a proper translation.

Second, the selected category is omitted from the proper fields by excessive restriction. There are cases where an unsuitable subcategory is selected due to the restriction even if its upper category suits the proper field. In this case, it is difficult to acquire proper translations.

To summarize this point, restricting the target fields of the query is effective for acquiring a suitable translation of a query. However, excessive restriction causes a decline in retrieval effectiveness. Thus, it is necessary to find an appropriate level of merged categories in a Web directory in order to resolve any translation disambiguity. However, a heavier negative influence becomes more serious than the restriction effect in lower categories that are too low.

8.5 Conclusion

In this paper, we analyzed what levels in Web directories are effective linguistic resources. We conducted retrieval experiments using the NTCIR3 test collection and verified that the proposed method is effective enough for cross-language information retrieval. We found that it is effective to restrict the search to the target fields of a query using lower-level merged categories in order to acquire a suitable translation of the query. However, excessive restriction has the possibility of causing a decline in the retrieval effectiveness. In this experiment, 3-lv had the best precision of the four levels. This means that 3-lv best balances the merit of restriction also showing the demerits of excessive restriction.

In future work, we need to detect the most suitable level of using merged categories in order to acquire more proper translations of query terms. In addition, we have to consider the use of Yahoo! category as a linguistic resource. There is also the possibility of improving the retrieval precision. However, it is difficult for lower categories to perform category matching among the different language categories. Lower categories also have a problem of an insufficient amount of Web documents. Thus, we have to consider using a suitable linguistic resource for the Yahoo! category (e.g. Wikipedia).

References

1. Kishida, K., Kando, N., and Chen, K.-H. (2004). Two-stage Refinement of Transitive Query Translation with English Disambiguation for Cross- Language Information Retrieval: A Trial at CLEF 2004, Working Notes for the CLEF 2004 Workshop, pp. 135–142.
2. Sakai, T., Manabe, T., Kumano, A., Koyama, M., and Kokubu, T. (2005). Toshiba BRIDJE at NTCIR-5 CLIR: Evaluation using Geometric Means, Proceedings of NTCIR-5 Workshop Meeting, pp. 56–63.
3. Davis, M. (1996). New experiments in cross-language text retrieval at NMSU's computing research lab. In the fifth text retrieval conference (TREC-5).
4. Hull, D. A. and Grefenstette, G. (1996). Querying across languages: a dictionary-based approach to multilingual information retrieval. the 19th ACM SIGIR conference (SIGIR '96).
5. Eichmann, D., Ruiz, M. E., and Srinivasan, P. (1998) .Cross-language information retrieval with the UMLS metathesaurus. The 21st ACM SIGIR conference (SIGIR '98).
6. Seo, H. C., Kim, S. B., Rim, H. C., and Myaeng, S. H. (2005). Improving Query Translation in English-Korean Cross- Language Information Retrieval. Information Processing and Management, 41(3), pp. 507–522.
7. Gonzalo, J., Verdejo, F., Peters, C., and Calzolari, N. (1998). Applying EuroWordNet to Cross-Language Text Retrieval, Computers and the Humanities, 32(2.3), pp. 185–207.
8. Deerwester, S., Dumais, S. T., Furnas, T. K., Landauer, G. W., and Harshman, R. A. (1990). Indexing by latent semantic analysis, journal of the American Society for Information Science, 46(1), pp. 391–407,
9. Dumais, S. T. (1995). Using LSI for information filtering: TREC-3 experiments. In The Third Text Retrieval Conference (TREC3), pp. 219–230,
10. Rehder, B., Littman, M.L., Dumais, S., and Landauer, T.K. (1997). Automatic 3-Language Cross- Language Information Retrieval with Latent Semantic Indexing, Proceedings of the Sixth Text Retrieval Conference (TREC-6), pp. 233–239.
11. Jansen, B. J., Spink, A., and Saracevic, T. (2000). Real life, real user queries on the Web. Information Processing & Management, 36(2), pp. 207–227.
12. Resnik, P., and Smith, N. A. (2003). The Web as a Parallel Corpus, Computational Linguistics, 29(3), pp. 349–380.
13. Hull, D. A. (1997). Using structured queries for disambiguation in cross-language information retrieval. Electronic Working Notes of the AAAI Symposium on Cross-Language Text and Speech Retrieval.
14. Lin, C. J., Lin, W. C., Bian, G. W., and Chen, H. H. (1999). Description of the NTU Japanese-English cross-lingual information retrieval system used for NTCIR workshop. First NTCIR Workshop on Research in Japanese Text Retrieval and Term Recognition, pp. 145–148.
15. Kimura, F., Maeda, A., Yoshikawa, M., and Uemura, S. (2003). Cross-Language Information Retrieval using Web Directories. Proceedings of IEEE Pacific Rim Conference on Communications, Computers and Signal Processing (PACRIM '03), pp. 911–914.
16. Cho, K. and Kim, J. (1997). Automatic Text Categorization on Hierarchical Category Structure by using ICF (Inverted Category Frequency) Weighting, Proc. of KISS Conference, pp. 507–510.
17. Frakes, W. and Baeza-Yates, R. (1992). Information Retrieval: Data Structures and Algorithms, chapter 7. Prentice-Hall, Englewood Cliffs.

Chapter 9
Ontology Based Improvement of Response Time and Reliability of Web Service Repositories

Viji Gopal and N.S. Gowri Ganesh

Abstract Web services allow clients to invoke procedures, functions, and methods on remote objects using an XML-based protocol. SOAP, UDDI and WSDL are the three core elements of a web service. The client queries a UDDI registry (Chiusano, March 2003, http://lists.ebxml.org/archives/ebxml-dev/200304/msg00000.html; Noy and McGuinness, 2001, http://protege.stanford.edu/publications/ontology_development/ontology101.html) for the service. To improve the response time and reliability of web service repositories, we introduce a new strategy based on ontologies. Ontology based Service Index Annotator (OSIAN) is a module that can work in association with a search engine (Gopal and Gowri Ganesh, Proc Int Multi Conf Eng Comput Scientists, 1:2008, 766–770). It acts as a mediator between the user and the search engine. Its main purpose is to ease the job of the search engine and to give quick results to the user.

Keywords Web service repositories · Semantic search engine · Service retrieval · Availability checking · Ontology

9.1 Introduction

A Web Service is a software component that is described via WSDL and is capable of being accessed via standard network protocols. It can support interoperable machine-to-machine interaction over a network. Web services allow clients to invoke procedures, functions, and methods on remote objects using an XML-based protocol. SOAP, UDDI and WSDL are the three core elements of a web service. The client queries a UDDI registry [1, 2] for the service either by name, category, identifier, or specification supported. Once located, the client obtains information about the location of a WSDL document from the UDDI registry.

V. Gopal (✉)
R.M.K Engineering College, Chennai, India

S.-I. Ao et al. (eds.), *Trends in Communication Technologies and Engineering Science*, Lecture Notes in Electrical Engineering 33, DOI 10.1007/978-1-4020-9532-0_9,

Repositories are a basic part of Web Services. They make it possible to find Web Services. Once a Web Service is found in a repository, the repository also describes how to use the Web Service. The owners of Web Services publish them to the UDDI registry. Once published, the UDDI registry maintains pointers to the Web Service description and to the service. The UDDI allows clients to search this registry, find the intended service and retrieve its details. These details include the service invocation point as well as other information to help identify the service and its functionality. Also UDDI enables companies to advertise the business products and services they provide, as well as how they conduct business transactions on the Web.

9.1.1 Why a Semantic Search Engine?

In a normal keyword search the search results are based on match between keywords present in the description of the published services and the search string. Pure keyword based search fails to retrieve services which are described using synonyms of the search string. Moreover, singular/plural word forms used in the service description also affect the search result. The result will contain all the services that contain the word in their interfaces. This becomes a stumbling stone in bringing out meaningful results. So using concepts instead of words for matching seems to be a better idea. A semantic search engine works on this principle.

9.1.2 Traditional System

In a traditional system (Fig. 9.1), service providers describe the interface to their web service using WSDL, and publish their services in a UDDI repository by providing appropriate "meta data" such as provider identity (white pages), a categorization of the provider's industry (yellow pages) and technical information necessary to invoke the service (green pages). Developers interested in using web services are then meant to be able to find components suitable for their needs by browsing the registry or using the keyword-based UDDI search facilities.

To make the search much context specific and meaningful, semantic techniques can be made use of. A web crawler collects semantic web service descriptions from the ontology-oriented UDDI registries, Web sites hosting these services etc and creates and maintains a semantic web service repository. Once a service request is received from a client, the broker in the system can invoke a matching algorithm and a set of web service references are returned to the user.

9.1.3 Semantic Web

Current web which can be assumed to be the biggest global database lacks the existence of a semantic structure to keep the interdependency of its components and as a result the information available on web is mostly human understandable. Semantic web provides some languages that express information in a machine process-able

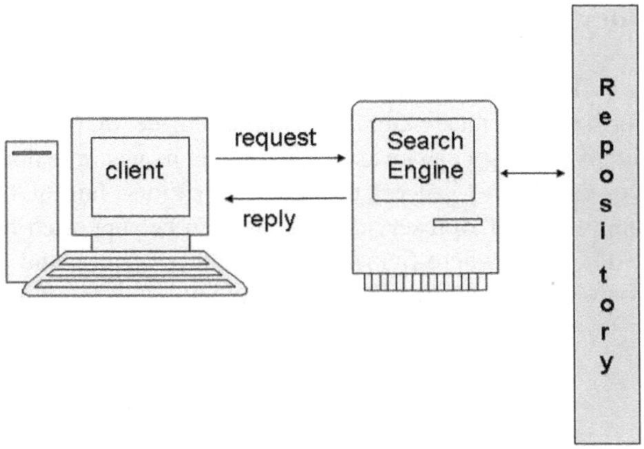

Fig. 9.1 Architecture of a traditional system

format. This implies that we can take more benefit from their processing power. A huge amount of data is conceptually related, but much of these relationships still have to be kept in human memory and not stored in an understandable way for machines. Ultimate goal of Semantic Web is to create some smarter content which could be understood by machines. When the content is understood by machine, some assertions may come out of the content and new pieces of information will be produced.

A search engine handles queries to retrieve web services (Fig. 9.2). It takes as input a set of input parameters of the web services and a set of output parameters of the web service. The search engine contains four parts: a Crawler, an Index, a query interface, and a result interface. The retrieved services are ranked and presented to the user via the result interface. The crawler discovers, analyzes and indexes semantic descriptions of Web services. The structure of the index allows the questions above to be answered, by indexing services according to the concepts they relate to, and according to their relations with other services.

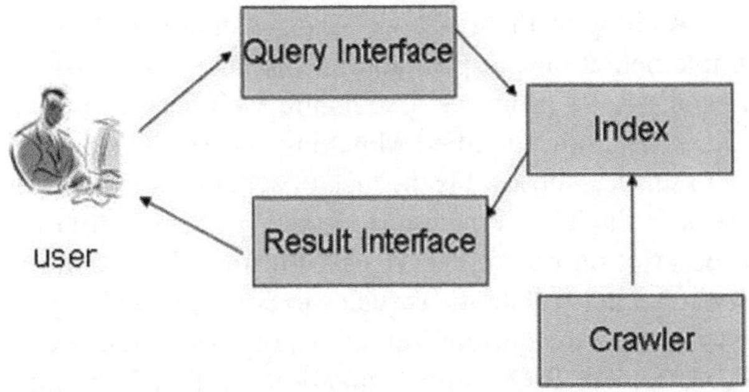

Fig. 9.2 The architecture of a search engine for web services

9.1.4 Ontology

Ontologies are building blocks of Semantic Web based systems. Creating ontologies is not an easy task and obviously there is no unique correct ontology for any domain. The real quality of ontology can be assessed only by its use in real applications.

Web services enable us to access relevant applications, but the discovery, invocation and composition of web services still need to be supported by human interaction. This is the point where Semantic Web comes to play and to support web services with ontologies as an added value. Combining the strength of web services and the added value of Semantic Web will result in a concrete base for enterprise applications.

9.2 Previous Works in This Direction

This section deals with the research work carried out by the various researchers in the related area of the initial phase of software development. The problems encountered by various authors in their work and the proposed methods for the identification of those problems are discussed.

Colin Atkinson et al. [3] argues that one of the fundamental pillars of the web service vision is a brokerage system that enables services to be published to a searchable repository and later retrieved by potential users. This is the basic motivation for the UDDI standard. But this technology was not successful enough, and the few websites that today attempt to provide a web service brokerage facility do so using a simple cataloguing approach rather than UDDI.

Atkinson and Stoll [4] states that Service Oriented Architecture depends on the availability of accurate and universally understandable specifications of services. WSDL [5] is a mechanism which gives the description of the "procedures" that a service offers. The information to invoke a service is available from its WSDL file. OWL-S [6, 7] and some other languages define the semantics of services. But they do not change the underlying WSDL service specification. This principle has a drawback. Much of the information we need is context specific where WSDL specification is not context independent. Therefore it is good to specify a service using different abstractions which give a more precise description of context sensitive information. As long as there is a well defined mapping from the specification to one of the implementations, a specification can take any form.

Eran Toch et al. [8, 9] proposes a semantic web service search engine called Object-PrOcedure-SemanticS Unified Matching (OPOSSUM). It is shown that the main challenge in service-retrieval is the lack of semantics in their interface description for precise search. The semantic approach to service-retrieval is based on expanding the description of Web services with formal semantic models, such as OWL-S [2]. These models relate the services to concepts making it easy to retrieve them. In order to address the current limitations of service retrieval, they had developed OPOSSUM (Object-PrOcedure-SemanticS Unified Matching). It is a Web-based search engine that uses semantic methods for precise and efficient retrieval of

Web services, based on their WSDL descriptions. OPOSSUM crawls the Web for WSDL descriptions, transforming them into ontological-based models of the Web services.

9.3 The New Enhanced Search Engine

To improve the response time and reliability of web service repositories, we introduce a new strategy based on ontologies. Ontology based Service Index Annotator (OSIAN) is a module that can work in association with a search engine [10]. It acts as a mediator between the user and the search engine. Its main purpose is to ease the job of the search engine and to give quick results to the user. OSIAN comprises of two parts:

- an availability checker
- RTub (Recent Tub)

The availability checker checks whether a web service is still alive. If yes, it adds the service to the RTub. The web services which are in the RTub as well as in the search engine output will not be checked by the availability checker. RTub is a tub of web services which have been recently checked for availability. RTub is an ontology which has different classes for various domains, for example travel, conferences, pay rules etc. Based on the user input, the services from the corresponding domain are forwarded to the user.

Once a service is added to the RTub, it will automatically removed from the tub after a specified amount of time that is considered to be a reasonable time in which chances are very less for a web service to shut down. This time interval is called the *lifetime of a service entry* in the RTub. If at all a web service shuts down just after it has been entered in the RTub, it will remain there only for this time interval and it will be quickly removed when its life time expires.

OSIAN will act as a front end of the search engine. It acts as a mediator between the user and the search engine. But OSIAN will use the input user interface and output user interface of the search engine. Client posts the search criteria to the interface from where OSIAN absorbs it. Now OSIAN searches its RTub for matches. User can specify the input concept and/or output concept of the services that he needs. In RTub services are associated with their input and output concepts or ontologies. A search on the basis of the associated ontology is performed. If matches are available, it is returned to the output interface.

Else the search engine is invoked to search for matching web services in its repository. The search gives an output which is the input of OSIAN. Now OSIAN checks for the availability using its availability checker. The available web services are inserted in the RTub under the proper category. Now they are returned to the output interface so that the user can view them and make a selection from among the set of web services that he/she is presented with. Figure 9.3 shows how Ontology based Service Index Annotator connects the client to the search engine.

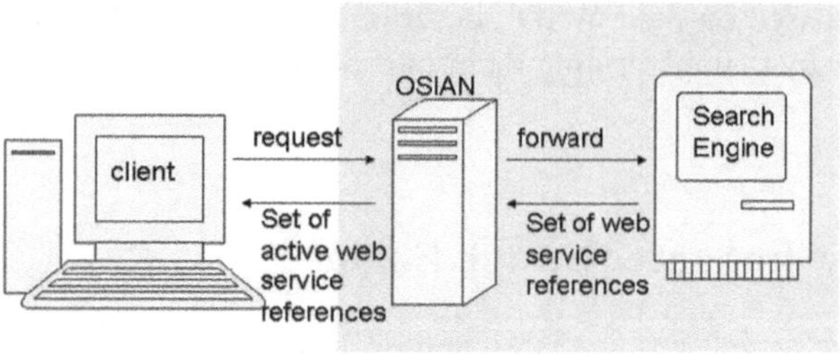

Fig. 9.3 The role of OSIAN in the web service retrieval system

OSIAN has the following advantages:

- The system is able to automatically check the availability of the web services in the repository without user intervention
- Thesystem is able to provide a cache effect when users access the services from the repository, making the response very fast
- The system is able to reduce the use of resources like processor time etc. by reducing the number of disk access needed
- The system is able to increase the overall performance of the search engine using it.

Figure 9.4 shows the structure of Ontology based Service Index Annotator system. We can test the availability of a potential web service by invoking one of its methods using randomly generated test data. This can be stored in a database and used for later periodic re-evaluation of the web service's availability. The receipt of any valid SOAP response can be interpreted as an indication that the service is at least responding and can be regarded as being "live" [3].

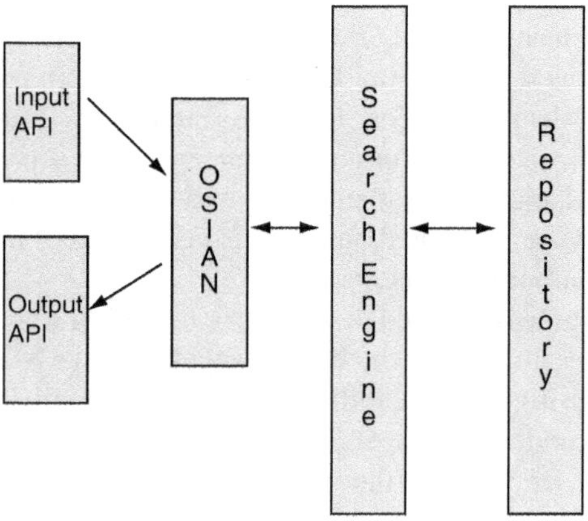

Fig. 9.4 Block diagram of the structure of OSIAN

9.4 Structure of Major Subsystems

9.4.1 Service Retrieval Subsystem

As shown in Fig. 9.5, once a query has been given to the system, it checks its RTub. If matching services are found, it is immediately returned to the user. If matching services are not found, it goes and checks the tables on the disk (repository). The retrieved services are checked for availability and available services are added in the RTub. Next time when same query comes, it is directly answered back from the fast memory which stores the services for later use.

RTub contains instances of the class *RTubEntry*. Each instance has a unique ID, a service name, a concept associated with it, a type number showing if it's the input concept/output concept of the service and a timestamp which indicates the time of the last availability checking. The value of the timestamp is updated every time an availability check is performed.

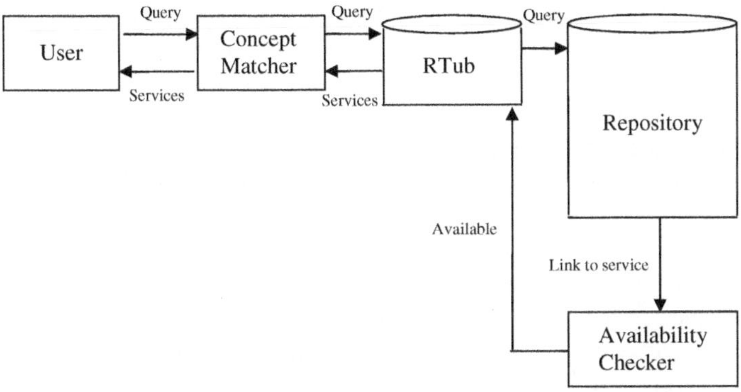

Fig. 9.5 Service retrieval

9.4.2 Periodic Availability Checking

As shown in Fig. 9.6, the purpose of this feature is to detect the non-working services in the repository and remove them. Input is the link to the WSDL of the service and based on the output, service is retained in RTub or removed from RTub. *TubConnector* is a module that initiates *Updator* in a periodic basis. *Updator* module retrieves services in the RTub and invokes the *AvailabilityChecker* module. Available services are retained in the RTub, others are deleted.

9.5 Experimental Evaluation

In this section, we evaluate our approach in two ways

1. Improvement in response time
2. Automatic availability checking

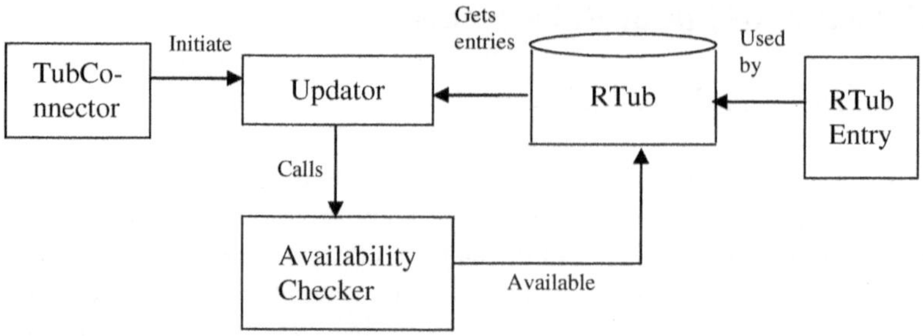

Fig. 9.6 Availability checker

9.5.1 Improvement in Response Time

Evaluation was based on an implementation of OSIAN using Java and a MySQL server. A dedicated personal computer running Windows XP with 1.256 GB RAM was used for all the experiments. The services and ontologies are all locally stored in the disk. When launched in World Wide Web, the time for network traffic will be added up. But this will be same for both the benchmark product and our product. So we can neglect it without affecting the performance comparison result.

We compared the response time values of OSIAN with those of OPOSSUM by running several queries. Our results show that we succeeded in improving the response time of the search engine compared to that of basic OPOSSUM. The basic search engine and the enhanced search engine vary considerably in query response time. Table 9.1 presents a comparison of response time of OSIAN and OPOSSUM to different queries. The results clearly show the benefits of a direct memory storage mechanism for answering repeated queries, which improves the performance of the query evaluation and response time.

From Table 9.1, it is clear that once result of a query has been stored in the fast primary memory then there needs no more disk reading or availability checking for the same query or for a similar query. The first time a query is given, OSIAN takes some time to check for the availability of those services which are in the result set,

Table 9.1 The query response time of OSIAN vs. OPOSSUM (measured in milliseconds)

Query	No. of time the query is evaluated	OPOSSUM	OSIAN
Input: quotes and Output : greet	1	14	40
Input: quotes	1	8	1
Output : greet	1	6	1
Greet	1	12	1
Quotes	1	9	1
Greet	2	15	1
Quotes	2	13	1
Input: quotes and Output : greet	2	14	1

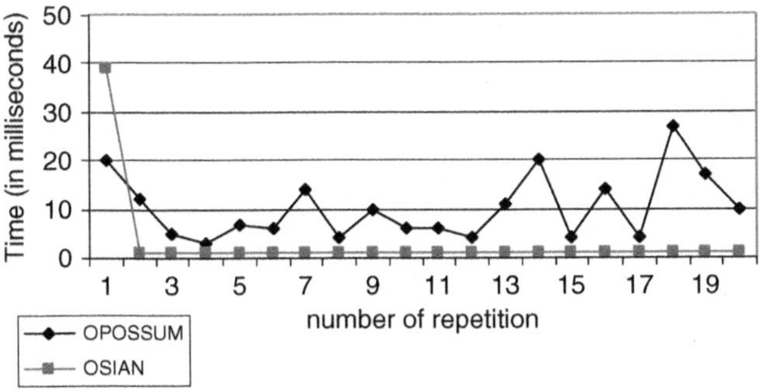

Fig. 9.7 Response time comparison for multiple executions of single query

but not in the RTub. So it will be good for the service provider to run a query which will retrieve the newly deployed service. Once it has taken its place in the RTub, next time when a user gives a similar query, OSIAN will skip the time for availability checking. Of course all the above specified data is not constant whenever you run the machine. It changes slightly depending on the machine's current load. Nevertheless, OSIAN's response time has shown no changes around 95% of the time we ran the test (except for a query's first evaluation).

A service's availability is checked again in a specified interval when its *safe life span* is over. This is done by the availability checker. This process can be scheduled for some low traffic time like midnight to avoid unnecessary delay.

A query searching for all services which give *quotes*as output is given to both OPOSSUM and OSIAN. Figure 9.7 shows how each of them respond to the same query each time.

Figure 9.8 shows the response time of both on giving different queries searching for entirely different result sets. Five random queries were run 20 times each, the response time at each time was noted down and their average was calculated. The experiments show that OSIAN is particularly useful in databases which are queried for similar services. As number of repetitions increase, its efficiency increases.

9.5.2 Automatic Availability Checking

Automatic availability checking needed no comparison because we could not find any product in the market which does the same. So, only a console output of the background work has been shown here just to illustrate its working. It shows the console output when user gave the query "quotes" two times in a row. In real, this part is not visible to the user.

QUERY FROM PARSER : quotes

The query took 39 milliseconds

final content given is : QuoteOf Day

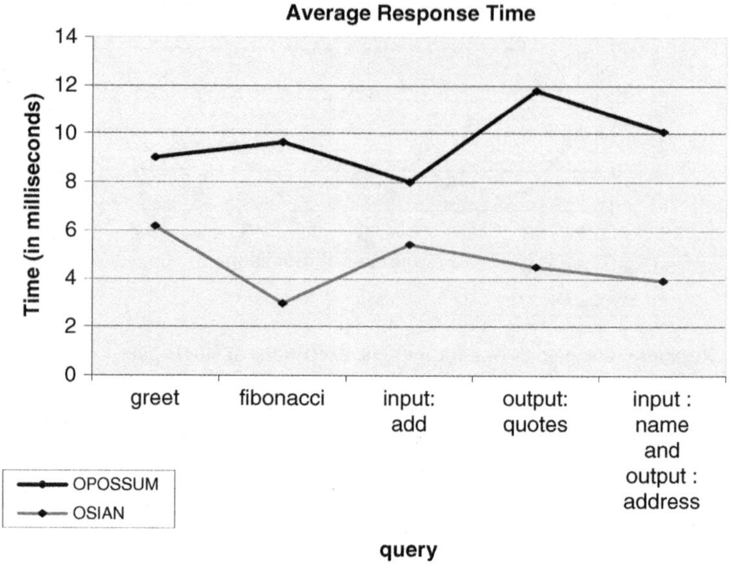

Fig. 9.8 Average response time

final content given is : publish quotes
final content given is : NiceQuoteService
****
Initiating the scheduler
****

- Job execution threads will use class loader of thread: http-8080-Processor25
- Quartz Scheduler v.1.6.0 created.
- RAMJobStore initialized.
- Quartz scheduler 'DefaultQuartzScheduler' initialized from default resource file in Quartz package: 'quartz.properties'
- Quartz scheduler version: 1.6.0
- Scheduler DefaultQuartzScheduler_$_NON_CLUSTERED started.

Entered the scheduled class
Services in RTub : 3
service : QuoteOf Day lifetime : 763
service : publish quotes lifetime : 760
service : NiceQuoteService lifetime : 760

QUERY FROM PARSER : quotes

The query took 1 millisecond

final content given is : QuoteOf Day

final content given is : publish quotes
final content given is : NiceQuoteService

Entered the scheduled class
Services in RTub : 3
service : QuoteOf Day lifetime : 45743
lifetime > safe life span
now deleting : RTubEntry@103c29b
Service available..!!!
service : publish quotes lifetime : 45740
lifetime > safe life span
now deleting : RTubEntry@1dd7bc5
Service available..!!!
service : NiceQuoteService lifetime : 45740
lifetime > safe life span
now deleting : RTubEntry@1e8c585
Service available..!!!

9.6 Conclusion

There are many service search engines available to search and retrieve web services. The traditional system depends on keyword matching of service descriptions which has been proved to be very inefficient for service retrieval. Those who use semantic strategies have clearly shown an improvement in their precision and recall. OPOSSUM is one such search engine which has been selected as a platform for this project OSIAN. None of the present service search engines do any availability checking which is an important factor that decides the reliability of the repository and thereby user satisfaction. OSIAN is a module that works with a search engine improving its reliability and response time by using an availability checker and a fast memory which keeps track of recently used services. It ensures that the number of disk accesses is minimum, which saves a lot of time while searching for services. Similar queries have been run in both OPOSSUM and OSIAN and the results were compared. Our results show that we succeeded in improving the response time performance compared to those of OPOSSUM. Also OSIAN's availability checker checks for the availability of services before they are delivered to the user. This ensures that all the services reaching the user's hands are in proper working condition. It improves user satisfaction which is an important criterion of evaluating today's competitive business world. Increased user trust on the search engine justifies the overhead of adding a verifier. It is proved that OSIAN can improve the performance by a great deal by giving a cache effect to the retrieval part of the search engine and by automatically checking the availability of the services.

Appendix: Ontology Building Tools

The following are two common tools that can be used for the creation of ontologies.

Protégé

Web Ontology Language (OWL) is the most recent development in standard ontology languages, endorsed by the World Wide Web Consortium (W3C). An OWL ontology may include descriptions of classes, properties and their instances. The Protégé-OWL API is an open-source Java library for OWL and RDF(S). The API provides classes and methods to load and save OWL files, to query and manipulate OWL data models, and to perform reasoning based on Description Logic engines. Furthermore, the API is optimized for the implementation of graphical user interfaces. The API is designed to be used in two contexts:

- For the development of components that are executed inside of the Protégé-OWL editor's user interface
- For the development of stand-alone applications (e.g., Swing applications, Servlets, or Eclipse plug-ins)

The Protégé platform supports modeling of ontologies. When you use Protégé to create and edit ontologies you will generate at least 2 files:

A *project file* (with the extension .pprj). The project file stores information relating to any interface customizations or editor options you have selected.

A *source file* (with the extensions .owl, .rdfs, or .rdf). This file is where your ontology classes, individuals and properties are defined.

OntoBuilder

OntoBuilder is an efficient tool, which allows users to drive the ontology creation process using a user-friendly and intuitive interface. It was created by Israel Institute of Technology. Once ontologies are created, computer-supported tools can utilize them as part of the information seeking process. Mainly two issues come up when one try to create ontologies. The initial creation of ontologies is a tedious, time-consuming process. Another issues is the rapid evolution of ontologies. OntoBuilder let the machine generate the ontologies. Four constructs should be considered before extracting ontology from HTML pages: Terms (things), Values, Composition, Precedence.

References

1. Chiusano, J. (Booz|Allen|Hamilton)., UDDI and ebXML Registry: A Co-Existence Paradigm. March 2003. Available at http://lists.ebxml.org/archives/ebxml-dev/200304/msg00000.html

2. Noy, N.F. and McGuinness D.L. (2001) "Ontology Development 101: Guide to Creating Your First Ontology" http://protege.stanford.edu/publications/ontology˙development/ontology101.html
3. Colin, A., Philipp, B., Oliver, H. and Dietmar, S. A practical approach to web service discovery and retrieval. IEEE International Conference on Web Services (ICWS 2007). 9–13 July 2007. pp. 241–248
4. Atkinson, C. and Stoll, D. and Acker, H. and Dadam, P. and Lauer, M. and Reichert, M.U. (2006) *Separating Per-client and Pan-client Views in Service Specification.* In: Proceedings of the 2006 International Workshop on Service-oriented Software Engineering, 27–28 May 2006, Shanghai, China. pp. 47–53.
5. Christensen, E., Curbera, F., Meredith, G., and Weerawarana, S. Web Services Description Language (WSDL) 1.1. W3C, Accessed W3C Note. 15 March 2001. Available from http://www.w3.org/TR/2001/NOTE-wsdl-20010315
6. Bechhofer, S., van Harmelen, F., Hendler, J., Horrocks, I., McGuinness, D., Patel Schneider, P., and Stein, L. 2004. OWL web ontology language reference. W3c candidate recommendation, W3C
7. Martin, D., Paolucci, M., McIlraith, S., Burstein, M., McDermott, D., McGuinness, D., Parsia, B., Payne, T.R., Sabou, M., Solanki, M., Srinivasan, N., and Sycara, K. (SRI, CMU, Univ. Toronto). Bringing Semantics to Web Services: The OWL-S Approach. First International Workshop on Semantic Web Services and Web Process Composition (SWSWPC 2004), 6 July 2004, San Diego, California, USA. pp. 26–42
8. Toch, E., Reinhartz-Berger, I., Gal, A., and Dori, D., OPOSSUM: Bridging the Gap between Web Services and the Semantic Web, proceedings of NGITS 2006, pp. 357–358
9. Toch, E., Gal, A., and Dori, D. 2005. Automatically grounding semantically-enriched conceptual models to concrete web services. In Proceedings of the International Conference on Conceptual Modeling (ER'05). Lecture Notes in Computer Science, vol. 3716. pp. 304–319.
10. Gopal, V. and Gowri Ganesh, N.S. Ontology based Service Index Annotator. In Proceedings of the International MultiConference of Engineers and Computer Scientists. (Vol 1) 19–21 March 2008. pp. 766–770

Chapter 10
Quad-Tree Based Adaptive Wavelet Packet Image Coding

Tze-Yun Sung and Hsi-Chin Hsin

Abstract In wavelet domain, if a wavelet coefficient is insignificant, the spatially related wavelet coefficients taken from all the successively higher frequency subbands of the same orientation are likely to be insignificant. This chapter will present quad-tree based adaptive wavelet packet transform in conjunction with the above-mentioned rearrangement scheme to generate adaptive wavelet packet trees (AWPT). We particularly combine SPIHT with AWPT, which is named set partitioning in adaptive wavelet packet trees (SPIAWPT), for image compression.

Keywords Quad-tree · Adaptive wavelet packet transform · Adaptive wavelet packet trees · Image compression · Image coding

10.1 Introduction

With the rapid growth of modern communications, multimedia applications, and computer technologies, image compression continues to be in great demand [1–4]. The discrete cosine transform (DCT) based Joint Photographic Experts Group (JPEG) standard shows good compression results at moderate to high rates of bits per pixel [5]. Wavelet transform provides many desirable properties and has drawn a lot of attention to the image compression applications [6]. The wavelet based multiresolution representation matches the Human Visual System [7], in which the higher detailed information is projected onto the shorter basis function with higher spatial resolution; the lower detailed information is projected onto the larger basis function with higher spectral resolution.

In wavelet domain, if a wavelet coefficient is insignificant, the spatially related wavelet coefficients taken from all the successively higher frequency subbands of the same orientation are likely to be insignificant. Shapiro first introduced this property called the self similarity of wavelet coefficients, and proposed the embedded

T.-Y. Sung (✉)
Department of Microelectronics Engineering, Chung Hua University, Hisnchu, Taiwan, ROC

S.-I. Ao et al. (eds.), *Trends in Communication Technologies and Engineering Science*, Lecture Notes in Electrical Engineering 33, DOI 10.1007/978-1-4020-9532-0_10,

zero-tree wavelet (EZW) algorithm [8]. The improved version known as the set partitioning in hierarchical trees (SPIHT) algorithm has been considered a benchmark [9]. Mukherjee proposed a vector extension of SPIHT, in which wavelet coefficients are first grouped into small vectors and then coded by SPIHT [10].

In addition to the above-mentioned quad-tree algorithms, there are some algorithms based on the energy clustering of wavelet coefficients. The amplitude and group partitioning (AGP) algorithm divides regions of high energy wavelet coefficients into small blocks by quad-tree partitioning [11]. The SWEET algorithm uses octave-band partitioning to exploit the pyramid structure of wavelet transform [12]. The set partitioning embedded block (SPECK) algorithm uses both quad-tree and octave-band partitioning to divide wavelet coefficients [13]. Taubman proposed the embedded block coding with optimized truncation (EBCOT) algorithm [14], which has been adopted by the JPEG2000 standard [15]. EBCOT is a two-tier algorithm: tier-1 performs embedded block coding with arithmetic coding; tier-2 features a rate distortion control using the post compression rate distortion (PCRD) algorithm, in which a large amount of memory space is required to store all the code streams of code blocks with their respective rate distortion information. Moreover, every code block must be processed in the tier-1 of EBCOT; however, some code blocks may not be needed to reconstruct the optimal decoded image at a given bit rate. Fang proposed a pre-compression rate distortion optimization to avoid unnecessary computations [16]. The computational complexity of EBCOT is highly intense in comparison with SPIHT.

For images with textures, many high-frequency wavelet coefficients are likely to become significant after several code passes of SPIHT, which leads to a dramatic deterioration in coding performances. As discussed in [17], a great diversity of basis functions is desirable for decomposing the significant wavelet subbands of an image. Wavelet packet transform provides a much larger family of basis functions in contrast to wavelet transform [18–19]. Ramchandran [20] and Xiong [21] proposed rate-distortion based wavelet packet selection to suit the various space-frequency properties of images; however, this approach is computationally intensive. A generalized tree structure proposed by Rajpoot [22] and the extended version proposed by Sprljan [23] aim at constructing adaptive wavelet packet trees for zero-tree image coding. In which, a set of rules is used to solve the so-called parental conflict problem: the spatial resolution of a parent in a tree is finer than that of the children. Specifically, a coefficient that has more parents than one on the next higher tree level can be reassigned a new single parent located at the same relative position on a much higher tree level in order to solve parental conflict. The idea behind this approach is to generalize parent-child relationships for wavelet packets. The resulting tree structures are variable and may not be of quad-tree structure. Thus, the widely used quad-tree based SPIHT algorithm would have to be modified to code trees with arbitrary structures. In [24], a simple scheme has been proposed to rearrange the wavelet packet coefficients of an image such that the constructed hierarchical trees are still of quad-tree structure and therefore can be efficiently coded by using the SPIHT algorithm directly. This chapter will present quad-tree based adaptive wavelet packet transform in conjunction with the above-mentioned rearrangement scheme to generate adaptive wavelet packet trees (AWPT) [25]. We particularly combine SPIHT with AWPT, which is named set partitioning in adaptive wavelet packet trees (SPIAWPT), for image compression.

10.2 Background: Wavelet Transform and Wavelet Packet Transform

Let $L^2(\Re)$ be a set of finite energy 1-D signals. There exists a function: $\psi(t) \in L^2(\Re)$ called mother wavelet, which can be used to span $L^2(\Re)$ by dilating and translating itself as follows:

$$L^2(\Re) = \text{span}\{\psi_{mn}(t) \equiv 2^{-m/2}\psi(2^{-m}t - n);\ m, n \in Z\} \tag{10.1}$$

where m and n denote the scale index and translation index, respectively. In this definition, a large m means a coarse resolution. The detail space W_m and approximation space V_m are given by

$$W_m = \text{span}\left\{\psi_{mn}(t);\ n \in Z\right\}, \text{ and} \tag{10.2}$$

$$V_m = \bigoplus_{j=m+1}^{\infty} W_j \tag{10.3}$$

respectively, where $\oplus$ denotes the direct sum. Note that V_m can be spanned by $\{\phi_{mn}(t) \equiv 2^{-m/2}\phi(2^{-m}t - n);\ n \in Z\}$ with the corresponding scaling function: $\phi(t)$, and thus provides a multiresolution analysis for $L^2(\Re)$. A finite energy continuous signal $s(t)$ can be written by

$$s(t) = \sum_n s_J(n)\phi_{Jn}(t) + \sum_{m=-\infty}^{J}\sum_n d_m(n)\psi_{mn}(t) \tag{10.4}$$

where $s_J(n)$ and $d_m(n)$ are the scaling coefficient at the coarsest resolution J and the wavelet coefficient at resolution m, respectively.

The wavelet transform (WT) of $s_\ell(n)$, which is a 1-D (discrete) signal at resolution ℓ, is given by

$$\begin{aligned} s_{\ell+1}(n) &= \sum_k s_\ell(k)\, h(2n-k) \\ d_{\ell+1}(n) &= \sum_k s_\ell(k)\, g(2n-k) \end{aligned} \tag{10.5}$$

where $s_{\ell+1}(n)$ denotes the approximation at the next coarser resolution $\ell + 1$, $d_{\ell+1}(n)$ denotes the detail between resolutions ℓ and $\ell + 1$, $h(n) =< \phi, \phi_{1,n} >$, $g(n) =< \psi, \phi_{1,n} >$, and $< \cdot, \cdot >$ is the inner product operator. $s_\ell(n)$ can be exactly reconstructed by using the following inverse wavelet transform:

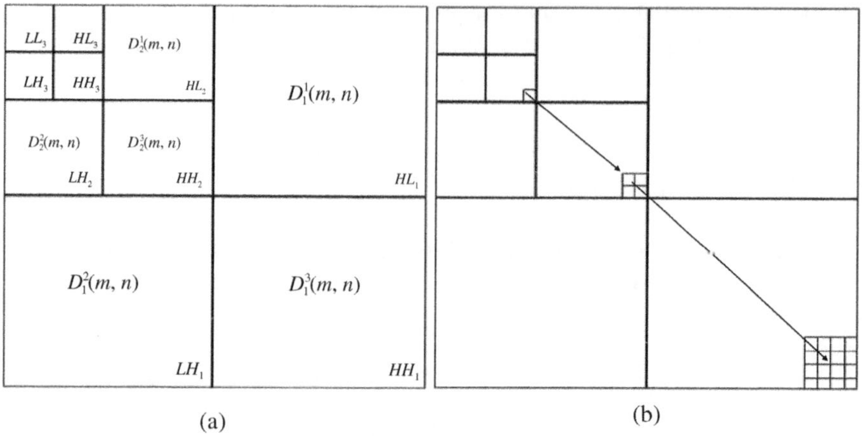

Fig. 10.1 (**a**) a 3-level wavelet transform; (**b**) a wavelet tree in the diagonal orientation

$$s_\ell(n) = \sum_k s_{\ell+1}(k)\,\tilde{h}(n-2k) + \sum_k d_{\ell+1}(k)\,\tilde{g}(n-2k) \tag{10.6}$$

where, $\tilde{h}(n) = h(-n)$ and $\tilde{g}(n) = g(-n)$ for orthogonal wavelets.

2-D WT can be obtained by the tensor product of 1-D WT. Figure 10.1(a) shows a 3-level 2-D WT, where HL_ℓ, LH_ℓ and HH_ℓ denote the subbands composed of wavelet coefficients $D_\ell^1(m,n)$, $D_\ell^2(m,n)$ and $D_\ell^3(m,n)$ representing the detail information at resolution ℓ in the horizontal, vertical and diagonal orientations, respectively. LL_3 denotes the approximation at the coarsest resolution 3. The original image is usually taken as the scaling coefficients, $S_0(m,n)$, at the finest resolution 0. $S_\ell(m,n)$ can be decomposed into $S_{\ell+1}(m,n)$, $D_{\ell+1}^1(m,n)$, $D_{\ell+1}^2(m,n)$ and $D_{\ell+1}^3(m,n)$ by using 2-D WT.

Wavelet transform is focused on the low frequency decomposition, i.e. only the scaling coefficients are successively decomposed. As a result, wavelet transform may not be suitable for images with textures. As both scaling and wavelet coefficients can be successively decomposed, the so-called wavelet packet transform (WPT) will result. Similarly, 2-D WPT can be obtained by the tensor product of 1-D WPT. Figure 10.2 shows a 3-level 2-D WPT, where the wavelet coefficients: $D_1^3(m,n)$ and $D_2^3(m,n)$ are further decomposed into wavelet packets. Note that WPT provides a much larger family of subband decompositions.

10.3 Wavelet Image Coding

By wavelet transform, an image can be decomposed into subbands with orientation selectivity. The spatially related wavelet coefficients taken from all the subbands of the same orientation can be grouped to form hierarchical wavelet trees. The tree

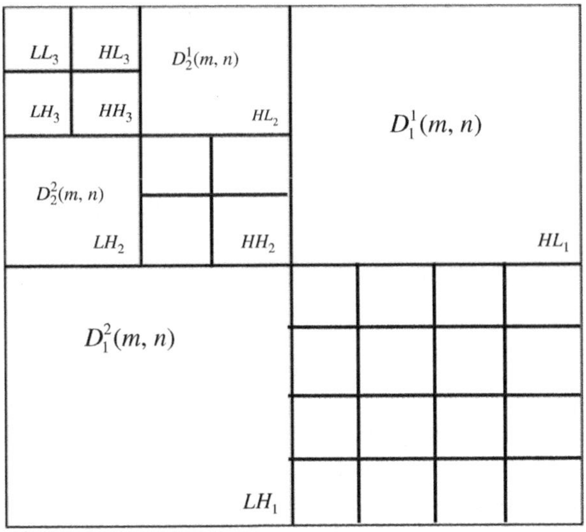

Fig. 10.2 A 3-level wavelet packet transform

hierarchy is based on the resolution level. A wavelet coefficient called parent node at a resolution level has four corresponding wavelet coefficients called children nodes at the next finer resolution level. The root node is at the coarsest resolution level, and the leaf nodes are at the finest resolution level. Figure 10.1(b) shows a wavelet tree in the diagonal orientation. Most of the significant wavelet coefficients of homogeneous image regions are at the coarser resolution levels. If a parent node is insignificant with respect to a given threshold, all the descendant nodes are likely to be insignificant with respect to the same threshold; thus, these nodes form a zero-tree and therefore can be efficiently coded.

10.3.1 The SPIHT Algorithm

The wavelet based SPIHT algorithm has received a lot of attention since its introduction [9]. In SPIHT, two code passes, namely sorting pass and refinement pass are processed, which can be combined to form a single scan pass. Three symbols are used, namely zero tree (ZT), insignificant pixel (IP), and significant pixel (SP), which are stored in their respective lists called list of insignificant sets (LIS), list of insignificant pixels (LIP), and list of significant pixels (LSP). The following is SPIHT presented in steps.

Step 1: Initialization: Compute $N = \lfloor \log_2(\max_{(m,n)} |c_{m,n}|) \rfloor$, where $c_{m,n}$ is the wavelet coefficient at position (m, n); Set the initial threshold $T_b = 2^N$; $b = 1$.

Step 2: Sorting pass: Identify coefficients with $T_b < |c_{m,n}| \leq 2T_b$; Output their respective signs and positions.
Step 3: Refinement pass: Output the bth bit of the significant coefficients with $|c_{m,n}| > 2T_b$, which have been identified in the previous passes.
Step 4: Increase b by one, divide the threshold by 2, and go to Step 2.

The scan pass, i.e. step 2 followed by step 3, is performed repeatedly until the bit budget reaches. In the sorting pass, tree nodes stored in LIS and LIP are evaluated as follows. For a node with magnitude greater than the current threshold, it becomes significant and then is stored in LSP. For an insignificant node with respect to the current threshold, if all the descendants are also insignificant, it is stored in LIS; otherwise, it is stored in LIP. After the bth sorting pass, the newly significant nodes with magnitude in the range of $(T_b, T_{b-1}]$ for $b > 1$ (or $(T_1, 2T_1]$ for $b = 1$) will be stored in LSP with one bit per node to code their respective signs. In the refinement pass, the significant nodes are refined with one bit per node to update their respective magnitudes. The basic idea behind SPIHT is as follows: If a parent node is insignificant, all the descendants are likely to be insignificant and therefore can be efficiently coded by a single symbol: ZT.

10.4 Adaptive Wavelet Packet Image Coding

In the framework of SPIHT with wavelets, the spatially related wavelet coefficients are organized into hierarchical quad-trees. Typically, most of images' energy is concentrated in the lower frequency subbands. For images with textures, however, many significant coefficients are found in the high frequency subbands. Thus, wavelet packets seem to be more suitable for coding these images.

10.4.1 Rearrangement of Wavelet Packet Coefficients

In wavelet domain, it is straightforward to construct hierarchical trees because of the pyramid structure of WT. As the construction of wavelet trees is based on the spatial position of wavelet coefficients, a spatial position based rearrangement scheme has been proposed to organize wavelet packet coefficients into hierarchical trees [24]. Figure 10.3 depicts the block diagram, in which a sequence of wavelet coefficients is decomposed into two sub-sequences of wavelet packet coefficients, and then multiplexed into a single sequence by interleaving one with the other.

For image applications, a separable 2-D rearrangement can be obtained by the tensor product of 1-D rearrangements. More specifically, wavelet coefficients are first decomposed into wavelet packet coefficients, and then rearranged by using the 1-D rearrangement scheme horizontally followed by vertically, or vice versa. Figure 10.4 shows the differences between the Sprljan's algorithm [23] and the rearrangement scheme [24] in terms of parent-child assignments. As shown in

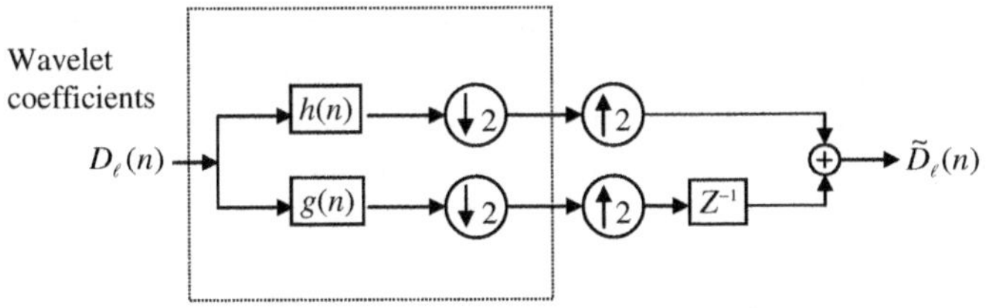

Fig. 10.3 Rearrangement of 1-D wavelet packet coefficients [24]

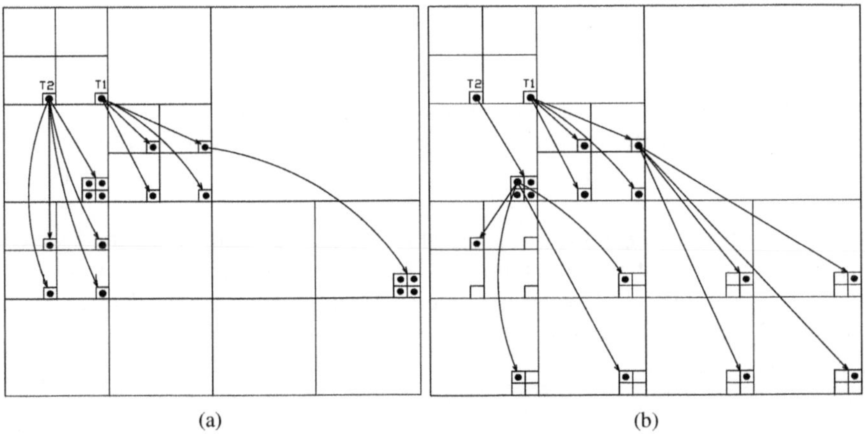

Fig. 10.4 Differences of parent-child assignments using (**a**) the algorithm proposed by Sprljan et al. [23] and (**b**) the rearrangement scheme [24]

Fig. 10.4(a), the wavelet packet tree: T2 obtained by using the Sprljan's algorithm is not a quad-tree, and therefore can not be coded by using the SPIHT algorithm directly. In contrast, the wavelet packet tree: T2 (shown in Fig. 10.4(b)) obtained by using the rearrangement scheme is still a quad-tree, which can be efficiently coded by using SPIHT at no extra cost of coding complexity.

10.4.2 Quad-Tree Based Adaptive Wavelet Packet Tree

In SPIHT, if a zero-tree becomes significant, each offspring node (of the tree) is coded as a single pixel, but the non-offspring nodes are expected to be insignificant and then coded in a group manner. Thus, the desirable quad-trees with adaptive wavelet packets should be the one with as many insignificant non-offspring nodes as possible. Specifically, after ℓ-level WT, an image is decomposed into multiresolution wavelet subbands (from the coarsest resolution ℓ to the finest resolution 1). For each wavelet subband at resolution k ($\leq \ell - 2$), the wavelet packet coefficients obtained by p-level WPT are rearranged; notice that wavelet packet coefficients are actually wavelet coefficients while$p = 0$. Together with the subbands at coarser resolutions: $k+1$ and $k+2$, a collection of 3-level quad-trees can be obtained, in which the rearranged wavelet packet coefficients (obtained from the wavelet subband at resolution$k$) are the leaf nodes. If any descendant nodes of a tree become significant, it is called a broken tree. For computational simplicity, the following measure indicating a weighted number of broken (3-level) quad-trees due to leaf nodes is used.

$$S_p = \sum_{t \in B_p} N(t) \cdot Sig(t) \tag{10.7}$$

where B_p denotes the set of the constructed wavelet packet trees with leaf nodes obtained by p-level WPT, $Sig(t) = 1$ if t is a broken tree; otherwise $Sig(t) = 0$, and $N(t)$ is the number of significant groups of 4 neighboring leaf nodes in a broken tree: t.

In view of zero-tree coding, it is desirable that the sum of the measures over each individual wavelet subband of an image is minimized. For the sake of simplicity, an efficient algorithm is used to construct hierarchical quad-trees with adaptive wavelet packets. The pseudo-code is as follows:

Step 1: Decompose the input image using ℓ-level WT. Set $k = \ell - 2$.
Step 2: For each wavelet subband at resolution k, set $p = 1$.
Loop:
  Compute the wavelet packet coefficients using p-level WPT.
  Rearrange these wavelet packet coefficients.
  Together with the subbands (of the same orientation) at resolutions: $k+1$ and $k+2$, construct 3-level quad-trees, and compute S_p (defined in equation (10.7))
  If $\frac{S_{p-1} - S_p}{S_{p-1}} \times 100\% < T_r$, exit the loop; otherwise increase p by 1, and continue the loop until $p = \ell - k$.
Step 3: Decrease k by 1. If $k \geq 1$, go to step 2.

As the numbers of the low frequency wavelet coefficients at resolutions ℓ and $\ell - 1$ are quite small, no further decomposition is performed. Recall that wavelet packet coefficients with $p = 0$ are actually wavelet coefficients. Figure 10.5 shows flowchart of the quad-tree based adaptive wavelet packet decomposition (AWPD) for zero-tree image coding.

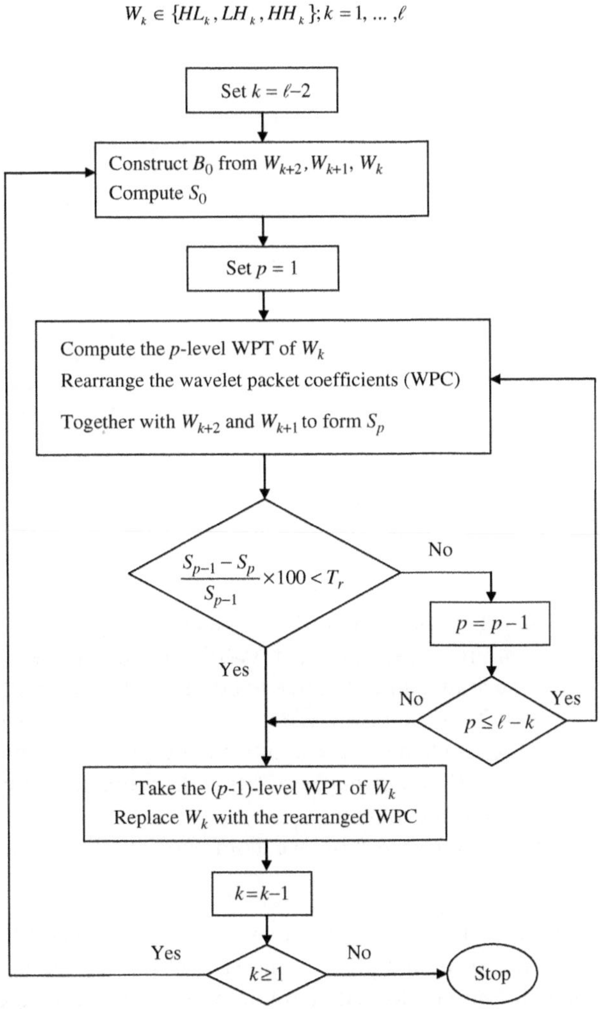

Fig. 10.5 Flowchart of the quad-tree based adaptive wavelet packet decomposition (AWPD) algorithm

10.4.3 The SPIAWPT Algorithm

As the adaptive wavelet packet trees (AWPT) of an image are always quad-trees, an efficient image coder can be obtained by applying SPIHT to AWPT, which is called set partitioning in adaptive wavelet packet trees (SPIAWPT). The following is SPIAWPT presented in steps:

Fig. 10.6 Block diagram of the SPIAWPT algorithm

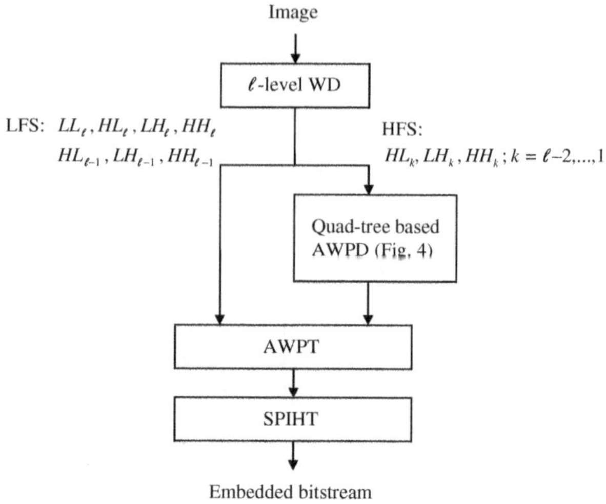

Step 1: Initialization: Decompose an image using the AWPD algorithm (shown in Fig. 10.5), construct AWPT, and determine an initial threshold: T_b; $b = 1$ such that all of the tree nodes are in the range of $[-2T_1, 2T_1]$.

Step 2: Sorting pass: Identify significant tree nodes in comparison with T_b. If a tree node becomes significant, output its sign bit.

Step 3: Refinement pass: Output the refinement bits of the significant tree nodes found previously.

Step 4: Increase b by one, compute$T_b = T_{b-1}/2$, and go to step 2.

Figure 10.6 shows the block diagram. To reconstruct the image, the received code stream is first parsed, the wavelet packet coefficients are rearranged back to the conventional format, and finally the decoded image can be obtained by inverse wavelet packet transform [24].

10.5 Experiments of the Coding Algorithms

The coding algorithms have been evaluated on grayscale images. The coding results of images namely Barbara, Earth, and Fingerprint are presented in this chapter. The original images with the subband decompositions obtained by using AWPD are shown in Figs. 10.7, 10.8 and 10.9, respectively. The bi-orthogonal 9/7 wavelet is used. The number of decomposition levels is 5. The threshold: T_r is set to 2. The compression rate is measured in bits per pixel (bpp). The distortion defined by the

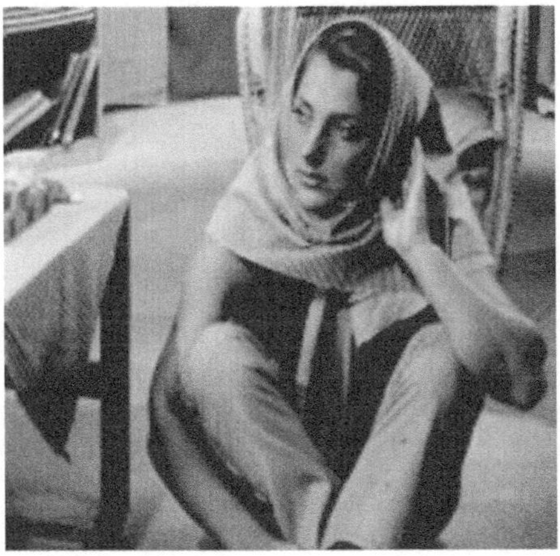

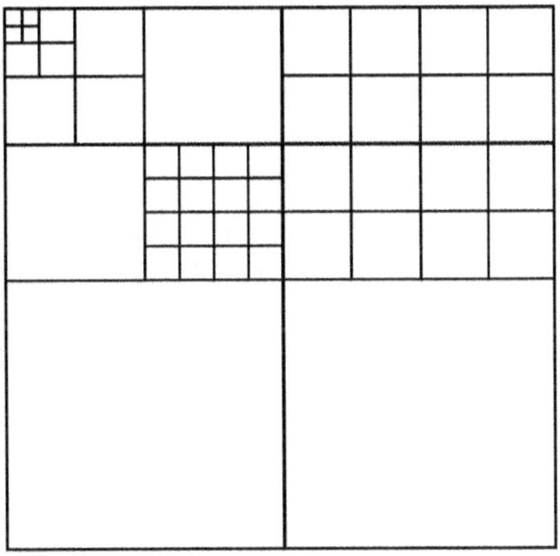

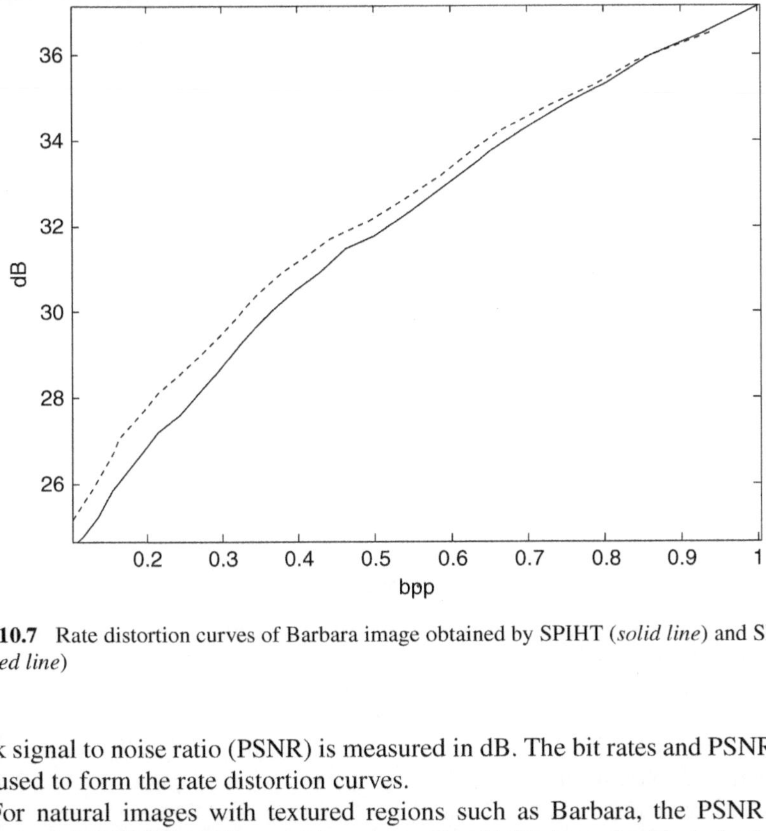

Fig. 10.7 Rate distortion curves of Barbara image obtained by SPIHT (*solid line*) and SPIAWPT (*dotted line*)

peak signal to noise ratio (PSNR) is measured in dB. The bit rates and PSNR values are used to form the rate distortion curves.

For natural images with textured regions such as Barbara, the PSNR performance of SPIAWPT (without arithmetic coding (AC)) is preferable to both SPIHT (without AC) and CZQ-WP (without AC) at low bit rates (< 0.5 bpp) as shown in Table 10.1; SPIAWPT is comparable to SPIHT but outperforms CZQ-WP at high bit rates (> 0.6).

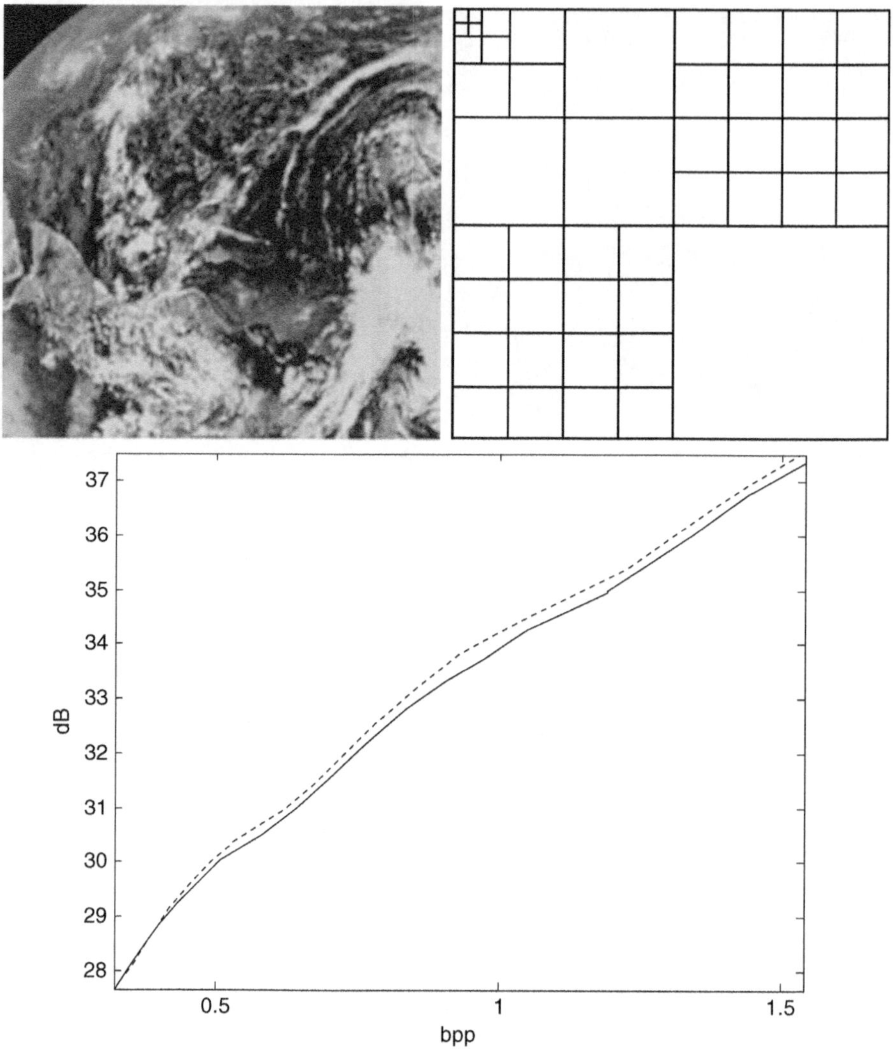

Fig. 10.8 Rate distortion curves of Earth image obtained by SPIHT (*solid line*) and SPIAWPT (*dotted line*)

Figures 10.7, 10.8 and 10.9 show comparisons between SPIAWPT with AC and SPIHT with AC in terms of the rate distortion curves, where the horizontal and vertical axes are the bit rates and PSNR values, respectively. For Barbara image, as there are large areas of high detail textures, lots of middle and high frequency wavelet coefficients will become significant in the early code passes of SPIHT. Figure 10.7 shows that SPIAWPT with AC outperforms SPIHT with AC at low-middle bit rates (<0.8). For the satellite image, Earth, SPIAWPT with AC is preferable to SPIHT with AC at middle-high bit rates (>0.5) as shown in Fig. 10.8.

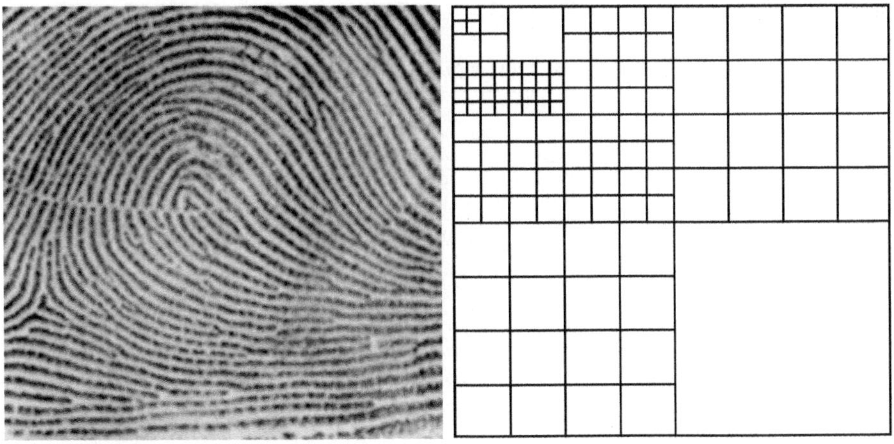

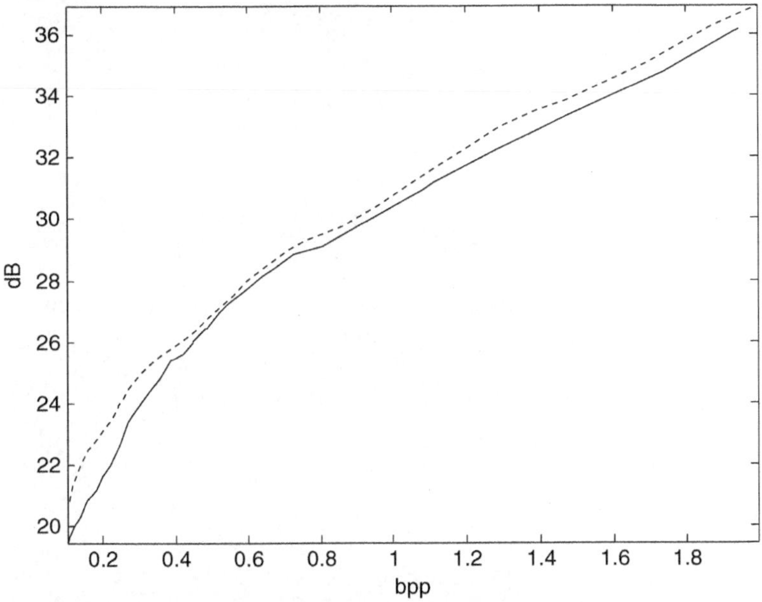

Fig. 10.9 Rate distortion curves of Fingerprint image obtained by SPIHT (*solid line*) and SPIAWPT (*dotted line*)

Fingerprints image compression is an important issue that demands the best solution. Fingerprints images appear semi-regular oscillatory patterns in all directions. As one can expect, further decompositions of wavelet subbands will be needed. We compare SPIAWPT with AC to SPIHT with AC both visually and in terms of the PSNR values. We notice in Fig. 10.9 that a significant improvement can be obtained by using SPIAWPT. Figure 10.10 shows visual quality of the decoded Fingerprint images at bit rates 0.05 and 0.1 bpp.

Table 10.1 PSNR comparison of Barbara image

	PSNR (dB)		
Bit rate (bpp)	SPIHT	CZQ-WP	SPIAWPT
1.0	36.41	36.15	36.26
0.8	34.66	33.91	34.59
0.7	33.75	33.21	33.76
0.6	32.53	32.56	32.70
0.5	31.40	31.60	31.59
0.4	30.10	29.85	30.43
0.3	28.56	28.77	28.91
0.25	27.58	28.12	28.14
0.2	26.65	26.64	27.14
0.1	24.26	24.27	24.78

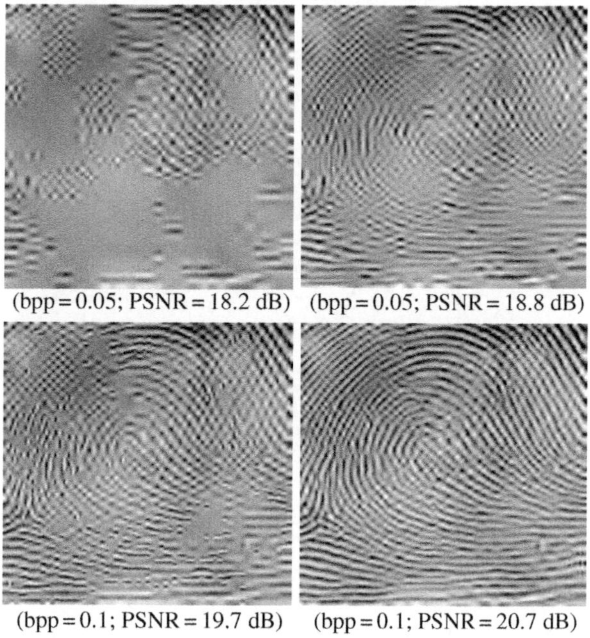

Fig. 10.10 Visual quality of the decoded Fingerprint images obtained by SPIHT (*left column*) and SPIAWPT (*right column*)

10.6 Conclusion

Wavelet transform has drawn a lot of attention to image compression. In SPIHT, the wavelet trees of an image can be efficiently coded by exploiting the self similarity property across wavelet subbands. However, for texture-rich images, many middle and high frequency wavelet coefficients are likely to become significant after a few code passes of SPIHT. Thus, it is desirable that each individual wavelet subband can be further decomposed by wavelet packet transform. To incorporate with the framework of SPIHT, the quad-tree based adaptive wavelet packet decomposition (AWPD) algorithm can be used to construct adaptive wavelet packet trees (AWPT), which can be efficiently coded by using SPIHT at no extra cost of coding complexity. Experimental results show that set partitioning in adaptive wavelet packet trees (SPIAWPT) is superior to set partitioning in wavelet trees, especially for fingerprints images.

References

1. H. G. Musmann, P. Pirsch, and H. J. Grallert, "Advances in Picture Coding," Proc. IEEE, vol. 73, pp. 523–548, April, 1985.
2. R. J. Clarke, Transform Coding of Images, New York: Academic Press, 1985.
3. O. J. Kwon and Rama Chellappa, "Region Adaptive Subband Image Coding," IEEE Trans. On Image Processing, vol. 7, no. 5, pp. 632–648, May, 1988.
4. K. R. Rao and J. J. Hwang, Techniques and Standards for Image, Video and Audio Coding, Englewood Cliffs: Prentice Hall, 1996.
5. W. B. Pennebaker and J. L. Mitchell, JPEG Still Image Data Compression Standards, New York: Van Nostrand, 1993.
6. M. Antonini, M. Barlaud, P. Mathieu, and I. Daubechies, "Image Coding Using Wavelet Transform," IEEE Trans. On Image Processing, vol. 1, pp. 205–220, April, 1992.
7. G. Strang and T. Nguyen, "Wavelets and Filter Banks," Wellesley-Cambridge, Wellesley, MA., USA, 1996.
8. J. M. Shapiro, "Embedded Image Coding Using Zero-Trees of Wavelet Coefficients," IEEE Trans. On Signal Processing, vol. 40, pp. 3445–3462, 1993.
9. A. Said and W. A. Pearlman, "A New, Fast, and Efficient Image Codec Based on Set Partitioning in Hierarchical Trees," IEEE Trans. On Circuits Syst. Video Tech. vol. 6, pp. 243–250, 1996.
10. D. Mukherjee and S. K. Mitra, "Vector SPIHT for Embedded Wavelet Video and Image Coding," IEEE Trans. On Circuits Syst. Video Tech. vol. 13, pp. 231–246, March, 2003.
11. A. Said and W. A. Pearlman, "Low Complexity Waveform Coding via Alphabet and Sample-Set Partitioning," Proc. SPIE Visual Communications and Image Processing, vol. 3024, pp. 25–37, Feb., 1997.
12. J. Andrew, "A Simple and Efficient Hierarchical Image Coder," Proc. IEEE Int. Conf. Image Processing (ICIP), vol. 3, pp. 658–661, Oct., 1997.
13. W. A. Pearlman, A. Islam, N. Nagaraj, and A. Said, "Efficient, Low Complexity Image Coding With a Set-Partitioning Embedded Block Coder," IEEE Trans. On Circuits Syst. Video Tech. vol. 14, pp. 1219–1235, Nov., 2004.
14. D. Taubman, "High Performance Scalable Image Compression with EBCOT," IEEE Trans. On Image Processing, vol. 9, pp. 1158–1170, July, 2000.
15. A. Skodras, C. Christopoulos, and T. Ebrahimi, "The JPEG 2000 still image compression standard," IEEE Signal Process. Mag., vol. 18, pp. 36–58, September, 2001.
16. H.-C. Fang, Y.-W. Chang, T.-C. Wang, C.-T. Huang, and L.-G. Chen, "High-Performance JPEG 2000 Encoder with Rate-Distortion Optimization," IEEE Trans. On Multimedia, vol. 8, no. 4, pp. 645–653, August. 2006.
17. F. G. Meyer, A. Z. Averbuch, and J.-O. Stromberg," Fast Adaptive Wavelet Packet Image Compression," IEEE Trans. Image Processing, vol. 9, pp. 792–800, 2000.
18. D. Engle, A. Uhl, "Adaptive Object Based Image Compression Using Wavelet Packets," VIPromCom-2002, 4th EURASIP, IEEE Region 8 International Symposium on Video/Image Processing and Multimedia Communications, pp. 183–187, 2002.

19. N. M. Rajpoot, R. G. Wilson, F. G. Meyer and R. R. Coifman, "Adaptive Wavelet Packet Basis Selection for Zerotree Image Coding," IEEE Trans. On Image Processing, Vol. 12, pp. 1460–1472, 2003.
20. K. Ramchandran and M. Vetterli "Best Wavelet Packet Bases in A Rate Distortion Sense," IEEE Trans. On Image Processing, vol. 2, pp. 160–175, 1993.
21. Z. Xiong, K. Ramchandran, and M. Orchard, "Wavelet Packet Image Coding Using Space-Frequency Quantization," IEEE Trans. On Image Processing, vol. 7, pp. 892–898, 1998.
22. N. M. Rajpoot, R. G. Wilson, F. G. Meyer and R. R. Coifman, "Adaptive Wavelet Packet Basis Selection for Zerotree Image Coding," IEEE Trans. On Image Processing, vol. 12, pp. 1460–1472, 2003.
23. N. Sprljan, S. Grgic and M. Grgic, "Modified SPIHT Algorithm for Wavelet Packet Image Coding," Real-Time Imaging, vol. 11, pp. 378–388, 2005.
24. T.-Y. Sung and H.-C. Hsin, "An Efficient Rearrangement of Wavelet Packet Coefficients for Embedded Image Coding Based on SPIHT Algorithm," IEICE Trans. Fundamentals, vol. E90-A, no. 9, pp. 2014–2020, 2007.
25. H.-C. Hsin and T.-Y. Sung, "Image Coding with Adaptive Wavelet Packet Trees," IMECS 2008, Hong Kong, (The Certificate of Merit).

Chapter 11
Plaque Boundary Extraction in Intravascular Ultrasound Image in Conjunction with Image Separability and Fuzzy Inference

Ryosuke Kubota, Shohei Ichiyama, Noriaki Suetake, Eiji Uchino, Genta Hashimoto, Takafumi Hiro and Masunori Matsuzaki

Abstract In the conventional boundary extraction methods, the medical doctors have to put the several seed points on the IVUS image. However, those methods need some iterative processes and take big computing time. To meet with the practical demand in clinic, a quick boundary extraction by a small number of seed points is definitely necessary. In this paper, to cope with those problems, we propose a new fuzzy rule-based boundary extraction method, which achieves a fast, precise and robust extraction of plaque. In the proposed method, a boundary to be extracted is piecewise approximated by polynomial series inferred by fuzzy rules. The proposed method employs not only the information of the seed points itself but also the separability information of the IVUS image. The validity and the effectiveness of the proposed method have been verified by applying it to the practical problem of plaque boundary extraction in clinic.

Keywords Plaque boundary extraction · Intravascular ultrasound · Fuzzy rule-based · Fuzzy inference · Image separability

11.1 Introduction

A backscattered intravascular ultrasound (IVUS) method is one of the tomographic imaging techniques [1, 2]. The IVUS method, which provides a real-time cross sectional image of a coronary artery in vivo, is employed for visualization of an atherosclerotic plaque [3–5] for a diagnosis of the acute coronary syndromes (ACS) [6]. In the diagnosis of ACS, a precise boundary extraction of a plaque is required. However, it's image is very grainy due to speckle noise.

R. Kubota (✉)
Ube National College of Technology, Department of Intelligent System Engineering, Ube, Yamaguchi 755-8555, Japan
e-mail: kubota@ube-k.ac.jp

S.-I. Ao et al. (eds.), *Trends in Communication Technologies and Engineering Science*, Lecture Notes in Electrical Engineering 33, DOI 10.1007/978-1-4020-9532-0_11,

Medical doctors are evaluating manually the area of plaque in hundreds of images. The extraction of plaque boundary involves a great deal of task and time for the medical doctors. Therefore, an automatic precise boundary extraction of plaque is strongly required.

In the conventional boundary extraction methods, the medical doctors have to put the several seed points on the IVUS image. Those points are then interpolated smoothly by a spline function [7, 8]. These methods can reduce slightly the workload of medical doctors for plaque boundary extraction task. However, its precision is considerably affected by a number of the seed points and/or a distance between those points. For a precise boundary extraction, a lot of seed points with even interval are required. On the other hand, the methods which are based on automatic search algorithms, e.g., an active contour (snake) model, genetic algorithm, and so on, have been proposed so far [9–11]. However, those methods need some iterative processes and take big computing time. To meet with the practical demand in clinic, a quick boundary extraction by a small number of seed points is definitely necessary.

In this paper, to cope with those problems, we propose a new fuzzy rule-based boundary extraction method, which achieves a fast, precise and robust extraction of plaque. In the proposed method, a boundary to be extracted is piecewise approximated by polynomial series inferred by fuzzy rules [12, 13]. The series coefficients are determined by the weighted least square method using the separability which is a kind of statistical discriminant measure for the edge extraction of image [14].

The proposed method employs not only the information of the seed points itself but also the separability information of the IVUS image. The proposed method does not include any time-consuming iterative processes as in the conventional methods, e.g., snake model.

The validity and the effectiveness of the proposed method have been verified by applying it to the practical problem of plaque boundary extraction in clinic.

11.2 Intravascular Ultrasound (IVUS) Image and Conventional Boundary Extraction Method

11.2.1 IVUS Image

The IVUS method uses an ultrasound probe mounted on a catheter as shown in Fig. 11.1. The radiofrequency (RF) signal is emitted from the probe. The transmitting frequency is 40 MHz. The probe rotates at the tip of the catheter inserted in the arterial lumen. The probe also receives an RF signal reflected from the plaque tissue and the vascular wall. The RF signal is sampled at 200 or 400 MHz.

Figure 11.2 shows a sample of the IVUS image. The IVUS image is constructed by the RF signal. The sampled RF signal is transformed into intensity, and the intensities in all radial directions are formed as a tomographic cross-sectional image of a coronary artery as shown in Fig. 11.2. This image is called a "B-mode image."

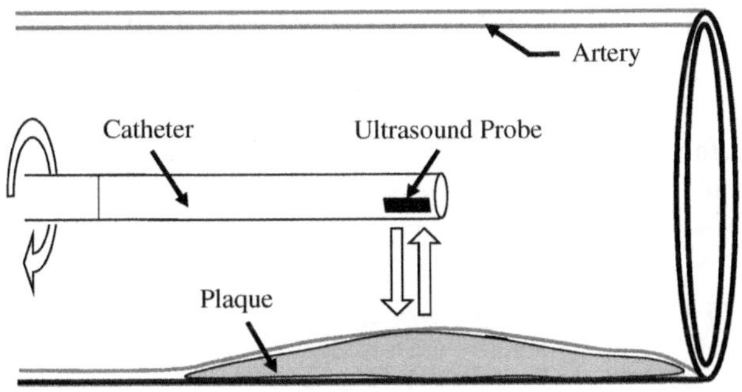

Fig. 11.1 An ultrasound probe mounted on a catheter

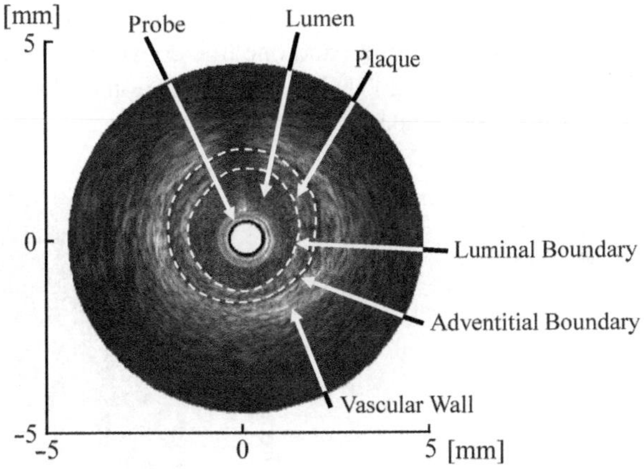

Fig. 11.2 A sample of IVUS B-mode image. The *dotted lines* show luminal and adventitial boundaries, respectively

11.2.2 Conventional Boundary Extraction Method by Using Spline Function

The following two boundaries are extracted. One is a luminal boundary between the lumen and the plaque, and the other is an adventitial boundary between the plaque and the vascular wall as shown in Fig. 11.2.

The conventional methods using spline function need several seed points. The seed points are marked directly on the IVUS image by a medical doctor. Those points are then interpolated smoothly by a parametric spline function [8]. Interpolation is done based on the marked seed points and its marking order. However, its precision is considerably affected by a number of the seed points and/or a distance between those points. Furthermore it is difficult to mark the seed points precisely on the boundary.

11.3 Proposed Boundary Extraction Method Based on Fuzzy Inference

In this paper, we propose a novel fuzzy rule-based boundary extraction method by using a separability measure of the IVUS image.

11.3.1 B-Mode Image

In the present method, the IVUS image of Fig. 11.2 is transformed into the Cartesian system of Fig. 11.3. The horizontal axis corresponds to the angle in clockwise radial direction. The angle in Fig. 11.3 starts from the 3-o'clock direction of the IVUS image of Fig. 11.2. The vertical axis corresponds to the distance from the probe located at the center of the IVUS image of Fig. 11.2. Figure 11.3 consists of 256 lines in radial direction, and one line consists of 1,024 pixels in case where the sampling frequency is 200 MHz. The resolution of distance and angle are 3.91 μm/pixel and 1.41°/line, respectively.

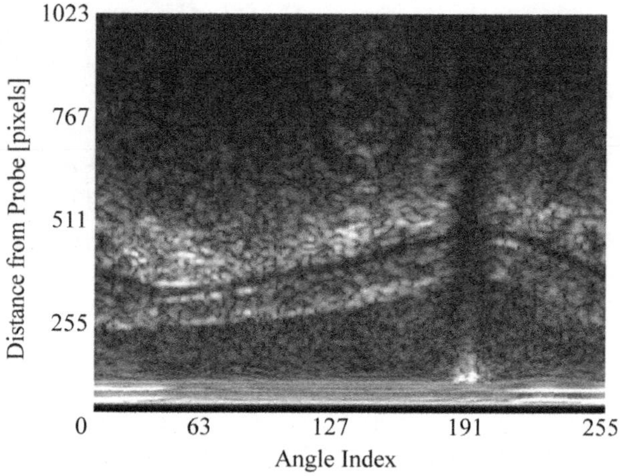

Fig. 11.3 Transformed image of Fig. 11.1 into the Cartesian system

11.3.2 Separability for Boundary Extraction

The present method employs a statistical discriminant measure of separability of image [14]. The separability η_{ij} for the pixel (i, j) of Fig. 11.4(a) is defined by:

$$\eta_{ij} = \frac{n_A(\bar{I}_A - \bar{I}_m)^2 + n_B(\bar{I}_B - \bar{I}_m)^2}{\sum_{k=1}^{S}(I_k - \bar{I}_m)^2}, \tag{11.1}$$

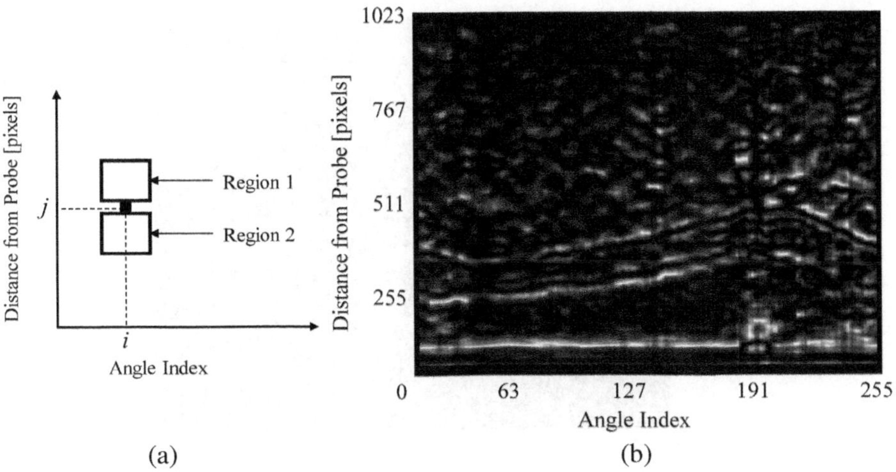

Fig. 11.4 (**a**) Area for evaluating separability. (**b**) Discriminant image of Fig. 11.3 by using separability

where n_A and n_B represent a number of the pixels in regions A and B, respectively. $\bar{I}_A$ and $\bar{I}_B$ are averages of the intensities in each regions. $\bar{I}_m$ stands for the average of the intensities in the combined two regions A and B. S and I_k are a number of the pixels and a intensity of the k-th pixel in the combined regions A and B. η_{ij} is $0 \leq \eta_{ij} \leq 1$, and takes a large value when two regions are separated with each other.

Figure 11.4(b) shows a discriminant image of Fig. 11.3 by using η_{ij}. Each pixel of its image is the separability η_{ij}. When η_{ij} is large, the brightness assigned to its corresponding pixel becomes high. That is, a chain of white pixels with high brightness can then be a candidate of a boundary.

11.3.3 Proposed Boundary Extraction by Applying Fuzzy Inference

In this paper, a plaque boundary is approximated by a piecewise polynomial series inferred by fuzzy rules. The coefficients of its series are adjusted optimally by a weighted least square method (WLSM) using a separability.

In the first step, the seed points given by a medical doctor are linearly interpolated on the Cartesian system to get primitive boundary shown by the solid line in Fig. 11.5(a). The accurate boundary is found by starting from this primitive boundary. The specified boundary search area is also shown in Fig. 11.5(a) by the dotted lines.

The boundary to be extracted is piecewise approximated by the polynomials in the Cartesian system by a series of the following fuzzy if-then rules:

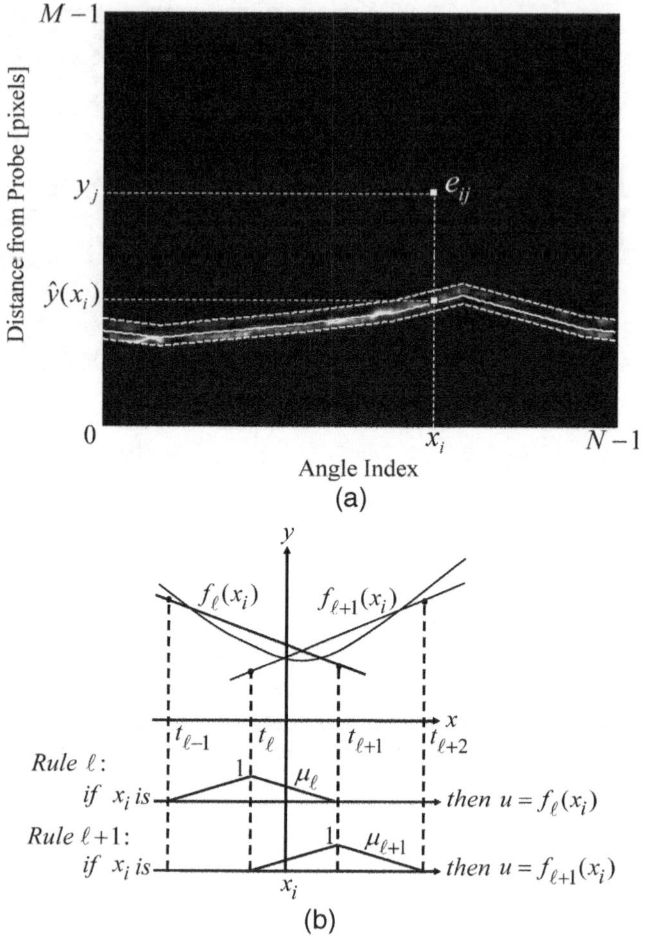

Fig. 11.5 Polynomial assignment by fuzzy rules. (**a**) A primitive boundary and a specified boundary search area. (**b**) Membership functions assigned to each $f_\ell(x_i)$, and the fuzzy rules for the interval of x_i

$$\begin{aligned}
\text{Rule } 1 &: \text{ if} x_i \text{ is } A_1 \text{ then } u = f_1(x_i),\\
&\vdots\\
\text{Rule } \ell &: \text{ if } x_i \text{ is } A_\ell \text{ then } u = f_\ell(x_i),\\
&\vdots\\
\text{Rule } L &: \text{ if} x_i \text{ is } A_L \text{ then } u = f_L(x_i),
\end{aligned}$$

where A_ℓ is a fuzzy set whose membership function is μ_ℓ as shown in Fig. 11.5(b). x_i corresponds to the angle index, and u corresponds to the location of the boundary from the probe in the Cartesian system. $f_\ell(x_i)$ is a polynomial series defined by:

$$f_\ell(x_i) = \sum_{m=0}^{p} a_{(\ell,m)} x_i^m. \tag{11.2}$$

The ℓ-th rule thus stands for the piecewise approximation of boundary by polynomial in the interval $[t_{\ell-1}, t_{\ell+1}]$.

The inferred value $\hat{y}(x_i)$ is given by [13]:

$$\hat{y}(x_i) = \frac{\sum_{\ell=1}^{L} \mu_\ell(x_i) f_\ell(x_i)}{\sum_{\ell=1}^{L} \mu_\ell(x_i)}. \tag{11.3}$$

When the complementary triangular functions are employed for the membership functions as shown in Fig. 11.5(b), Eq. (11.3) is simplified as:

$$\hat{y}(x_i) = \mu_\ell(x_i) f_\ell(x_i) + \mu_{\ell+1}(x_i) f_{\ell+1}(x_i), \tag{11.4}$$

when x_i is in the interval $[t_\ell, t_{\ell+1}]$.

The optimum coefficients $a^*_{(\ell,m)}$ of Eq. (11.2) are determined with use of the weighted least square method (WLSM) so as to minimize the following weighted error function:

$$E = \sum_{j=0}^{M-1} \sum_{i=0}^{N-1} \eta_{ij}^2 e_{ij}^2, \tag{11.5}$$

where e_{ij} is an error defined by:

$$e_{ij} = \{y_j - \hat{y}(x_i)\}. \tag{11.6}$$

In this method η_{ij} only inside the specified search area, enclosed by the dotted lines in Fig. 11.5(a), are used as the weights of WLSM.

The following is a boundary extraction procedure of the present method.

Step 1: The seed points are marked by a medical doctor on the B-mode image. The image with the marked seed points is transformed into the Cartesian system.
Step 2: The primitive boundary is given by a linear interpolation of the seed points. Search area is specified around this primitive boundary.
Step 3: The discriminant image is obtained by using the separability .
Step 4: The objective boundary is inferred by Eq. (11.4) with use of the WLSM of Eq. (11.5).

The present method performs a quick and robust boundary extraction of plaque without any time-consuming iterative processes.

11.4 Experimental Results

The proposed method is applied to the practical boundary extraction problem of plaque. Figure 11.6, which is a clipped and enlarged image of Fig. 11.2, is the IVUS image (Image 1) employed for the experiments. The extraction results by the proposed method are compared with those by the conventional method using the parametric spline-interpolation.

For the boundary extraction in Image 1, the membership functions of the fuzzy rules are assigned with even interval. The luminal and the adventitial boundaries shown by the white lines in Fig. 11.8(a) are independently extracted by the proposed method. The number of the seed points is 3 for the extraction of each boundary. In each boundary extraction, the number of the fuzzy rules is $L = 4$, and the order of the polynomial series is $p = 1$. Figure 11.7 shows the assigned membership functions for the extraction of the luminal boundary in Image 1. The horizontal and vertical sizes of each evaluation region of Fig. 11.4(a) are 7 and 30, respectively.

Figure 11.8(b) and (c) shows the extraction results by each method for the IVUS image of Fig. 11.8(a). Moreover, the proposed method, by employing the separability of image, is robust for the luminal boundary extraction in vivo corrupted by the

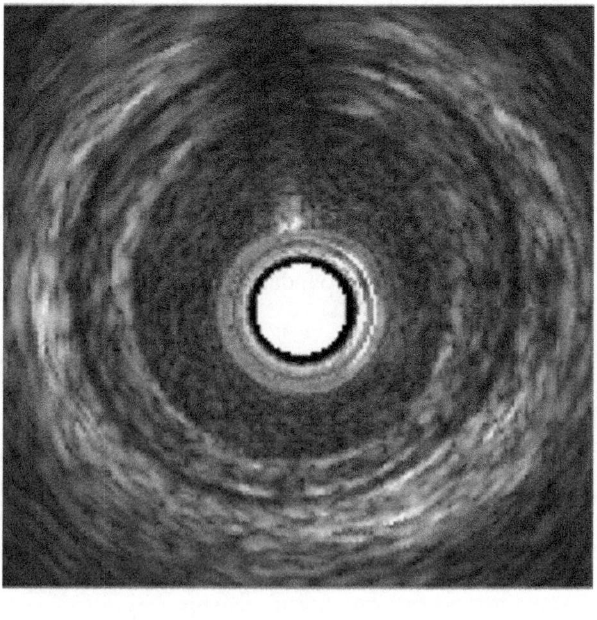

Fig. 11.6 Enlarged IVUS image of Fig. 11.2 (Image 1)

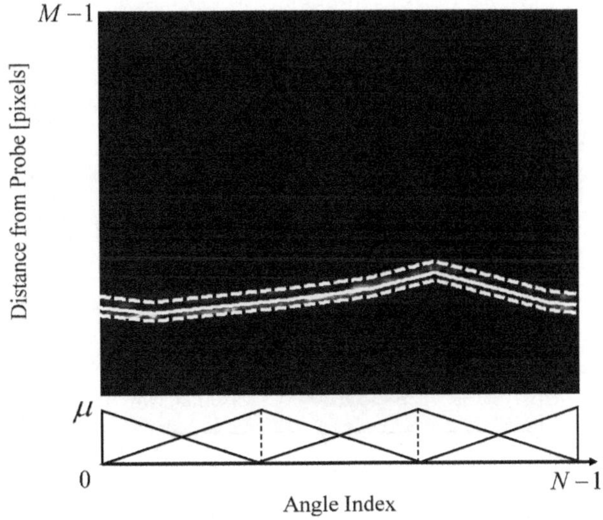

Fig. 11.7 Assigned membership functions for the luminal boundary extraction in Image 1

blood speckles. These results show that the proposed boundary extraction method works very well compared with the conventional method.

The proposed method is also applied to another boundary extraction problem of IVUS image (Image 2) shown in Fig. 11.9. There are two growing plaques in this image. In this case, the extraction of the luminal boundary shown by the dotted lines in Fig. 11.9 is difficult, because the shape of the boundary is complicated.

In this extraction, the number of the assigned membership functions is $L = 11$ for the luminal boundary extraction as shown in Fig. 11.10. The numbers of the seed points for the extraction of the luminal and the adventitial boundaries are 7 and 4. The other parameters of each method are the same as the experiments for Fig. 11.8.

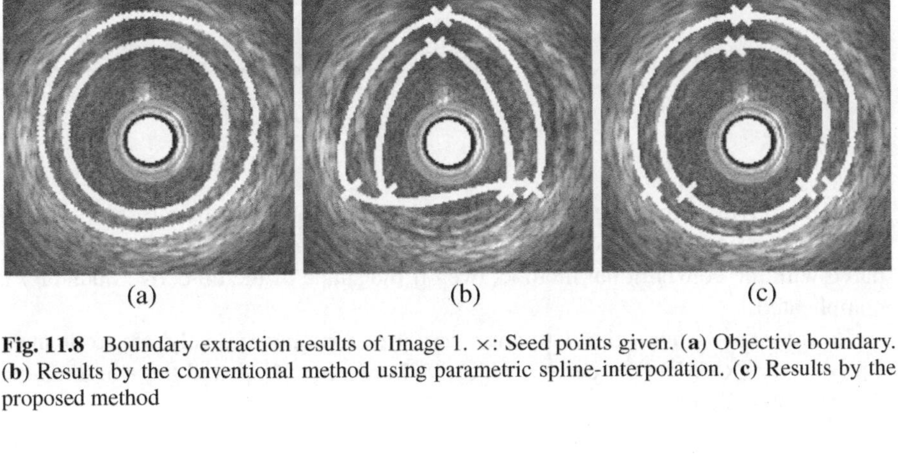

Fig. 11.8 Boundary extraction results of Image 1. ×: Seed points given. (**a**) Objective boundary. (**b**) Results by the conventional method using parametric spline-interpolation. (**c**) Results by the proposed method

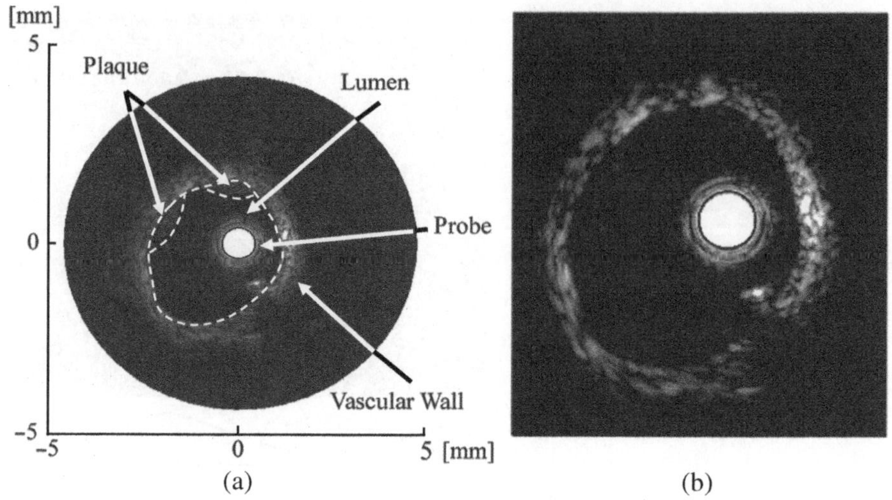

Fig. 11.9 An IVUS image of another coronary artery including the growing plaques. (**a**) B-mode image. (**b**) Enlarged image of (**a**) (Image 2)

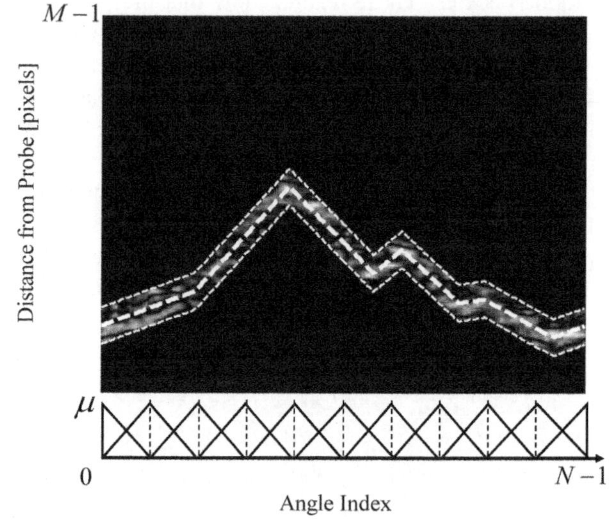

Fig. 11.10 Assigned membership functions for Image 2

Figure 11.11 shows the objective boundary and the extraction results by each method. These results show that the proposed method performs fine extraction compared with the conventional method, even if the shape of the objective boundary is complicated.

In order to evaluate the extraction performance quantitatively, the root mean square error (RMSE) between the objective boundary $y^*(x_i)$ and the extracted one $\hat{y}(x_i)$ is calculated by:

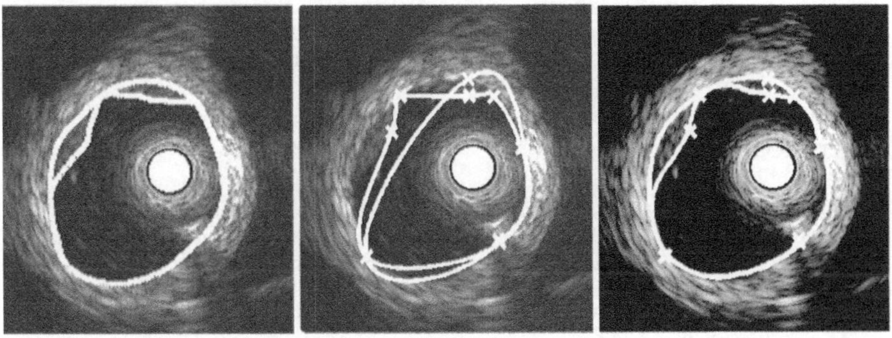

(a) (b) (c)

Fig. 11.11 Boundary extraction results of Image 2. ×: Seed points given. (**a**) Objective boundary. (**b**) Extraction results by the conventional method using parametric spline-interpolation. (**c**) Extraction results by the proposed method

Table 11.1 RMSE for boundary extraction results. One pixel corresponds to 3.91 μm in the actual size

Method	Image	Luminal boundary	Adventitial boundary
Conventional method	1	67.9 (pixels)	96.1 (pixels)
	2	60.4 (pixels)	187.4 (pixels)
Proposed method	1	9.6 (pixels)	13.0 (pixels)
	2	26.5 (pixels)	23.6 (pixels)

$$RMSE = \frac{1}{N}\sum_{i=0}^{N-1} \|y^*(x_i) - \hat{y}(x_i)\|,$$

where N is the number of the angle indexes. The objective boundary is calculated by the parametric spline-interpolation with a lot of seed points.

Table 11.1 shows the RMSE between the objective boundary and the extracted one. It is observed that the proposed method gives better extraction results than the conventional method.

11.5 Conclusion

We have proposed a new fuzzy rule-based boundary extraction method for a plaque, which employs the separability information of the IVUS image. In the proposed method, the objective boundary to be extracted is piecewise approximated by the polynomial series inferred by the fuzzy rules. The present method performs a quick, precise and robust extraction of the boundary of a plaque.

Future work is an automatic assignment of optimum membership functions and the applications of the proposed method to other object extraction problems.

Acknowledgment This work was supported in part by "Knowledge Cluster Initiative" funded for 2004-2009 by Japan Ministry of Education, Culture, Sports, Science and Technology.

References

1. Hodgson, J.B., Graham, S.P., Savakus, A.D., Dame, S.G., Stephens, D.N., Dhillon, P.S., Brands, D., Sheehan, H., Eberle, M.J.: Clinical percutaneous imaging of coronary anatomy using an over-the-wire ultrasound catheter system. Int. J. Cardiac Imaging **4**, 187–193 (1989)
2. Potkin, B.N., Bartorelli, A.L., Gessert, J.M., Neville, R.F., Almagor, Y., Roberts, W.C., Leon, M.B.: Coronary artery imaging with intravascular high-frequency ultrasound. Circulation **81** 1575–1585 (1990)
3. Nissen, S.E., Gurley, J.C., Grines, C.L., Booth, D.C., McClure, R., Berk, M., Fischer, C., DeMaria, A.N.: Intravascular ultrasound assessment of lumen size and wall morphology in normal subjects and patients with coronary artery disease. Circulation **84** 1087–1099 (1991)
4. Tobis, J.M., Mallery, J., Mahon, D., Lehmann, K., Zalesky, P., Griffith, J., Gessert, J., Moriuchi, M., McRae, M., Dwyer, M.L.: Intravascular ultrasound imaging of human coronary arteries in vivo: analysis of tissue characterizations with comparison to in vitro histological specimens. Circulation **83** 913–926 (1991)
5. Nicholls, S.J., Tuzcu, E.M., Sipahi, I., Schoenhagen, P., Nissen, S.E.: Intravascular ultrasound in cardiovascular medicine. Circulation **114** 54–59 (2006)
6. Falk, E., Shah, P.K., Fuster, V.: Coronary plaque disruption. Circulation **92** 657–671 (1995)
7. Ahlberg, J.H., Nilson, E.N., Walsh, J.L.: The Theory of Splines and Their Applications. Academic Press, New York (1967)
8. Ferguson, J.: Multivariable curve interpolation. J. Assoc. Comput. Mach. **11** 221–228 (1967)
9. Sonka, M., Zhang, X., Siebes, M., Bissing, M.S., DeJong, S.C., Collins, S.M., McKay, C.R.: Segmentation of intravascular ultrasound images: A knowledge-based approach. IEEE Trans. Med. Imaging **14** 719–732 (1995)
10. Shekhar, R., Cothren, R.M., Vince, D.G., Chandra, S., Thomas, J.D., Cornhill, J.F.: Three-dimensional segmentation of luminal and adventitial borders in serial intravascular ultrasound images. Comput. Med. Imaging Graph. **23** 299–309 (1999)
11. Klingensmith, J.D., Vince, D.G., Kuban, B.D., Shekhar, R., Tuzcu, E.M., Nissen, S.E., Cornhill, J.F.: Assessment of coronary compensatory enlargement by three-dimensional intravascular ultrasound. Int. J. Cardiac Imaging **16** 87–98 (2000)
12. Takagi, T., Sugeno, M.: Fuzzy identification of systems and its applications to modeling and control. IEEE Trans. Syst., Man Cybernet. **15** 116–132 (1985)
13. Sugeno, M.: Fuzzy Control. Nikkan Kogyo (in Japanese), Tokyo (1988)
14. Fukui K.: Edge extraction method based on separability of image features. IEICE Trans. Inf. Syst. **E78–D** 1533–1538 (1995)

Chapter 12
Dedicated Hardware for Hybrid Evolutionary Computation

Masaya Yoshikawa, Hironori Yamauchi and Hidekazu Terai

Abstract This paper discusses new dedicated hardware architecture for hybrid optimization based on Genetic algorithm (GA) and Simulated Annealing (SA). The proposed hardware achieves high speed processing comparison with software processing. Moreover, the hybrid optimization of GA and SA in the proposed architecture realized searching not only globally but also locally. To keep general purpose, self-control processing by a handshake system is introduced. By introducing this type of handshake system, the proposed hardware can be applied to various combinatorial optimization problems by only changing encoder, decoder, and evaluation circuit. Furthermore, the proposed architecture is flexible for many genetic operations on GA and achieves high speed processing by adopting dedicated hardware. Simulation results prove the effectiveness of the proposed hardware architecture.

Keywords Genetic algorithm · Simulated annealing · Dedicated hardware · High speed processing · Hybrid optimization

12.1 Introduction

Many combinatorial optimization problems are known to be non-deterministic polynomial-time (NP)-hard, and no polynomial-time algorithm is predicted to exist. Therefore, it is impossible to obtain the optimal solution by enumerating all of the combinations of discrete values. However, these problems can be applied to various fields. Even though no actual optimality is guaranteed, many cases are satisfied if a solution with sufficiently high accuracy can be obtained. Simulated Annealing (SA) [1, 2] and Genetic Algorithm (GA) [3,4] are the algorithms that can be used to obtain this type of semi-optimal solution.

M. Yoshikawa (✉)
Meijo university, Nagoya, Japan

S.-I. Ao et al. (eds.), *Trends in Communication Technologies and Engineering Science*, Lecture Notes in Electrical Engineering 33, DOI 10.1007/978-1-4020-9532-0_12,

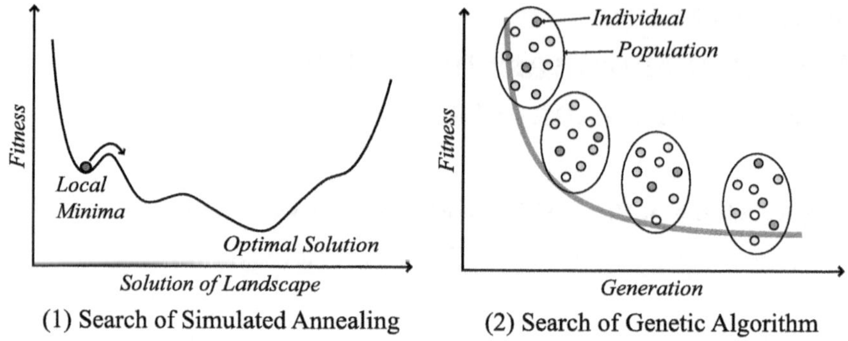

Fig. 12.1 Comparison GA and SA

Although local search methods, represented by SA, are superior in local searching ability, they are inferior in global searching ability as shown in Fig. 12.1 (1). On the other hand, GA, a population based algorithm, is superior in global searching ability, because it generates new search points mainly by a crossover operation. However, it is inferior in systematic search near a good solution as shown in Fig. 12.1 (2).

A hybrid search algorithm, in which such complementary algorithms are combined, would therefore be expected to exhibit excellent solution searching ability. Hence, several hybrid algorithms, in which global search is performed by GA and systematic search near a good solution is performed by SA, have been proposed, and excellent results have been reported [5–7]. However, GA and SA have the inherent problem of substantial processing time, because they require a large number of repetitive operations to find semi-optimal solution. That is, reducing processing time is the most important priority, if GA and SA are applied to practical applications.

In this paper, we propose a dedicated hardware of hybrid optimization based on GA and SA to achieve high speed processing. Moreover, simulation result using benchmark data proves the effectiveness of the proposed hardware.

12.2 Preliminaries

12.2.1 Genetic Algorithm and Simulated Annealing

Genetic Algorithm (GA) was proposed by Holland as an algorithm for probabilistic search, learning, and optimization, and is based in part on the mechanism of biological evolution and Darwin's theory of evolution. It repeatedly applies three genetic operations of selection, crossover, and mutation to a population.

On the other hand, Simulated Annealing (SA) is an algorithm developed based on technological modeling of the annealing process, i.e., a process in which a substance is cooled and its state transformed from the dissolved state to the crystallized state. SA essentially performs a sequential local search. When a solution is improved by the local search, SA accepts the improved solution. Even if a solution is deteriorated, SA accepts the deteriorated solution at the probability shown in equation (12.1).

$$\begin{cases} P = \exp\left[-\frac{\Delta E}{T}\right] & (\Delta E > 0) \\ P = 1 & (\Delta E \leq 0) \end{cases} \tag{12.1}$$

Where, delta E represents the difference of energy, and T represents temperature. In SA, the acceptance of the deteriorated solution enables to escape from local optimal solution.

12.2.2 Previous Work

Examples of dedicated GA hardware to reduce processing time have been reported by Scott et al. [8], Graham et al. [9], and Imai et al. [10]. Scott et al. developed a hardware-based GA and demonstrated its superiority to software in speed and solution quality. Imai et al. proposed a processor element constituting parallel GA, and achieved the parallelism due to the number effect.

However, most of these previous works deal with small-scaled problems and limit to some fixed genetic operations. Thus, no studies have been performed to develop a dedicated hardware to realize hybrid searches that combine both GA and SA.

12.3 Architecture Methodology

The proposed hardware has several merits: (1) realization of high general-purpose properties, (2) introduction of a hardware-oriented hybrid algorithm, and (3) adoption of an architecture combining GA and SA.

Regarding (1), in the proposed hardware, self-control processing by a handshake system is introduced. In this self-control processing, each circuit has request signal and acknowledge signal as enable signal. By introducing this type of handshake system, the timing design of each circuit becomes independent, and consequently, high general-purpose properties can be realized. Thus, the proposed hardware can be applied to various combinatorial optimization problems by only changing encoder, decoder, and evaluation circuit as shown in Fig. 12.2. Regarding genetic operations, roulette wheel selection, rank selection, and tournament selection are realized as the selection operator, and swap operation and inversion operation are realized as the mutation operator, in order to realize general-purpose properties similar to software processing. Consequently, the evolution strategy most suitable for various problems can be selected.

Regarding (2), a new SA is introduced, in which calculation cost is reduced maintaining the accuracy of the solution. The proposed algorithm adopts a new hardware-oriented calculation instead of the calculation of exponential function which is used in conventional SA. Thus, the proposed hardware achieves not only high speed processing, but also reduction of calculation steps.

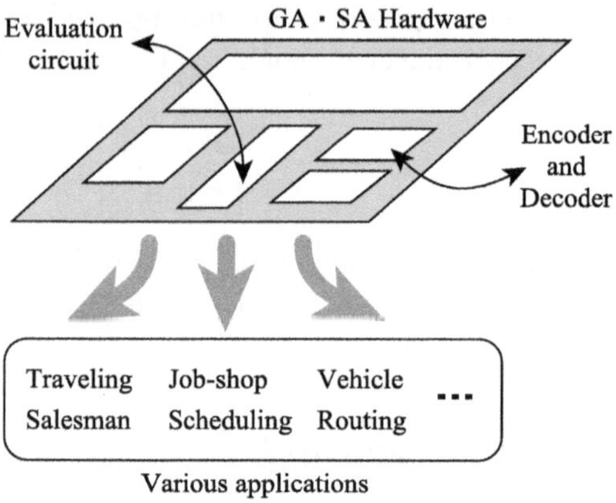

Fig. 12.2 Flexibility of the proposed hardware architecture

Regarding (3), the proposed hardware adopts a combined architecture which shares a circuit to perform local search processing of SA and a circuit to perform mutation processing of GA. By adopting this new architecture, the circuit scale can be reduced. Figure 12.3 shows the processing flow of the proposed hardware.

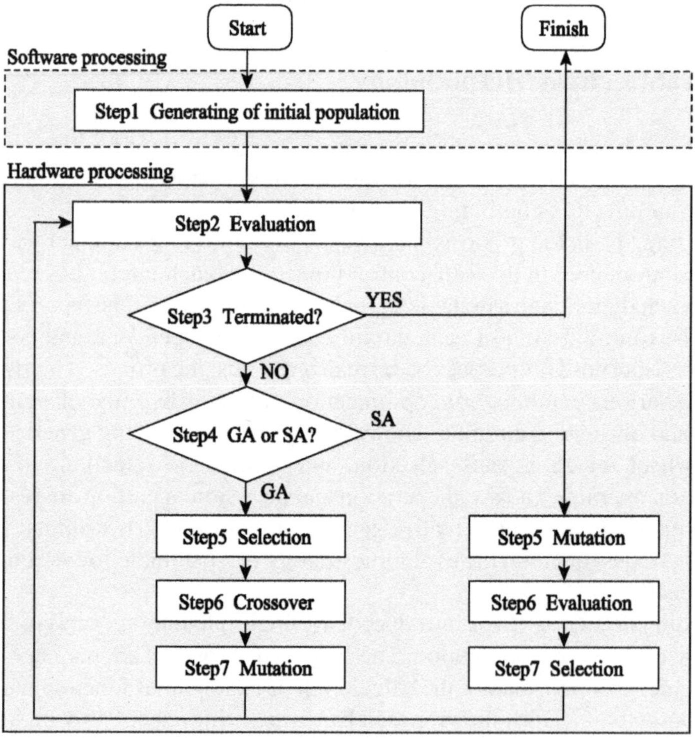

Fig. 12.3 Flow of the hybrid optimization procedure

12.4 Circuits

12.4.1 Hardware of Genetic Algorithm

12.4.1.1 Selection Operator

Regarding the selection operation, the method to implement the roulette wheel selection in the hardware is described here.

The rank selection can be realized by changing the selection probability of the roulette wheel selection. The tournament selection can be realized using a random number generator and a comparator of fitness (evaluation value).

The roulette wheel selection uses the ratio of each individual's evaluation value to the sum of all individual's evaluation values as each individual's selection probability. If the evaluation value of the *i*-th individual is denoted as $f(i)$, a roulette wheel of which the selection probability of the each individual correspond to the angle of the fan-shaped part is required as shown in Fig. 12.4 (1).

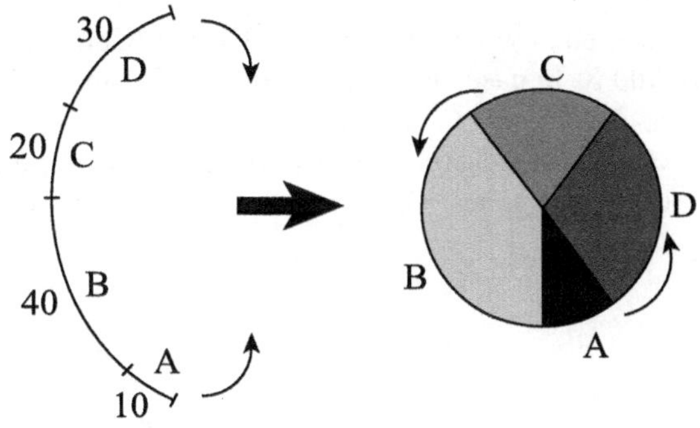

(1) Roulette wheel

Individual	A	B	C	D
Fitness	10	40	20	30

(2) Fitness of each individual

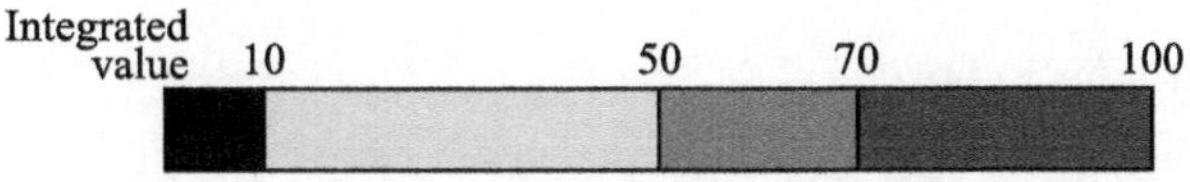

(3) Calculation of integrated value

Fig. 12.4 Example of roulette wheel selection in hardware

Consider the example with the evaluation value of each individual as shown in Fig. 12.4 (2). First, the integrated value of the individual's evaluation values is calculated in alphabetical order, as shown in Fig. 12.4 (3). Next, random numbers in the range between zero and the integrated value are generated. In the case of Fig. 12.4 (3), when the generated random numbers are between 0 and 9, individual "A" will be selected. When the generated random numbers are between 10 and 49, individual "B" will be selected.

12.4.1.2 Crossover Operator

In this paper, the proposed hardware targeted Traveling Salesman Problem (TSP) as an example. Regarding TSP, given a set of n cities and the travel cost between each pair of cities (usually distance), the salesman is to visit each once, and finally return to the start city. Therefore, a solution is the order of visiting the cities. Figure 12.5 shows an example of coding in TSP.

Here, if one-point crossover and multi-point crossover are adopted in this coding, it generates many lethal genes. Figure 12.6 shows an example of the one-point crossover. And Fig. 12.7 shows the tours of offspring generated by the crossover. Regarding one offspring (Child X), genes of "B" and "C" are overlapped, and conversely genes of "D" and "F" are lacked in this example. The overlap and lack indicate lethal genes. The other offspring (Child Y) is also the lethal genes as well as Child X. In case of Child Y, genes of "D" and "F" are overlapped, and conversely genes of "B" and "C" are lacked.

As a result, several crossover operators were proposed in order to prevent generating lethal genes. Grefenstette et al. [11] proposed a coding technique using a

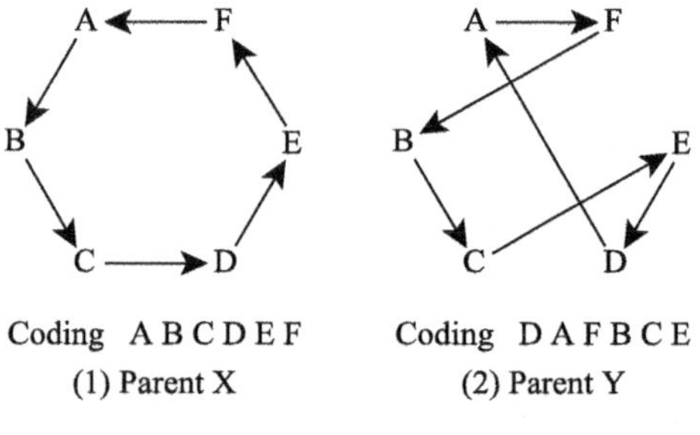

Fig. 12.5 Example of coding in TSP

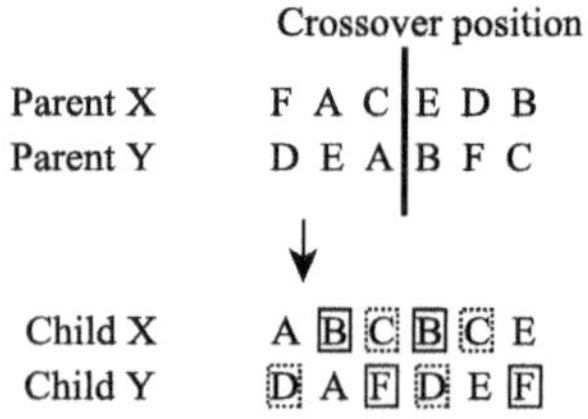

Fig. 12.6 Example of one-point crossover

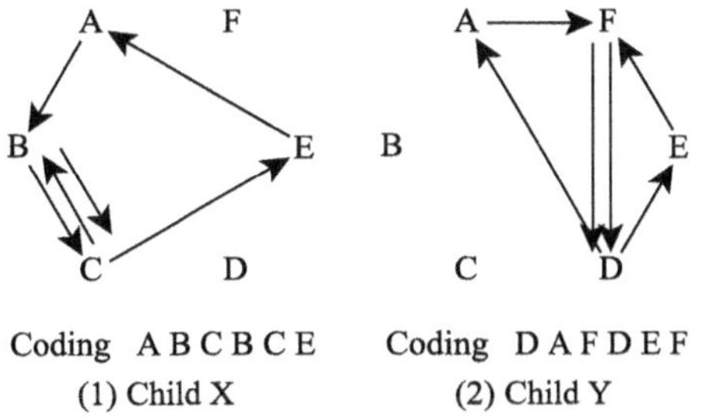

Fig. 12.7 Offspring's tours

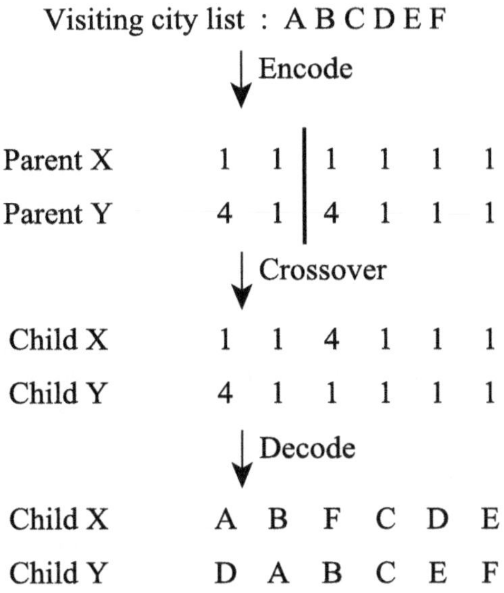

Fig. 12.8 Example of Grefenstette's approach

visiting city list which is created as preprocessing. An individual is represented by the order of the list instead of city name. Thus, it can prevent generating lethal genes if one-point crossover and multi-point crossover are adopted. Figure 12.8 shows an example of the Grefenstette's approach.

12.4.1.3 Mutation Operator

Inversion operation and swap operation are realized as the mutation operator, and each mutation technique is shown in Fig. 12.9. The inversion operation performs a role as the global search, and the swap operation performs a role as the local search.

In inversion operation, two loci are randomly selected using a random number generator and the order of the gene sequence between the loci is reversed. By this operation, the number of patterns (a pair of gene and locus) that do not exist in the

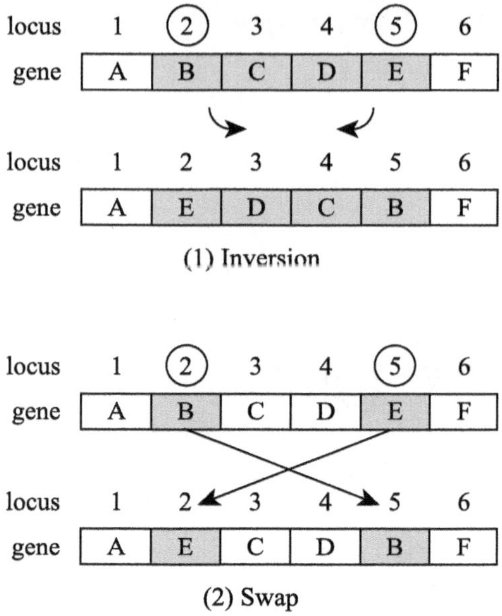

Fig. 12.9 Mutation operators

initial population can be drastically increased, and consequently, extension effect of the search space can be expected as shown in Fig. 12.9 (1).

On the other hand, swap operation can be realized by swapping genes at two arbitrary loci. By this operation, a gene that has become extinct during the evolution process may be recovered without a large change in a genotype as shown in Fig. 12.9 (2).

Thus, the proposed hardware achieves global search and local search efficiently by implementing these two mutation operations.

12.4.2 Hardware of Simulated Annealing

In a conventional SA, the calculation of exponential function is required for the acceptance criterion, as shown in the equation (12.1).

However, this calculation requires not only a very large number of calculation processing steps, but also a lot of hardware resources. Therefore, a new hardware-oriented SA algorithm is introduced in the proposed hardware. The equation for the acceptance criterion of the proposed hardware-oriented SA algorithm is as follows.

$$\begin{cases} P = \frac{1}{X+1} \\ P = \frac{\Delta E}{T} \end{cases} \tag{12.2}$$

As a result, the calculation processing step and the hardware resources can be reduced by adopting the new SA.

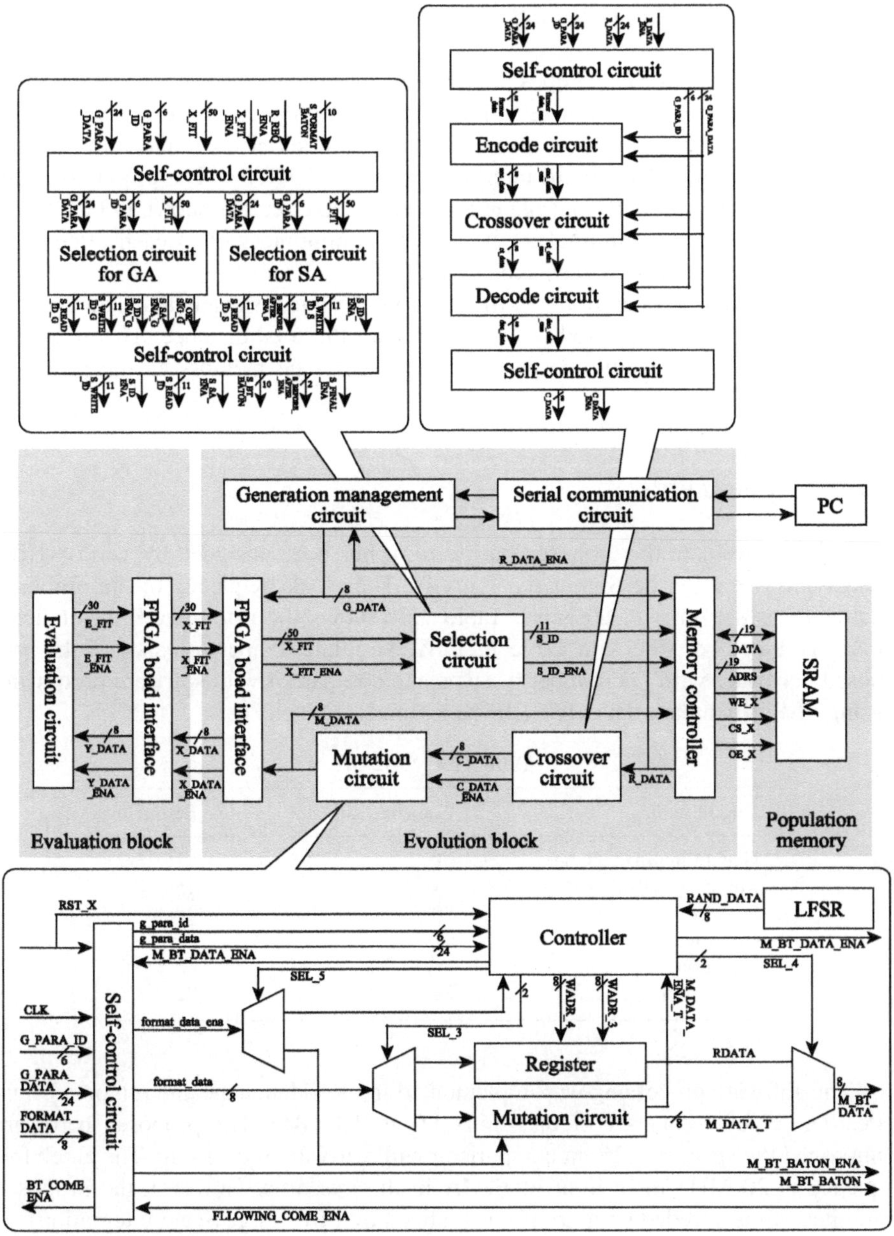

Fig. 12.10 Block diagram

12.4.3 GA and SA Architecture

The block diagram of the proposed architecture is shown in Fig. 12.10. The proposed architecture consists of selection circuit, mutation circuit, crossover circuit, memory controller, generation management circuit, serial communication circuit, evaluation circuit, FPGA board interface circuit, and SRAM. The proposed architecture introduces the self-control circuit in each sub circuit as shown in Fig. 12.10. It is required for the handshake type of data communication in each circuit and improves independence of each sub circuit.

In the evaluation operation of TSP, the total distance is evaluated. The distance between the cities is required in order to obtain the total distance. Therefore, the distance between the cities is previously saved in the memory in the proposed hardware.

12.5 Experiments

In order to evaluate the proposed hardware, it has been designed by Verilog-HDL and synthesized by the Synplicity Synplify. The clock frequency of the proposed hardware is set up with 20 MHz. Table 12.1 shows the gate size. ESBs indicate memory blocks of FPGA in Table 12.1. We simulate the performance of the proposed hardware with the number of steps and compare it with software processing using TSP benchmark data (TSP.LIB benchmark data: eil51).

Table 12.1 Gate size

Circuit name	Logic resource (%)	EBS (%)
Generating management	1	1
Memory controller	8	15
Selection	50	28
Crossover	7	5
Mutation	8	2
Evaluation	24	92

The software processing was implemented in the C language and run on a Linux PC. The clock frequency of the Linux PC is 2.4 GHz. The proposed hardware improved the speed of 22% in comparison with software processing. The clock frequency is 20 MHz for a limit of FPGA in this research. However, the proposed architecture is speeded up more if it is realized by LSI (Large Scale Integration).

12.6 Conclusion

In this paper, we proposed the new hardware architecture for hybrid optimization based on Genetic Algorithm and Simulated Annealing. The hybrid optimization of GA and SA in the proposed architecture realized searching not only globally but also locally. To keep general purpose, self-control processing by a handshake system is introduced. By introducing this type of handshake system, the proposed hardware can be applied to various combinatorial optimization problems by only changing encoder, decoder, and evaluation circuit. Furthermore, the proposed architecture is flexible for many genetic operations on GA and achieves high speed processing by adopting dedicated hardware. Simulation result evaluating the proposed architecture is shown to the effectiveness. Moreover, the gate size of the proposed architecture is reduced by adopting the combine architecture of GA and as SA. In relation to future works, we will expand the architecture to the artificial intelligence chip.

References

1. K. Kurbel, B. Schneider, and K. Singh, "Solving optimization problems by parallel recombinative simulated annealing on a parallel computer-an application to standard cell placement in VLSI design", IEEE Trans. on Systems, Man and Cybernetics, Part B, Vol. 28, No. 3, pp. 454–461, 1998.
2. V. Ravi, B.S.N. Murty, and J. Reddy, "Nonequilibrium simulated-annealing algorithm applied to reliability optimization of complex systems", IEEE Trans. on Reliability, Vol. 46, No 2, pp. 233–239, 1997.
3. J. Holland, Adaptation in Natural Artificial Systems, The University of Michigan Press (Second edition ; MIT Press), (1992)
4. D.E. Goldberg, Genetic algorithms in search optimization, and machine learning; Addison Wesley, New York (1989)
5. H. Chen, N.S. Flann, and D.W. Watson, "Parallel genetic simulated annealing: a massively parallel SIMD algorithm", IEEE Trans. on Parallel and Distributed Systems, Vol. 9, No. 2, pp. 126–136, 1998.
6. K.P. Dahal, G.M. Burt, J.R. McDonald, and S.J. Galloway, "GA/SA-based hybrid techniques for the scheduling of generator maintenance in power systems", Proc. of Congress on Evolutionary Computation, Vol. 1, pp. 567–574, 2000.
7. T. Kerdchuen and W. Ongsakul, "Optimal Measurement Placement for Power System State Estimation Using Hybrid Genetic Algorithm and Simulated Annealing", Proc. of International Conference on Power System Technology, pp. 1–5, 2006.
8. S.D. Scott, A. Samal and S. Seth, "HGA: A Hardware-Based Genetic Algorithm", Proceedings of International Symposium on Field-Programmable Gate Array, pp. 53–59, 1995.
9. P. Graham and B. Nelson, "Genetic Algorithm in Software and in Hardware A performance Analysis of Workstation and custom Computing Machine Implementations", FPGAs for Custom Computing Machines, pp. 216–225, 1996.
10. T. Imai, M. Yoshikawa, H. Terai, and H. Yamauchi, "VLSI Processor Architecture for Real-Time GA Processing and PE-VLSI Design", Proc. of IEEE International Symposium on Circuit and Systems, Vol. 3, pp. 625–628, 2004.
11. J. Grefenstette, R. Gopal, B. Rosmaita, and D. Van Gucht, "Genetic Algorithm for the Traveling Salesman Problem", Proc. of 1st Int. Conf. on Genetic Algorithms and their applications, pp. 160–168, 1985.

Chapter 13
Efficient Design of Arbitrary Complex Response Continuous-Time IIR Filter

Chi-Un Lei, Chung-Man Cheung, Hing-Kit Kwan and Ngai Wong

Abstract A continuous-time system identification technique, Vector Fitting (VF), is extended from symmetric functions, to asymmetrical cases and is used for complex infinite-impulse-response (IIR) continuous-time filter design. VF involves a two-step pole refinement process based on a linear least-squares solve and an eigenvalue problem. The proposed algorithm has lower complexity than conventional schemes by designing complex continuous-time filters *directly*. Numerical examples demonstrate that VF achieves highly efficient and accurate approximation to arbitrary asymmetric complex filter responses. The promising results can be realized for high dynamic frequency range networks. Robustness stability margin has also proposed for filter implementation robustness.

Keywords Complex filter · Vector fitting · Rational function approximation

13.1 Introduction

Complex signal processing has growing importance in high-speed communication systems [13, 14]. This can be seen, for example, by the majority of the LAN wireless transceivers papers in IEEE International Solid-State Circuits Conference (2001–2003) and recent papers [12, 16]. The topology of a complex signal processing block comprises two parallel paths, namely, an in-phase path and a quadrature path. The two cross-coupling paths have equal magnitude gain but a $\pi/2$ phase difference, as in Fig. 13.1. Modern wireless systems make use of this property in the design of quadrature receivers for image rejection in intermediate frequency (IF), and suppression of out-of-band signals and interferences [13]. This paper revisits and explores the design of complex continuous-time filters.

C.-U. Lei (✉)
Department of Electrical and Electronic Engineering, The University of Hong Kong
Pokfulam Road, Hong Kong
e-mail: culei@eee.hku.hk

S.-I. Ao et al. (eds.), *Trends in Communication Technologies and Engineering Science*,
Lecture Notes in Electrical Engineering 33, DOI 10.1007/978-1-4020-9532-0_13,

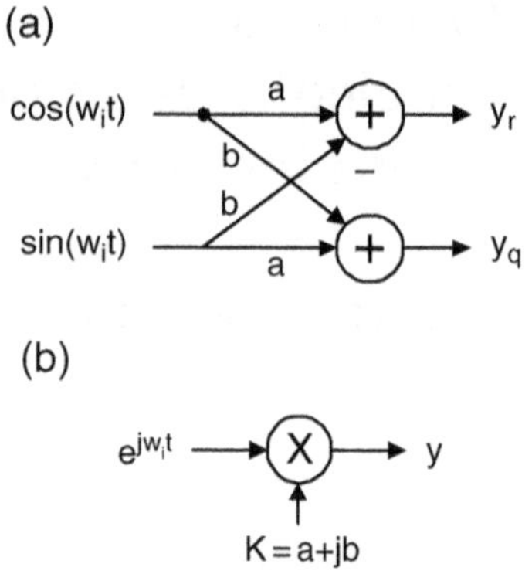

Fig. 13.1 (**a**) The real signal-flow graph (SFG) of a complex multiply and (**b**) the equivalent complex SFG

Existing complex infinite-impulse-response (IIR) continuous-time filter design algorithms include the first-order frequency transformation of a real lowpass filter [9], which usually results in a high-order filter due to unnecessary coefficient symmetry. In Martin [13] and Teplechuk and Sewell [18], algorithms are proposed to combine real filter design and conformal mapping to eliminate such transformation. However, both methods can only approximate a single passband response (but not arbitrary, say, multiple passband response) and do not consider phase matching and efficient design for advanced and adaptive applications.

On the other hand, a continuous-time frequency-domain system identification technique, known as Vector Fitting (VF) [6], is a robust macromodeling tool for fitting sampled frequency data with a rational function. Its extensive applications include power system modelling [4] and high-speed VLSI interconnect simulations [11]. Its counterpart, called discrete-time vector fitting (VFz), has been adapted to real digital filter design in Wong and Lei [19] with remarkable performance. VFz has been reformulated to discrete-time time-domain vector fitting (TD-VFz), adapted to high-speed VLSI interconnect simulations [10]. However, traditional VF has been restricted to symmetric functions. In this paper, VF is adapted to asymmetric functions for complex IIR continuous-time filter design.

The core of VF is a two-step process for refining the filter poles such that the prescribed response can be accurately reproduced with only a low-order rational function. VF enjoys simple coding and guarantees well-conditioned arithmetic by means of partial fraction basis. Flipping of unstable poles ensures system stability. Numerical examples then confirm the efficacy and accuracy of VF in complex IIR filter design over conventional schemes, especially in the high dynamic frequency range.

13.2 Vector Fitting

In complex filter design, the ultimate goal is to approximate a prescribed asymmetric frequency-domain response by a complex transfer function:

$$f(s) = H_0 \frac{\prod_{i=1}^{M} (s - \beta_i)}{\prod_{i=1}^{N} (s - \alpha_i)}, \tag{13.1}$$

where $H_0 \in \Re$ and $\alpha_i, \beta_i \in \mathbb{C}$. $f(s)$ is the desired frequency response, H_0, α_i, β_i, M and N represent gain, poles, zeros, the number of zeros and the number of poles, respectively.

To fit the continuous-time rational function, Vector Fitting (VF) [6] attempts to reformulate (13.1) with a partial fraction basis:

$$f(s) = \left(\sum_{n=1}^{N} \frac{c_n}{s - a_n} \right) + d + se, \tag{13.2}$$

to a set of calculated/sampled data points at frequencies $\{s_k\}$, for $k = 1, 2, \cdots, N_s$. In real response approximation of VF, the poles a_n and residues c_n are real or complex conjugate pairs. In complex filtering, this restriction is relaxed and VF readily handles complex poles and zeros arising in this stage. The constant term d and e are generally complex, too.

13.2.1 Problem Formulation

The fitting process begins with specifying an initial set of pole $\{\alpha_n^{(0)}\}$. By introducing of the scaling function $\sigma(s)$, a linear problem is set up for the ith iteration, namely

$$\underbrace{\left(\sum_{n=1}^{N} \frac{c_n}{s - \alpha_n^{(i)}} \right) + d + se}_{(\sigma f)(s)} \approx \underbrace{\left(\left(\sum_{n=1}^{N} \frac{\gamma_n}{s - \alpha_n^{(i)}} \right) + 1 \right)}_{\sigma(s)} f(s), \tag{13.3}$$

for $i = 0, 1, \cdots, N_T$, where N_T denotes the number of iterations when convergence is attained or when the upper limit is reached. The unknowns, c_n, d, e and γ_n, are solved through an overdetermined linear equation formed by evaluating (13.3) at the N_s sampled frequency points. One important feature in (13.3) is that $(\sigma f)(s)$ and $\sigma(s)f(s)$ are enforced to share the same set of poles, which in turn implies that the original poles of $f(s)$ are canceled by the zeros of $\sigma(s)$. And the relationship can be described as follows:

$$\underbrace{\frac{\prod_{n=1}^{N+1} \left(s - \beta_n^{(i)}\right)}{\prod_{n=1}^{N} \left(s - \alpha_n^{(i)}\right)}}_{(\sigma f)(s)} \approx \underbrace{\frac{\prod_{n=1}^{N} \left(s - \tilde{\beta}_n^{(i)}\right)}{\prod_{n=1}^{N} \left(s - \alpha_n^{(i)}\right)}}_{\sigma(s)} f(s), \tag{13.4}$$

$$\Rightarrow f(s) \approx \frac{(\sigma f)(s)}{\sigma(s)} = \frac{\prod_{n=1}^{N+1}\left(s - \beta_n^{(i)}\right)}{\prod_{n=1}^{N}\left(s - \widetilde{\beta}_n^{(i)}\right)}. \tag{13.5}$$

Subsequently, solving the zeros of $\sigma(s)$, in the least-squares sense, results in an approximation to the poles of $f(s)$, i.e., $\{\alpha_n^{(i+1)}\}$, without restriction of real coefficients. The new poles are then fed back to (13.3) as the next set of known poles for further refinement iteratively. In practical implementation, the fitting equation in (13.3) is linear in its unknowns c_n, d and γ_n. Therefore, the non-linear problem can be solved using over-determined equations and eigenvalue solving.

13.2.2 Pole Calculation

At a particular frequency point, with $e = 0$, (13.3) becomes

$$\left(\sum_{n=1}^{N} \frac{c_n}{s_k - \alpha_n^{(i)}}\right) + d - \left(\sum_{n=1}^{N} \frac{\gamma_n f(s_k)}{s_k - \alpha_n^{(i)}}\right) \approx f(s_k), \tag{13.6}$$

it can be reformulated into:

$$A_k x = b_k, \tag{13.7}$$

where $x = [c_1 \ \cdots \ c_N \ 1 \ \gamma_1 \ \cdots \ \gamma_N]^T$, $b_k = f(s_k)$ and

$$A_k = \left[\frac{1}{s_k - \alpha_1^{(i)}} \ \cdots \ \frac{1}{s_k - \alpha_N^{(i)}} \ d \ \frac{-f(s_k)}{s_k - \alpha_1^{(i)}} \ \cdots \ \frac{-f(s_k)}{s_k - \alpha_N^{(i)}}\right]$$

In the above expression, A_k and x are row and column vectors, respectively, while b_k is a scalar.

Equating (13.7) at all frequency samples, mathematically $s_k = j\Omega_k$, for $k = 1, 2, \cdots, N_S$, where $N_S > 2N + 1$, and by stacking A_k's and b_k's into a (tall) column matrix and a vector, an overdetermined equation for the ith iteration is obtained:

$$Ax = b. \tag{13.8}$$

The scaling function $\sigma(s)$ in (13.3) can be reconstructed from the last N entries obtained from the least-squares solution (i.e., $\gamma_1, \gamma_2, \cdots, \gamma_N$ in x), such that its zeros, $\{\alpha_n^{(i+1)}\}$, for $n = 1, 2, \cdots, N$, are taken as the new set of starting poles in the next VF iteration. Moreover, it can be shown that $\{\alpha_n^{(i+1)}\}$ are conveniently computed as the eigenvalues of $\Psi \in \mathbb{C}^{N \times N}$ where

$$\Psi = \begin{bmatrix} \alpha_1^{(i)} & & & \\ & \alpha_2^{(i)} & & \\ & & \ddots & \\ & & & \alpha_N^{(i)} \end{bmatrix} - \begin{bmatrix} 1 \\ 1 \\ \vdots \\ 1 \end{bmatrix} \begin{bmatrix} \gamma_1 \; \gamma_2 \cdots \; \gamma_N \end{bmatrix}. \tag{13.9}$$

Since the poles are not restricted to complex-conjugate pairs, Ψ is generally a complex matrix. Upon convergence, the update in $\alpha_n^{(i)}$ diminishes and $\sigma(s) \approx 1$.

To ensure system stability, it is necessary that $Re\left(\alpha_n^{(i+1)}\right) < 0$. If this is not the case, any unstable pole is flipped about the imaginary axis to the open left half plane $Re\left|\alpha_n^{(i+1)}\right| := -Re\left|\alpha_n^{(i+1)}\right|$. Such stability enforcement has the physical meaning of multiplying allpass filters onto the filter transfer function such that the magnitude response is preserved.

13.2.3 Building the Rational Function

Once a converged set of poles $\{\alpha_n^{(N_T)}\}$ is obtained, the final stage is to reconstruct the rational function complex IIR filter. Referring to (13.3), we should now have $\sigma(s) \approx 1$ so that

$$\left(\sum_{n=1}^{N} \frac{c_n}{s - \alpha_n^{(N_T)}}\right) + d \approx f(s_k), \tag{13.10}$$

for $k = 1, 2, \cdots, N_s$. The residues c_n of $F(s)$ are determined exactly in the same manner, except (13.7) is replaced as follows:

$$A_k x = b_k, \tag{13.11}$$

where $x = [c_1 \;\cdots\; c_N \;\; 1]^T$, $A_k = \left[\frac{1}{s_k - \alpha_1^{(i)}} \;\cdots\; \frac{1}{s_k - \alpha_N^{(i)}} \;\; d\right]$ and $b_k = f(s_k)$. The summation of partial fractions $\left(\sum_{n=1}^{N} \frac{c_n}{s - \alpha_n^{(N_T)}}\right) + d$ generates a rational function representing the complex IIR filter in the form of (13.2). By seeking a near-optimal fit to $f(s_k)$ [5], VF matches both magnitude and phase of $f(s_k)$ simultaneously.

In summary, comparing to VF, other complex IIR design algorithms such as the pole-placement algorithm in Martin [13] give similiar accuracy but has much higher computational requirement and does not emphasize the importance of phase linearity. For high-order filter design, it was claimed that 10–15 seconds are required on a 3-GHz computer [13], in great contrast to the 1–2 seconds by our design methodology implemented on a 1.8-GHz computer. In comparison to Martin [13] and other complex filter design algorithms, VF is superior in terms of the phase linearity in the passband and the much higher efficiency, as will be seen in our numerical examples.

13.3 Remarks

13.3.1 Numerical Consideration

Recent research has shown that VF is in fact a reformulation of the Sanathanan-Koerner (SK) iteration [7], an iterative continuous-time frequency-domain system identification technique [17], which minimizes the following objective function:

$$\sum_{k=0}^{N_s}\left|\frac{N^{(i)}(s_k)}{D^{(i-1)}(s_k)}-\frac{D^{(i)}(s_k)}{D^{(i-1)}(s_k)}f(s_k)\right|^2 = \sum_{k=0}^{N_s}\left|\sum_{n=1}^{N}\frac{c_n}{s_k+\alpha_n^{(i-1)}}-\left(1+\sum_{n=1}^{N}\frac{\gamma_n}{s_k+\alpha_n^{(i-1)}}\right)f(s_k)\right|^2, \tag{3.12}$$

where $N^{(i)}$, $D^{(i)}$ are the numerator and denominator determined during the ith iteration. In theory, the approximation converges for a noise-free model using (13.12), but numerical errors may still occur during iterative calculation [1]. Therefore, numerical considerations for improving the approximation accuracy are highlighted:

1. The approximation accuracy in noisy environment (e.g., finite precision) is affected by the equation normalization in the original VF. Furthermore, the normalization also gives a biased approximation in the low frequency region [5]. Recently, a relaxation is proposed to improve the normalization situation. Firstly, the unity basis in (13.3) is replaced by a variable and a relaxation constraint, namely,

$$Re\left\{\sum_{k=0}^{N_s}\left(1+\sum_{n=1}^{N}\frac{\gamma_n}{s_k+\alpha_n^{(i)}}\right)\right\} = N_s+1, \tag{13.13}$$

is introduced in (13.8) with a row weighting of $\|f(s)\|/(N_s+1)$. As shown in Deschrijiver et al. [2], this relaxation outperforms other relaxation approaches.
2. The conditioning and the approximant accuracy are affected by the choices of function basis and the location of initial-poles (i.e. $\alpha_n^{(0)}$ in (13.3)). In Gustavsen and Semlyen [6], it is recommended that the initial poles should be distributed over the frequency range of interest. As a rule of thumb in VF, the complex design algorithm starts with complex conjugate poles based on the following relationship:

$$a_n = -\alpha + j\beta, \quad a_{n+1} = -\alpha - j\beta, \tag{13.14}$$

where $\beta = 100\alpha$, α and β are the real part and the imaginary part of the initial poles, respectively. This selection criterion excludes the possibility of ill-conditioned computation in most practical cases.

This problem can also be alleviated by introducing orthonormal function basis [1] or digital partial fraction basis [15]. In this paper, column scaling technique from linear algerba [3] is used in system equation calculation to improve the numerical conditioning, and gives a similar accuracy as using the orthonormal basis.

13.3.2 Frequency Band Weighting

Besides L_2 norm minimization, some filter requires equiripple passband to minimize the maximum error of a filter. In IIR digital filter design, an equiripple filter can be designed by a weighted least-squares method with a suitable least-squares frequency response weighting function [8]. One of the choices is $w(s) = \sqrt{\frac{|f_{fit}(s)-f(s)|}{|f(s)|}}$. The frequency weighting can be included using row weighting in a particular frequency-data row of (13.7).

13.3.3 Stability Margin

An analytical stable filter can be obtained by unstable pole flipping during pole relocation. However, the filter can become unstable during non-ideal circuit implementation. Therefore, a stability margin is included to locate poles away from the imaginary axis. This is done by the explicit pole relocation in (13.9). Although the algorithm convergence may be affected, the poles are still guaranteed to converge, which is provided by the numerical examples.

13.4 Numerical Examples

The performance of VF in direct complex IIR filter design is illustrated by two examples run in the MATLAB 7.1 environment on a 512-RAM 1.8-GHz PC.

13.4.1 Example 1

In this example, we design a single-passband positive passband filter (PPF), which is widely used in communication systems (e.g. single-side-band communication [14]). The specification is extracted from Martin [13]. The response is sampled at 40 linearly spaced points in the passband, and 80 uniform sampling points in each stopband. The passband spans $[-10, 10]$.

Vector fitting constrains the numerator and denominator of the transfer function to have the same order. An 8-pole (8-zero) complex IIR filter was designed via VF to fit the prescribed response. It takes 0.9063 seconds for the system poles to attain convergence in 5 pole refining iterations. The frequency response and the passband group delay (the change of phase) are shown in Fig. 13.2. It can be observed that VF

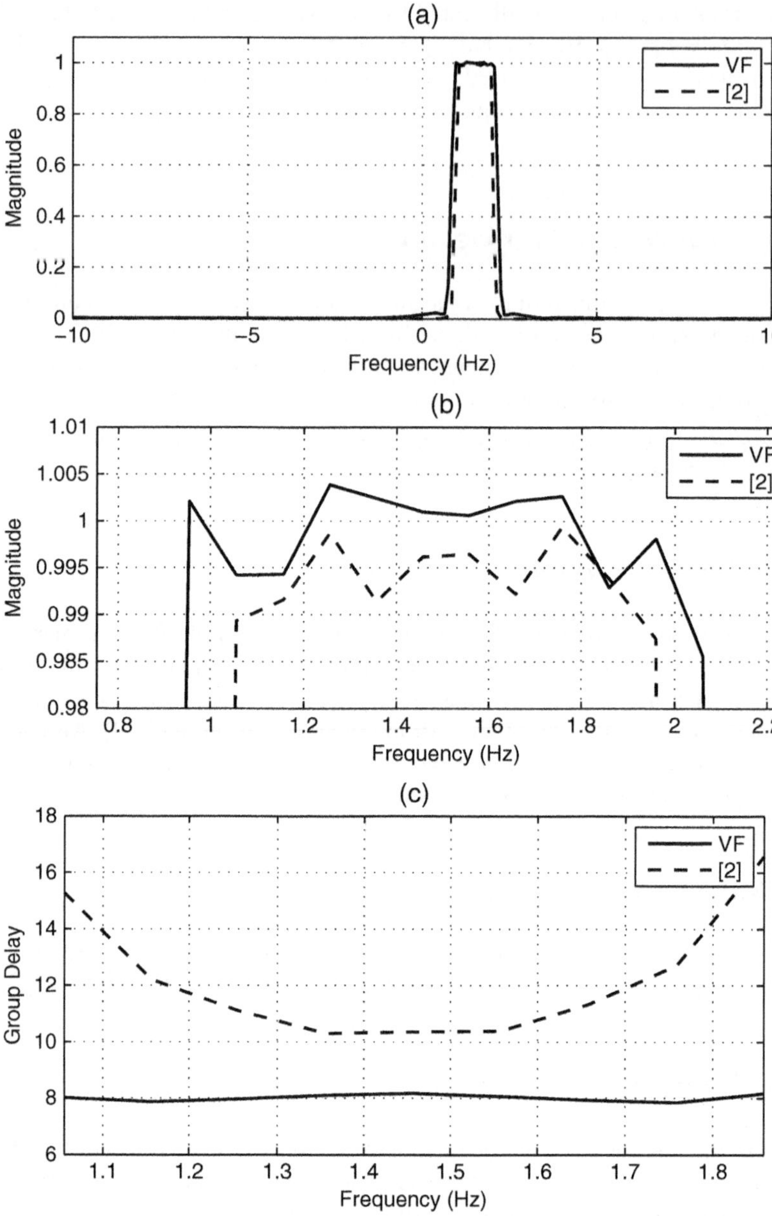

Fig. 13.2 Frequency response of Example 1. (**a**) Magnitude in the entire band, (**b**) magnitude in the passband and (**c**) group delay in the passband

Table 13.1 Approximation errors in the numerical examples

	Ex 1. (VF without pole constraint)	Ex 1. (VF with pole constraint)	Ex 1. ([Martin, 2005])	Ex. 2
L_2 passband	0.0124	0.0546	0.0234	0.0083
L_∞ passband	0.0071	0.0110	0.0126	0.0022
L_2 stopband	5.1452	5.1157	5.1953	0.0029
L_∞ stopband	0.9975	0.9987	1.0000	0.0068

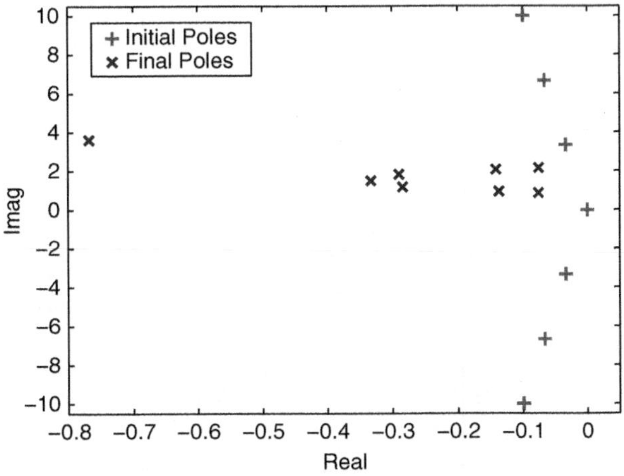

Fig. 13.3 Initial pole placement and final pole locations in Example 1

achieves an excellent fitting to both desired magnitude and phase response. The L_2 norm and L_∞ norm errors in magnitude of solution are summarized in Table 13.1. The numerical example is compared with [13], and is shown by Fig. 13.2. It can be concluded that the phase response matches the desired constant group delay (i.e. linear passband) with a significantly improved computation complexity and better approximation in the passband, with a slightly better stopband design. Figure 13.3 shows the initial and final system poles.

The same example is repeated with a constrained pole margin of ($\Re(\alpha) < -0.2$). The frequency response, passband group delay and approximation error details are shown in Fig. 13.4 and Table 13.1, respectively. Figure 13.5 shows the initial and final system poles. Two pairs of poles are closely packed together. The minimum real pole module is 0.2. The passband response of the robust design is slightly degraded, but with similar stopband response and improved robustness.

13.4.2 Example 2

The second example demonstrates the versatility of the VF algorithm through the design of a wide-frequency-range double-passband complex filter, which can be used in low-IF architecture [14]. The specifications are as follows:

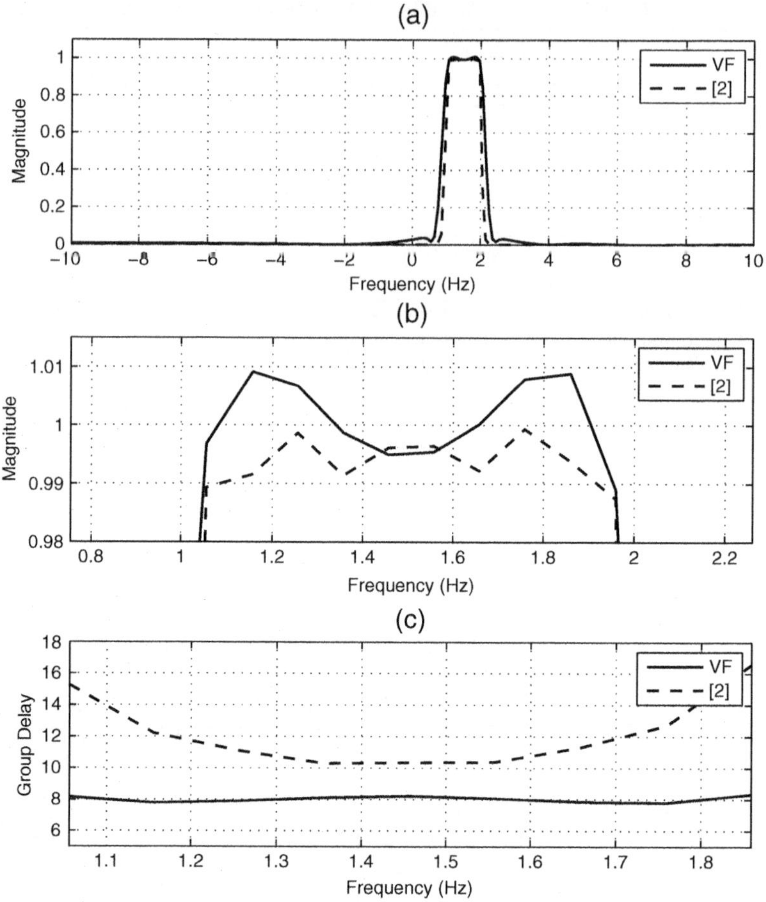

Fig. 13.4 Frequency response of Example 1 with pole constraint. (**a**) Magnitude in the entire band, (**b**) magnitude in the passband and (**c**) group delay in the passband

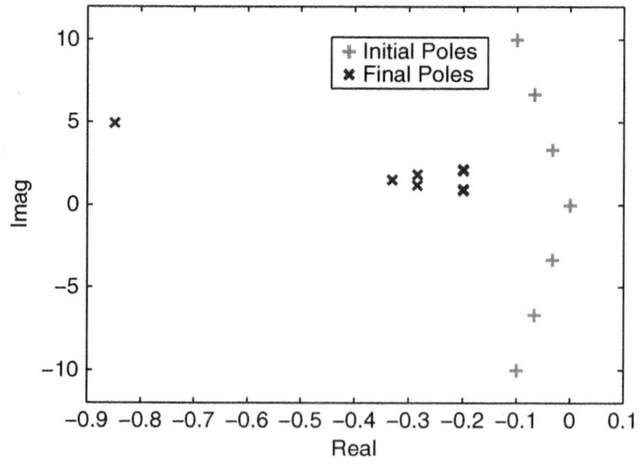

Fig. 13.5 Initial pole placement and final pole locations in Example 1 with pole constraint

$$H_d(e^{j\omega}) = \begin{cases} -60\,dBe^{j\omega D}, & (-10f_s \leq \omega \leq -7f_s) \\ 0\,dBe^{j\omega D}, & (-6f_s \leq \omega \leq -4f_s) \\ -60\,dBe^{j\omega D}, & (-3f_s \leq \omega \leq f_s) \\ 0\,dBe^{j\omega D}, & (2f_s \leq \omega \leq 6f_s) \\ -60\,dBe^{j\omega D}, & (7f_s \leq \omega \leq -10f_s) \end{cases} \tag{13.15}$$

where D is the group delay. To illustrate the capability of VF to design wide frequency range filters, we take f_s in the GHz. Uniform sampling points of 80 and 40 are used in each passband and stopband respectively, i.e., 280 sampling points in total. The transition bands are not sampled as a way of relaxation. We set the group delay to be 10 Mm in both the passband and stopband. To approximate the desired response, a $20th$-order IIR filter is designed via the VF algorithm with the numerical enhancements in Section 13.3. The set of system poles converges in only 4 iterations, taking 0.8063 seconds for computation. From Fig. 13.6 and Table 13.1,

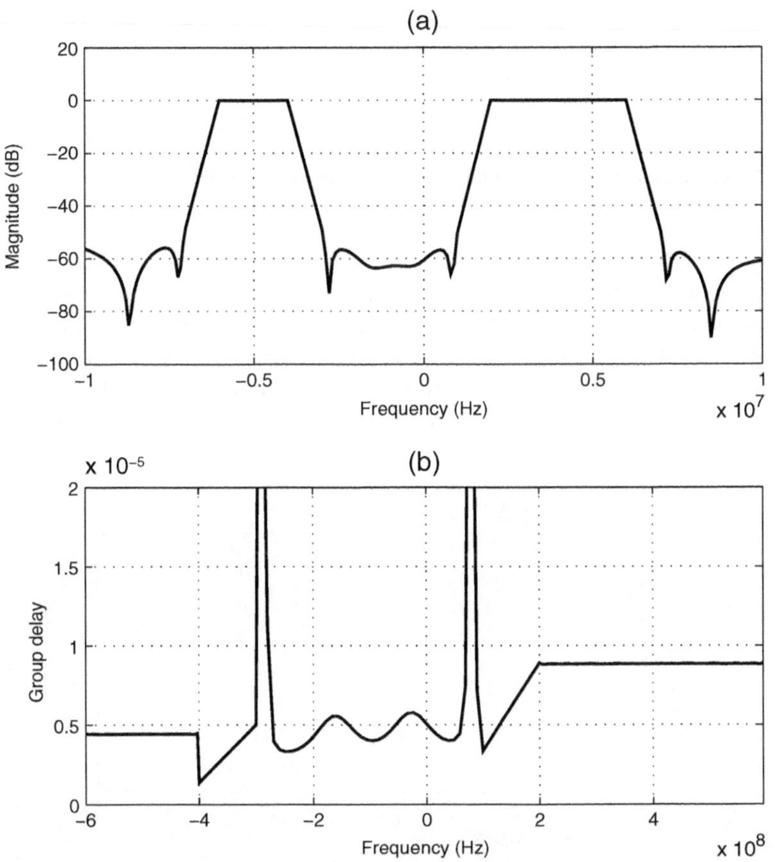

Fig. 13.6 Frequency response of Example 2. (**a**) Magnitude, (**b**) Group delay in passband

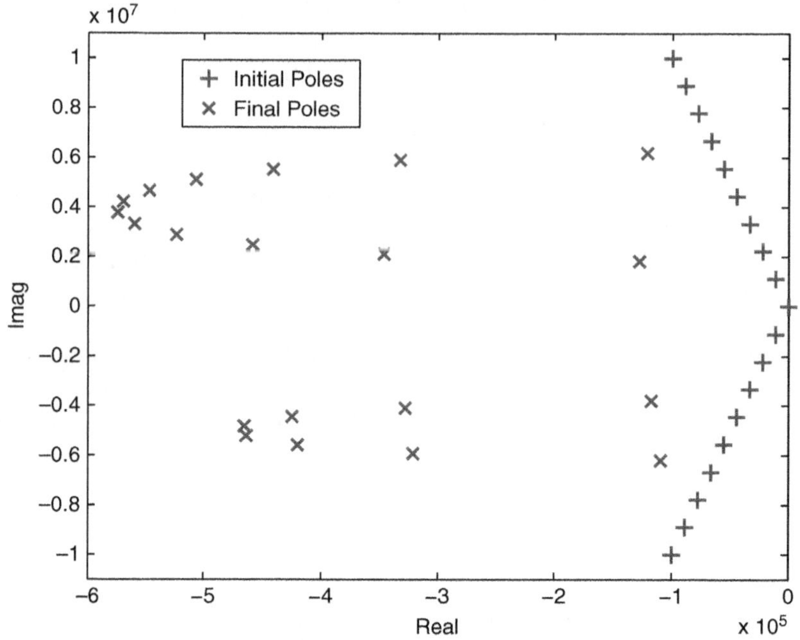

Fig. 13.7 Initial pole placement and final pole locations in Example 2

the complex IIR filter via VF matches the desired magnitude response and phase linearity well over a wide frequency range. The location of initial assigned poles and converged poles in the s-plane is plotted in Fig. 13.7. It is noticeable that the poles are roughly located separately into two elliptical regions, each of them corresponding to a passband. Figure 13.8 shows the L_2 error and the condition number in (13.3) during each iteration. Both quantities show a significant descending trend for

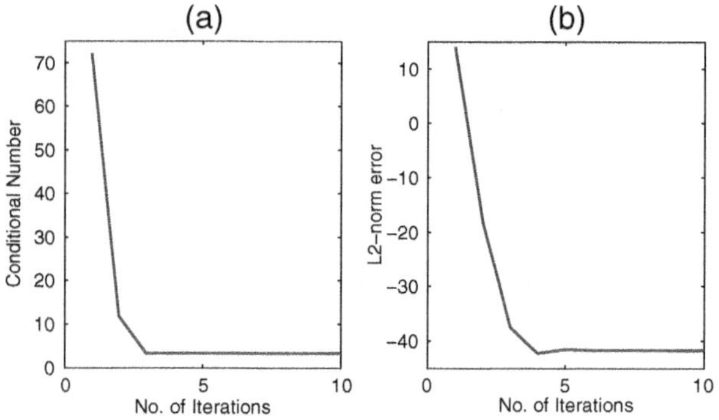

Fig. 13.8 Numerical information for each iteration in Example 2. (**a**) Conditional number and (**b**) L_2 error

each iteration and converge within a few iterations, and in fact converge much faster than the original VF. It also shows the numerical improvement using row scaling technique.

13.5 Conclusions

This paper has generalized VF to the design of arbitrary response complex IIR continuous-time filters. Without symmetry constraints, the proposed method demonstrates its efficacy in producing low-order approximation functions to the desired magnitude responses over a wide frequency range with matching of phase linearity. Arbitrary response approximation, passband phase matching, robust pole stability margin and the algorithm effectiveness are definite advantages over existing methods. Different enhancement techniques are applied in the numerical computation. Numerical examples have confirmed the superiority of VF over conventional complex IIR filter design algorithms.

Acknowledgment This work was supported in part by the Hong Kong Research Grants Council and the University Research Committee of The University of Hong Kong.

References

1. Deschrijver, D., Haegeman, B., and Dhaene, T. (2007). Orthonormal vector fitting: A robust macromodeling tool for rational approximation of frequency domain responses. *IEEE Trans. Adv. Packag.*, 30(2):216–225.
2. Deschrijver, D., Schoeman, M., Dhaene, T., and Meyer, P. (2007). Experimental analysis on the relaxation of macromodeling methods. In *Proc. IEEE Africon conference*.
3. Golub, G. H. and Loan, C. F. V. (1996). *Matrix Computations*. London: Johns-Hopkins, third edition.
4. Gustavsen, B. (2004). Wide band modeling of power transformers. *IEEE Trans. Power Delivery*, 19(1):414–422.
5. Gustavsen, B. (2006). Improving the pole relocating properties of vector fitting. *IEEE Trans. Power Delivery*, 21(3):1587–1592.
6. Gustavsen, B. and Semlyen, A. (1999). Rational approximation of frequency domain responses by vector fitting. *IEEE Trans. Power Delivery*, 14(3):1052–1061.
7. Hendrickx, W. and Dhaene, T. (2006). A discussion of "Rational approximation of frequency domain responses by vector fitting". *IEEE Trans. Power Syst.*, 21(1):441–443.
8. Kobayashi, T. and Imai, S. (1990). Design of IIR digital filters with arbitrary log magnitude function by WLS techniques. *IEEE Trans. Signal Processing*, 38(2):247–252.
9. Krukowski, A. and Kate, I. (2003). DSP system design: complexity reduced IIR filter implementation for practical application.
10. Lei, C. U. and Wong, N. (2008). Efficient linear macromodeling via discrete-time time-domain vector fitting. In *Proc. Intl. Conf. on VLSI Design*, pages 469–474.
11. Li, E. P., Liu, E. X., Li, L. W., and Leong, M. S. (2004). A coupled efficient and systematic full-wave time-domain macromodeling and circuit simulation method for signal integrity analysis of high-speed interconnects. *IEEE Trans. Antennas Propagat.*, 27(1):213–223.
12. Mahattanakul, J. and Khumsat, P. (2007). Structure of complex elliptic Gm-C filters suitable for fully differential implementation. *IEE Proceedings - Circuits, Devices and Systems*, 1(4):275–282.

13. Martin, K. (2005). Approximation of complex IIR bandpass filters without arithmetic symmetry. *IEEE Trans Circuits Syst. I*, 52(4):794–803.
14. Martin, K. W. (2004). Complex signal processing is not complex. *IEEE Trans. Circuits Syst. I*, 51(9):1823–1836.
15. Mekonnen, Y. S. and Schutt-Aine, J. E. (2007). Broadband macromodeling of sampled frequency data using z-domain vector-fitting method. In *Proc. IEEE workshop on sig. prop. on interconnects*. to appear.
16. Pandita, B. and Martin, K. W. (2007). Designing complex delta-sigma modulators with signal-transfer functions having good stop-band attenuation. In *Proc. IEEE Symp. Circuits and Systems*, pages 3626–3629.
17. Sanathanan, C. and Koerner, J. (1963). Transfer function synthesis as a ratio of two complex polynomials. *IEEE Trans. Automat. Contr.*, 8(1):56–58.
18. Teplechuk, M. A. and Sewell, J. I. (2006). Approximation of arbitrary complex filter responses and their realisation in log domain. *IEE Proceedings - Circuits, Devices and Systems*, 153(6):583–590.
19. Wong, Ngai and Lei, Chi-Un (2008). IIR approximation of FIR filters via discrete-time vector fitting. *IEEE Trans. Signal Processing*, 56(3):1296–1302.

Chapter 14
Fixed Structure Robust Loop Shaping Controller for a Buck-Boost Converter Using Evolutionary Algorithm

S. Kaitwanidvilai, A. Jangwanitlert, I. Ngarmroo, W. Khanngern and S. Karnprachar

Abstract In this paper, we propose a new technique used to design a robust controller that is not as high-order and complicated as the ones designed by conventional $\mathcal{H}_\infty$ loop shaping method. The proposed algorithm is called *Genetic Algorithm (GA) based fixed-structure H_∞ loop shaping control.* In the approach, GA is adopted to solve the H_∞ loop shaping design problem under a structure specified controller. The performance and robustness of the proposed controller are investigated in a buck-boost converter in comparison with the controllers designed by conventional H_∞ loop shaping and conventional ISE method. Results of simulations demonstrate the advantages of the proposed controller in terms of simple structure and robustness against plant perturbations and disturbances. Experiments are performed to verify the effectiveness of the proposed technique.

Keywords H_∞ loop shaping · Robust controller · Genetic algorithm · Buck-boost converter

14.1 Introduction

DC–DC converters have been widely used in computer hardware and industrial applications. Controlling of these converters is a challenging field because of their intrinsic nature of nonlinear, time-variant systems [1]. In previous research works, the linear models of these converters were derived by using linearization method [2, 3]. Some linear control techniques were applied to these converters based on the linear models [1, 4, 5]. Naim et al. [4] applied the $\mathcal{H}_\infty$ control to a boost converter. Three controllers; voltage mode, feed-forward and current mode control were investigated and compared the performance. G.C. loannidis and S.N. Manias [5] applied the $\mathcal{H}_\infty$ loop shaping control schemes for a buck converter. In their paper, the

S. Kaitwanidvilai (✉)
Electrical Engineering Department, Faculty of Engineering, King Mongkut's Institute of Technology, Ladkrabang 10520, Thailand

S.-I. Ao et al. (eds.), *Trends in Communication Technologies and Engineering Science*, Lecture Notes in Electrical Engineering 33, DOI 10.1007/978-1-4020-9532-0_14,

μ-analysis was used to examine the robust features of the designed controllers. Simone Buso [1] adopted the robust μ-synthesis to design a robust voltage controller for a buck-boost converter with current mode control. The parameter variations in the converter's transfer function were described in term of perturbations of linear fraction transformations (LFT) class.

The controllers in DC to DC converters are usually designed using analog circuit. Although the controllers designed by the techniques mentioned earlier are robust, and high performance, they are complicated and high order, thus they are difficult to be implemented in the converters. Nevertheless, the design of analog circuit for these controllers is not feasible. To solve this problem, fixed-structure controller is investigated. Fixed-structure robust controllers have become an interesting area of research because of their simple structure and acceptable controller order. However, it is difficult to design this controller using analytical method. Algorithms such as genetic algorithm, particle swarm optimization, and gradient method can be employed to simplify the design of this controller.

Several approaches to design a robust control for structure specified controller were proposed in Bor-Sen Chen and Yu-Min Cheng [6], Bor-Sen Chen [7], and Shinn-Jang Ho et al. [8]. In Bor-Sen Chen and Yu-Min Cheng [6], a robust $\mathcal{H}_\infty$ optimal control problem with structure specified controller was solved by using genetic algorithm (GA). As concluded in this paper, genetic algorithm is a simple and efficient tool to design a structure specified $\mathcal{H}_\infty$ optimal controller. Bor-Sen Chen et al. [7], proposed a PID design algorithm for mixed $\mathcal{H}_2/\mathcal{H}_\infty$ control. In their paper, PID controller parameters were tuned in the stability domain to achieve mixed $\mathcal{H}_2/\mathcal{H}_\infty$ optimal control. A similar work was proposed in Shinn-Jang Ho et al. [8] by using the intelligent genetic algorithm to solve the mixed $\mathcal{H}_2/\mathcal{H}_\infty$ optimal control problem.

The techniques mentioned above are based on the concept of $\mathcal{H}_\infty$ optimal control which two appropriate weights for both the uncertainty of the model and the performance are essentially chosen. A difficulty with the $\mathcal{H}_\infty$ optimal control approach is that the appropriate selection of close-loop objectives and weights is not straightforward. In robust control, $\mathcal{H}_\infty$ loop shaping which is a simple and efficient technique for designing a robust controller can be alternatively used to design the robust controller for the system. Uncertainties in this approach are modeled as normalized co-prime factors; this uncertainty model does not represent actual physical uncertainty, which usually is unknown in real problems. This technique requires only two specified weights, pre-compensator and post-compensator, for shaping the nominal plant so that the desired open loop shape is achieved. Fortunately, the selection of such weights is based on the concept of classical loop shaping which is a well known technique in controller design. By the reasons mentioned above, this technique is simpler and more intuitive than other robust control techniques. However, the controller designed by $\mathcal{H}_\infty$ loop shaping is still complicated and has high order. To overcome this problem, in this paper, we propose a fixed-structure $\mathcal{H}_\infty$ loop shaping control to design a robust controller for a buck-boost converter. In the proposed technique, the controller structure is first specified and the genetic algorithm is then used to evaluate the control's parameters. Simulation and experimental results show the advantages of simple structure, lower order and robustness of the proposed controller.

The remainder of this paper is organized as follows. Converter dynamics are described in Section 14.2. $\mathcal{H}_\infty$ loop shaping and the proposed technique are discussed in Section 14.3. Section 14.4 demonstrates the design example and results. Finally, Section 14.5 concludes the paper with some final remarks.

14.2 Converter Modeling

A typical circuit of buck-boost converter with current mode control is shown in Fig. 14.1. The dynamic model of this converter from the current reference (i_r) to output voltage (u_O) is given by Ridley [2] and Kassakian [3].

$$\frac{du_o}{di_r} = R_L \frac{V_i}{V_i + 2V_o} \frac{\left(1 - \frac{\mathrm{s} \cdot \mathrm{L}}{\mathrm{R_L}} \cdot \frac{\mathrm{V_o}}{\mathrm{V_i}} \cdot \frac{\mathrm{V_o} - \mathrm{V_i}}{\mathrm{V_o}}\right)}{\left(1 + \mathrm{s} \cdot \mathrm{C} \cdot \mathrm{R_L} \cdot \frac{\mathrm{V_o} + \mathrm{V_i}}{2\mathrm{V_o} + \mathrm{V_i}}\right)} \tag{14.1}$$

where R_L is the nominal load resistant, V_o is the nominal output voltage, V_i is the nominal input voltage, L is the inductance of an inductor used in the circuit, C is the capacitance, f_{sw} is the switching frequency. The accuracy of this model has been proved to be accepted, at least in frequency of interest in this application.

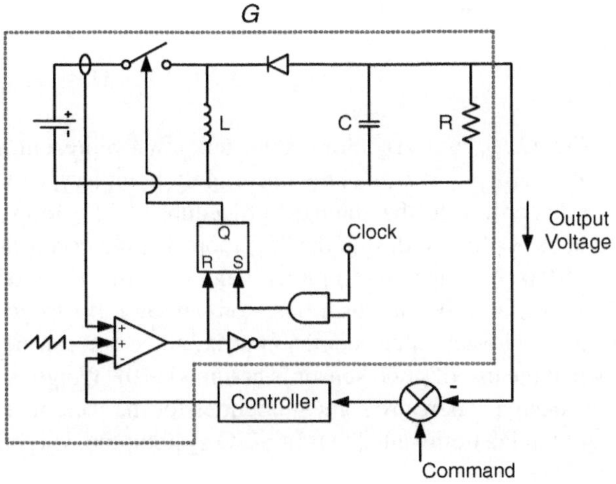

Fig. 14.1 Buck-boost converter with current mode control

14.3 $\mathcal{H}_\infty$ Loop Shaping Control and Proposed Technique

This section illustrates the concepts of the standard $\mathcal{H}_\infty$ loop shaping control and the proposed technique. Consider the system described by the block diagram (Fig. 14.2), where the plant G and the controller K are real rational and proper. y is the output, u is the control input, w is the vector signal including noises, disturbances, and reference signals, and z is the vector signal including all controlled signals and tracking errors. The $\mathcal{H}_\infty$ optimal control problem is to find admission controller $K(p)$ such that $||T_{zw}||_\infty$ is minimized, where $||T_{zw}||_\infty$ is the maximum norm of the transfer function from w to z, and the admission controller is the controller that internally stabilizes the system [9].

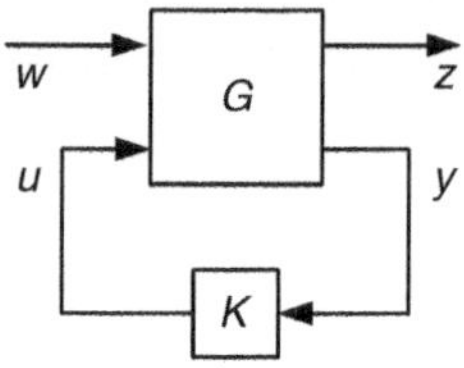

Fig. 14.2 General block diagram in robust control problem

14.3.1 *Standard $\mathcal{H}_\infty$ Loop Shaping*

$\mathcal{H}_\infty$ loop shaping control [9] is an efficient method to design a robust controller. This approach requires only a desired open loop shape in frequency domain. Two weighting functions, W_1 (pre-compensator) and W_2 (post-compensator), are specified to shape the original plant G. In this approach, the shaped plant is formulated as normalized coprime factor, which separates the plant G_s into normalized nominator N_s and denominator M_s factors. If the shaped plant $G_s = W_1 G W_2 = N_s M_s^{-1}$, then a perturbed plant can be written as

$$G_\Delta = (N_s + \Delta_{N_s})(M_s + \Delta_{M_s})^{-1} \tag{14.2}$$

Where Δ_{Ns} and Δ_{Ms} are stable, unknown representing the uncertainty satisfying $||\Delta_{Ns}, \Delta_{Ms}||_\infty \leq \varepsilon$, ε is the uncertainty boundary, called stability margin.

According to the standard procedure of $\mathcal{H}_\infty$ loop shaping, the following steps can be applied to design the $\mathcal{H}_\infty$ loop shaping controller.

Step 1 Shape the singular values of the nominal plant G_o by using a pre-compensator W_1 and/or a post-compensator W_2 to get the desired loop shape. W_2 can be chosen as an identity matrix, since we can neglect the sensor noise effect when the use of good sensor is assumed [10]. Weight selection is very important for the design. There are some guidelines for the weight selection in Siguard Skogestad and Ian Postlethwaite [11]. In SISO system, the weighting functions W_1 and W_2 can be chosen as

$$W_1 = K_W \frac{S+a}{S+b} \quad \text{and} \quad W_2 = \frac{c}{s+c} \tag{14.3}$$

where K_w , a, b and c are positive values.

Step 2 Minimize ∞ – norm of the transfer matrix T_{zw} over all stabilizing controllers K to obtain an optimal cost γ_{opt} , as

$$\gamma_{opt} = \varepsilon_{opt}^{-1} = \inf_{stabK} \left\| \begin{bmatrix} I \\ K \end{bmatrix} (I + G_s K)^{-1} M_s^{-1} \right\|_\infty \tag{14.4}$$

To determine ε_{opt} , there is a unique method explained in Siguard Skogestad and Ian Postlethwaite [11]. $\varepsilon_{opt} << 1$ indicates that W_1 or W_2 designed in step 1 are incompatible with robust stability requirement. If ε_{opt} is not satisfied ($\varepsilon_{opt} << 1$), then return to step 1, adjust W_1.

Step 3 Select $\varepsilon < \varepsilon_{opt}$ and then synthesize a controller K_∞ that satisfies

$$\|T_{zw}\|_\infty = \left\| \begin{bmatrix} I \\ K_\infty \end{bmatrix} (I + G_s K_\infty)^{-1} M_s^{-1} \right\|_\infty \leq \varepsilon^{-1} \tag{14.5}$$

Controller K_∞ is obtained by solving the optimal control problem. More detail is available in Siguard Skogestad and Ian Postlethwaite [11].

Step 4 Final controller (K) follows

$$K = W_1 K_\infty W_2 \tag{14.6}$$

14.3.2 Genetic Algorithm Based Fixed-Structure $\mathcal{H}_\infty$ Loop Shaping Optimization

Practical implementation of the controller derived from $\mathcal{H}_\infty$ loop shaping method is difficult because the order of the controller is quite high. In this paper, the genetic searching algorithm is adopted to solve this problem. Although the proposed controller is structured, it still retains the entire robustness and performance guarantee as long as a satisfactory uncertainty boundary ε is achieved. The proposed algorithm is explained as following.

Assume that the predefined structure controller *K(p)* has a satisfied parameters *p*. Based on the concept of $\mathcal{H}_\infty$ loop shaping, optimization goal is to find parameters *p* in controller *K(p)* that minimize infinity norm $\|T_{zw}\|_\infty$. In the proposed technique, the final controller *K* is defined as

$$K = K(p) W_2 \tag{14.7}$$

Assuming that W_1 is invertible, from (14.6) then it is obtained that

$$K_\infty = W_1^{-1} K(p) \tag{14.8}$$

In many cases, the weight W_2 is selected as identity matrix *I*. However, if W_2 is a transfer function matrix, then the final controller is the controller *K*(*p*) in series with the weight W_2. By Substitution of (14.8) into (14.5), then the ∞-norm of the transfer function matrix from disturbances to states, $||T_{zw}||_\infty$, which is subjected to be minimized can be written as

$$J_{\cos t} = \gamma = \|T_{zw}\|_\infty = \left\| \begin{bmatrix} I \\ W_1^{-1} K(p) \end{bmatrix} (I + G_s W_1^{-1} K(p))^{-1} M_s^{-1} \right\|_\infty \tag{14.9}$$

In this paper, GA is adopted to find the optimal control parameters p^* in the stabilizing controller *K(p)* such that the $||T_{zw}||_\infty$ is minimized.

14.3.2.1 Genetic Algorithms

GA is well known as a biologically inspired class of algorithms that can be applied to any nonlinear optimization problem. This algorithm applies the concept of chromosomes, and the operations of crossover, mutation and reproduction. At each step, called generation, fitness values of all chromosomes in population are calculated. Chromosome, which has the maximum fitness value (minimum cost value), is kept as a solution in the current generation and passed to the next generation. The new population of the next generation is obtained by performing the genetic operators such as crossover, mutation, and reproduction. Crossover randomly selects a site along the length of two chromosomes, and then splits the two chromosomes into two pieces by breaking them at the crossover site. The new chromosomes are then formed by matching the headpiece of one chromosome with the tailpiece of the other. Mutation operation forms a new chromosome by randomly changing value of a single bit in the chromosome. Reproduction operation forms a new chromosome by just copying the old chromosome. Chromosome selection in genetic algorithm depends on the fitness value; high fitness value means high chance to be selected. Operation type selection; mutation, reproduction, or crossover, depends on the pre-specified operation's probability. Chromosome in genetic population is coded as binary number. However, for the real number problem, decoding binary number to floating number is applied [12]. Our proposed algorithm is summarized as

Step 1 Shape the singular values of the nominal plant *G* by W_1 and W_2. Then evaluate the ε_{opt} using (14.4). If $\varepsilon_{opt} < 0.25$, then go to step 1 to adjust the weight W_1.

Step 2 Select a controller structure *K(p)* and initialize several sets of parameters *p* as population in the 1st generation. Define the genetic parameters such as initial population size, crossover and mutation probability, maximum generation, etc.

Step 3 Evaluate the cost function J_{cost} of each chromosome using (14.9). Assign $J_{cost} = 100$, or large number if *K(p)* does not meet the constraints in our optimization problem. The fitness value is assigned as $1/J_{cost}$. Select the chromosome with minimum cost function as a solution in the current generation. For the first generation, Gen = 1.

Step 4 Increment the generation for a step.

Step 5 While the current generation is less than the maximum generation, create a new population using genetic operators and go to step 3. If the current generation is the maximum generation, then stop.

Step 6 Check performances in both frequency and time domains. If the performance is not satisfied, such as too low ε (too low fitness function), then go to step 3 to change the control structure. Low ε indicates that the selected control structure is not suitable for the problem.

14.4 Simulation and Experimental Results

In this paper, a buck-boost converter designed for a photovoltaic system is studied. Converter's parameters and considered variation ranges used in this paper are given in Table 14.1.

By (14.1), the nominal transfer function is found to be

$$G = \frac{-0.0042\text{s} + 480}{0.7896\text{s} + 72} \tag{14.10}$$

Both $\mathcal{H}_\infty$ loop shaping control and our proposed technique are applied to this converter. Firstly, we design a controller by the conventional $\mathcal{H}_\infty$ loop shaping procedure. Based on the concept of $\mathcal{H}_\infty$ loop shaping, W_1 and W_2 can be selected as

$$W_1 = 30\frac{(\text{s} + 26.7)}{(\text{s} + 0.001)},\ W_2 = \frac{100000}{s + 100000} \tag{14.11}$$

Figure 14.4(a) shows the plot of open loop shape of nominal plant and shaped plant. As seen in this figure, the bandwidth of the nominal plant is about 600 rad/sec. With these weighting functions, bandwidth of the desired control system is increased to 20,000 rad/sec. Significant improvement in terms of performances and robustness is carried out by these weighting functions. The shaped plant is written as

$$G_s = W_1 G W_2 = 30\frac{(s + 26.7)}{(s + 0.001)}\frac{(-0.0042\text{s} + 480)}{(0.7896\text{s} + 72)}\frac{(100000)}{(s + 100000)} \tag{14.12}$$

Table 14.1 Converter's parameters and considered variation ranges

Parameter	Name	Nominal value
R_L	Load resistant	40 Ω
V_o	Output voltage	30 V
V_i	Input voltage	12 V
L	Inductance	100 μH
C	Capacitor	470 μF
f_{sw}	Switching frequency	100 kHz

By applying the $\mathcal{H}_\infty$ loop shaping method, the optimal stability margin (ε_{opt}) is founded at 0.612 ($\gamma_{opt} = 1.6338$). This value indicates that the selected weighting function is compatible with the robust stability requirement. The $\varepsilon = 0.590$ (γ=1.6949), which is less than the optimal stability margin, is chosen to synthesize the controller. Based on the conventional technique in Section 14.3, the conventional $\mathcal{H}_\infty$ loop shaping controller is synthesized as following

$$K(s) = W_1 K_\infty W_2 = 30\frac{(s+26.7)}{(s+0.001)} \frac{(261841)(s+1.002\times 10^5)(s+26.9)}{(s^2+3.265\times 10^5 s+3.608\times 10^{10})(s+26.7)} \times \frac{(100000)}{(s+100000)} \quad (14.13)$$

As shown in (14.13), the controller is 5th order controller resulting in a complicated structure. Next, PI controller is investigated as a fixed-structure controller. The controller structure is expressed in (14.14). K_p and K_i are parameters that will be evaluated.

$$K(p) = K_p + \frac{K_i}{s} \quad (14.14)$$

The controller parameters, their ranges, and genetic algorithms parameters are selected as follows: $K_p \in [0,200]$, $K_i \in [0,1000]$, population size = 100, crossover probability = 0.7, mutation probability = 0.25, and maximum generation = 30. An optimal solution is obtained after 18 generations. The optimal solution is shown in (14.15), which has stability margin (ε) of 0.586 ($\gamma = 1.7064$).

$$K(p)^* = 21.84 + \frac{597.6}{s} \quad (14.15)$$

The final controller (*K*) is shown in (14.16).

$$K = \left(21.84 + \frac{597.6}{s}\right)\left(\frac{100000}{s+100000}\right) \quad (14.16)$$

Figure 14.3 shows plots of convergence of cost function J_{cost} versus generations by genetic algorithm. As seen in this figure, the optimal fixed-structure controller provides the satisfied stability margin at 0.586 ($\gamma = 1.7064$). The open loop bode diagrams of the nominal and shaped plants are shown in Fig. 14.4(a). As shown in this figure, at low frequency, the open loop gain of shaped plant is much larger than that of the nominal plant. This makes the system good in term of performance tracking and disturbance rejection. Open loop bode diagrams are plotted in Fig. 14.4(b) to verify the proposed algorithm. It is clearly shown that the loop shapes of $\mathcal{H}_\infty$ control and proposed PI controller are close to the desired loop shape. Figure 14.4(c) shows the step responses of the optimal solutions from the proposed robust PI and the conventional $\mathcal{H}_\infty$ controllers. As shown in this figure, the settling time of all responses is about 500 µsec.

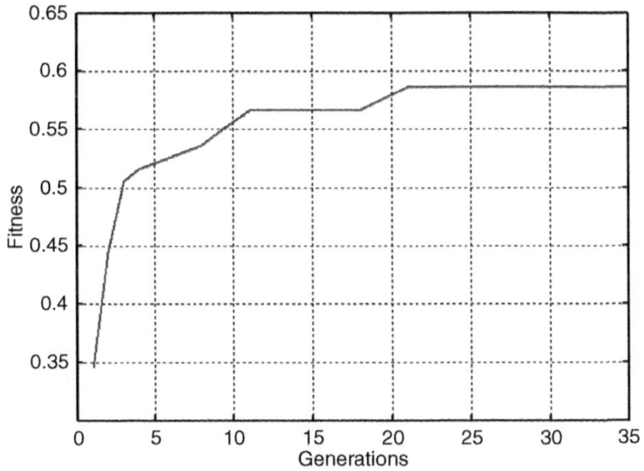

Fig. 14.3 Fitness functions versus iterations in genetic algorithm

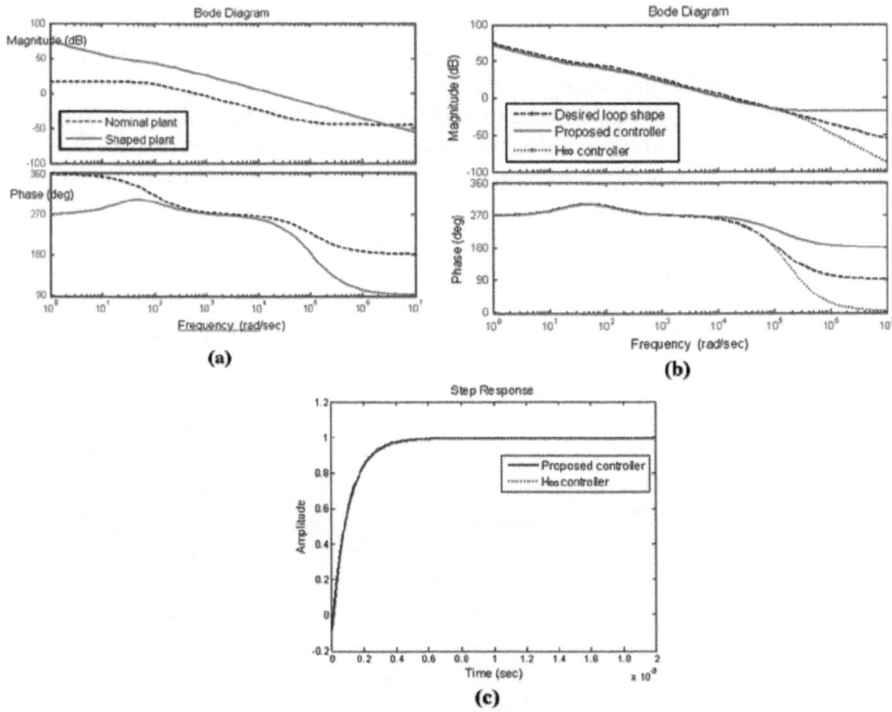

Fig. 14.4 (**a**) Bode plots of the nominal plant and the shaped plant (*desired loop shape*) (**b**) The desired loop shape and the loop shape by the conventional H_∞ loop shaping and the proposed PI, (**c**) Step responses by the proposed PI and H_∞ loop shaping controllers

In addition, in this paper, we apply the conventional PI controller based on the ISE method. In this method, the controller parameter is tuned in such a way that the integral of square error between output and desired response is minimized. However, to prevent the oscillations in the response, ISE with the model reference can be applied [13]. In this paper, we adopted the ISE with model reference to design a PI controller to make the settling time close to 500 μsec. The resulting controller is,

$$K_{ISE} = 19 + \frac{2800}{s} \tag{14.17}$$

The step responses of proposed and $\mathcal{H}_\infty$ controllers in the nominal plant and perturbed plant are shown in Fig. 14.5(a) and (b), respectively. As shown in the figures, the responses in perturbed plant are almost the same as the responses in the nominal plant with some different in the setting time. The results show that the designed system from the proposed controller and $\mathcal{H}_\infty$ loop shaping has a good performance and robustness. To verify the robust performance, we change the converter's parameters as: $R_L = 10\ \Omega$, $V_i = 10.8$ V, $L = 130\ \mu$H and $C = 2200\ \mu$F. The designed controllers in (14.16) and (14.17) are adopted to control the perturbed plant. Obviously, this condition (increase L and C and decrease the load resistance and input voltage) is worse than the nominal condition, because in this case the gain and phase of the plant are decreased in the crossover region.

Some experiments are performed to verify the effectiveness of the proposed controller. The nominal values in Table 14.1 are used to design a buck-boost converter with current mode control. A proposed robust PI controller in (14.16) and PI controller tuned by ISE method in (14.17) are used to control the converter. As seen in Figs. 14.5(a) and 14.6, the response of experimental result is almost the same as that of the simulation result.

To verify the robust performance of the system, an experiment is performed. The component values and operating point of the converter are changed to:

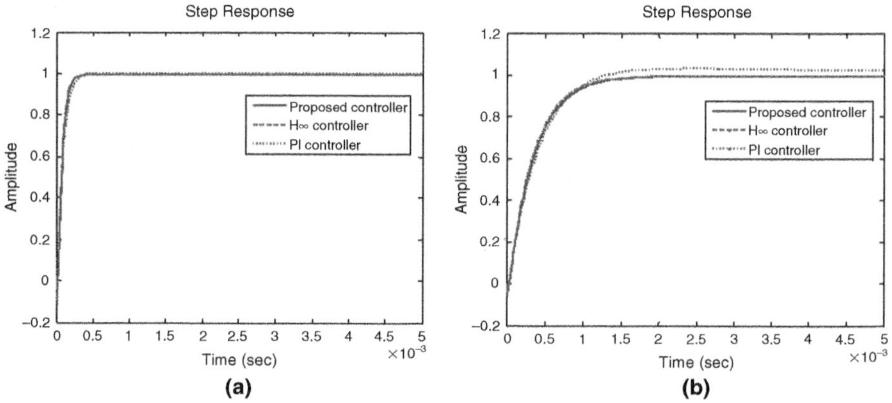

Fig. 14.5 (**a**) Step responses in the nominal plant. (**b**) Step responses in the perturbed plant

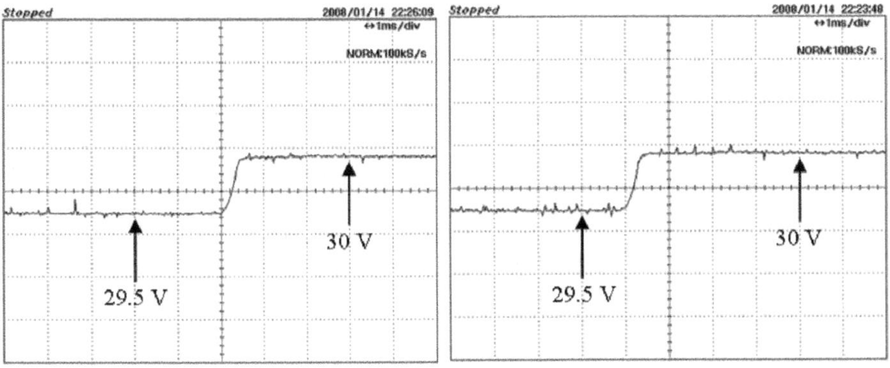

(a) (b)

Fig. 14.6 Step responses in nominal condition. (**a**) Proposed PI controller. (**b**) PI tuned by ISE method

$R_L = 10\ \Omega$, $V_i = 10.8$ V, $L = 130\ \mu$H and $C = 2200\ \mu$F. The performance is verified by using the step response. As shown in Fig. 14.7, the step response of the proposed controller is almost the same as the response in nominal conditions while the step response of the PI controller tuned by ISE method has an overshoot when the system parameters change. This can be verified that the robust performance of the proposed technique is better than that of PI controller by ISE method.

To verify the robust against the sudden change of load, an experiment were performed. As shown in Fig. 14.8, when the load is abruptly changed, the proposed controller can maintain the desired voltage.

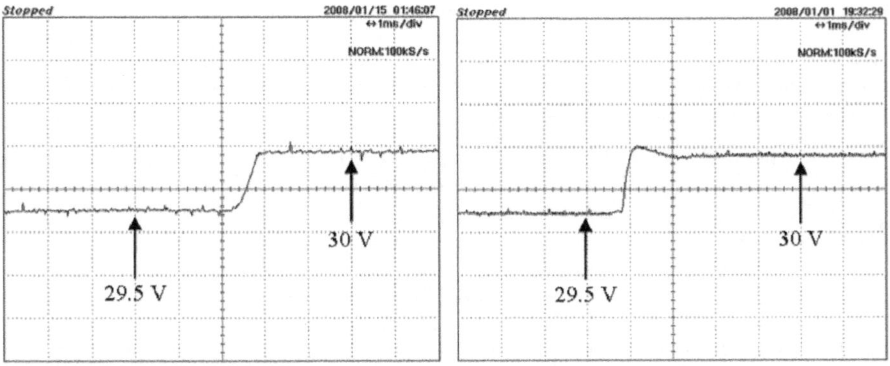

(a) (b)

Fig. 14.7 Step response in the closed loop in perturbed conditions. (**a**) Proposed PI controller. (**b**) PI tuned by ISE method

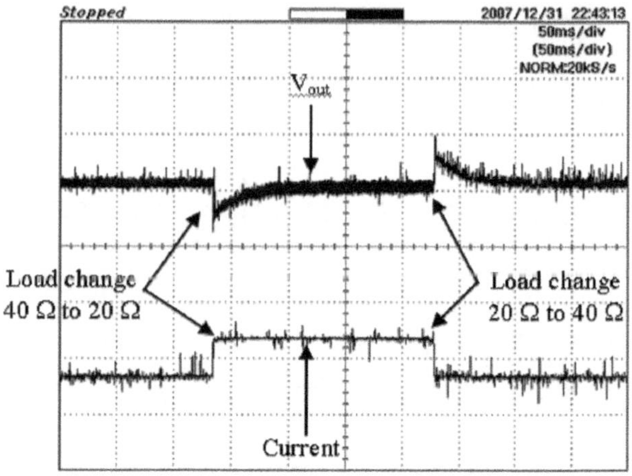

Fig. 14.8 Transient response of propose controller when the load change 40–20 Ω

14.5 Conclusion

Both $\mathcal{H}_\infty$ loop shaping and *GA based fixed-structure* $\mathcal{H}_\infty$ *loop shaping control*, are applicable in designing a robust controller for a current mode buck-boost converter. However, the proposed approach significantly improves the controller design in practical control viewpoint by simplifying the controller structure, reducing the controller order and retaining the robust performance. Structure of controller in the proposed technique is selectable. This is desirable, especially in the DC–DC converter which analog circuit is normally used to design the controller. In conclusion, by combining of these two approaches, genetic algorithms and $\mathcal{H}_\infty$ loop shaping; fixed-structure controller design can be achieved. Implementation in buck-boost converter assures that the proposed technique is valid and flexible.

Acknowledgments This research work is financially supported by the Thailand Research Fund (Project. No. MRG4980087) and the research fund from the Faculty of Engineering, King Mongkut's Institute of Technology Ladkrabang.

References

1. S. Buso, 1999, Design of a Robust Voltage Controller for a Buck-Boost Converter Using – Synthesis, *IEEE Trans. On Control Systems Technology*, 7(2): 222–229.
2. R.B. Ridley, 1989, A new continuous-time model for current-mode control, *Power Conversion Intell. Motion (PCIM) Conf. Proc.*, pp. 455–464.
3. J.G. Kassakian, M.F. Schlecht, and G.C. Verghese, *Principles of Power Electronics. Reading*, (Addison-Wesley, Boston, MA, 1991).
4. R. Naim, G. Weiss, and S. Ben-Yaakov, 1995, H_∞ control of boost converters: comparison to voltage mode, feed-forward and current mode controls, *PESC'95*, pp. 1327–1332.

5. G.C. Loannidis and S.N. Manias, 1999, H_∞ loop-shaping control schemes for the buck converter and their evaluation using μ-analysis, *IEE Proc.-Electr. Power Appl.*, 146(2).
6. B.-S. Chen and Y.-M. Cheng, 1998, A structure-specified optimal control design for practical applications: A genetic approach, *IEEE Trans. on Control System Technology*, 6(6).
7. B.-S. Chen, Y.-M. Cheng, and C.-H. Lee, 1995, A Genetic Approach to Mixed H_2/ H_∞ Optimal PID Control," *IEEE Trans. on Control Systems*, pp. 51–60.
8. S.-J. Ho, S.-Y. Ho, M.-H. Hung, L.-S. Shu, and H.-L. Huang, 1995 Designing structure-specified mixed H_2/ H_∞ optimal controllers using an intelligent genetic algorithm IGA, *IEEE Trans. on Control Systems*, 13(6): 1119–24.
9. D.C. McFarlane and K. Glover, 1992, A loop shaping design procedure using H_∞ synthesis, *IEEE Trans. On Automatic Control* AC-37 (6):759–769.
10. K. Zhou and J.C. Doyle, *Essential of Robust Control.* (Prentice-Hall, Englewood Cliffs, NJ, 1998), pp 315–327.
11. S. Skogestad and I. Postlethwaite, *Multivariable Feedback Control Analysis and Design.* (John Wiley & Son, New York, 1996), pp. 118, 376–380.
12. C. Houck, J. Joines, and M. Kay, 1995, A Genetic Algorithm for Function Optimization: A MATLAB Implementation by, NCSU-IE TR 95-09.
13. J. Lee and T.F. Edgar , 2004, ISE tuning rule revisited, *Automatica*, 40(8): 1455–1458.

Chapter 15
Ant Colony Optimization (ACO) Technique in Economic Power Dispatch Problems

Ismail Musirin, Nur Hazima Faezaa Ismail, Mohd. Rozely Kalil, Muhammad Khayat Idris, Titik Khawa Abdul Rahman and Mohd Rafi Adzman

Abstract Most of electrical power utilities in the world are required to ensure that electrical energy requirement from the customer is served smoothly in accordance to the respective policy of the country. Despite serving the power demands of the country, the power utility has also to ensure that the electrical power is generated within minimal cost. Thus, the total demand must be appropriately shared among the generating units with an objective to minimize the total generation cost for the system in order to satisfy the economic operation of the system. Economic dispatch is a procedure to determine the electrical power to be generated by the committed generating units in a power system so that the total generation cost of the system is minimized, while satisfying the load demand simultaneously. This paper presents the economic power dispatch problems solved using Ant Colony Optimization (ACO) technique. ACO is a meta-heuristic approach for solving hard combinatorial optimization problems. In this study, the proposed technique was tested using the standard IEEE 26-Bus RTS and the results revealed that the proposed technique has the merit in achieving optimal solution for addressing the problems. Comparative studies with other optimization technique namely the artificial immune system (AIS) were also conducted in order to highlight the strength of the proposed technique.

Keywords Ant colony optimization · Economic dispatch · Meta-heuristic · Power systems · Optimization

15.1 Introduction

The main role of electrical power utility is to ensure that electrical energy requirement from the customer is served. However in doing so, the power utility has also to ensure that the electrical power is generated with minimum cost. Hence, for economic operation of the system, the total demand must be appropriately shared

I. Musirin (✉)
The Faculty of Electrical Engineering, Universiti Teknologi MARA Malaysia
e-mail: i_musirin@yahoo.co.uk

S.-I. Ao et al. (eds.), *Trends in Communication Technologies and Engineering Science*,
Lecture Notes in Electrical Engineering 33, DOI 10.1007/978-1-4020-9532-0_15,

among the generating units with an objective to minimize the total generation cost for the system. Economic Dispatch is a procedure to determine the electrical power to be generated by the committed generating units in a power system so that the total generation cost of the system is minimized, while satisfying the load demand simultaneously. To address this problem, optimization is a priori in solving the cost minimization problems. Power system optimization is an important field in the operation, planning and control of power systems. Many modern heuristic techniques to the solution of complex power system optimization problems have been proposed, each differing in their method of representation, implementation and solution procedure [1]. Economic power dispatch is one of the fundamental problems in power system operation and planning [2]. It is defined as the process of allocating generation levels to the generating units so that the system load is supplied entirely and most economically [3]. It concerned on the minimization of an objective function, usually the total cost of generation, while satisfying both the equality and inequality constraints [4]. Depending on load variations, the output of generators has to be changed to meet the balance between loads and generation power to make the system efficient. Recently, a new global optimization techniques known as Ant Colony Optimization (ACO) has become a candidate for many optimization applications. The ACO has solved several combinatorial optimization traveling salesman (ATSP), quadratic assignment problem (QAP), optimal design and scheduling problem of thermal units [5]. ACO has been used to address various optimization problems as reported by [6–12]. This paper presents the application of Ant Colony Optimization (ACO) in solving economic power dispatch problems. The objective function is cost minimization considering loss of transmission implemented on a 6 generating unit system. Comparison was implemented using AIS.

15.2 Economic Dispatch Problem Formulation

Economic Dispatch problem can be solved by minimizing the cost of generation in the system. The solution gives the optimal generation output of the online generating units that satisfy the system's power balance equation under various system and operational constraints. The Economic Dispatch problem can be formulated mathematically as follows:

15.2.1 Objective Function

$$\text{minimize cost} = \sum_{i}^{Ng} \text{Fi(Pi)} \tag{15.1}$$

where *cost* is the operating cost of power system. N_g is the number of units. $F_i(P_i)$ is the cost function and P_i is the power output of the unit *i*. $F_i(P_i)$ is usually approximated by a quadratic function of its power output P_i as:

$$Fi(Pi) = a_i P_i^2 + b_i P_i + c \tag{15.2}$$

where a_i, b_i and c_i are the cost coefficients of unit i. Wire drawing effects occurs when each steam admission valve in a turbine starts to open, and at the same time a rippling effect on the unit curve is produced. To model the effects of "valve-points", a recurring rectified sinusoid contribution is added to the cost function. The result is:

$$Fi(Pi) = a_i P_i^2 + b_i P_i + c + g_i \; sin[h_i(P_i - P_i^{\min})] \tag{15.3}$$

where g_i and h_i are the valve-points coefficients. $P_i^{\min}$ is the lower generation limit of unit i. Ignoring valve point effects, some inaccuracy would be introduced into the resulting dispatch [3].

15.2.2 Constraint Equations

1. Unit Operation Constraints are given by:

$$P_i^{\min} \le P_i \le P_i^{\max} \qquad \mathrm{i} = 1, 2, \ldots, N_{\mathrm{g}} \tag{15.4}$$

   where $P_i^{\min}$ and P_i^{max} are the lower and upper generation limit of unit i.

2. Power Balance equation:

$$\sum_{i=1}^{N_g} P_i = P_L + P_D \tag{15.5}$$

   where P_D is the demand and P_L is transmission loss. The transmission loss can be calculated bye the B-coefficients method or power flows analysis.
   B-coefficients used in the power system is:

$$P_L = P^T BP + P^T B_0 + B_{00} \tag{15.6}$$

   where P is a – dimensional column vector of the power output of the units.

3. Line flow constraints:

$$|Lf_i| \le Lf_i^{\max} \qquad i = 1, 2, \ldots N_L \tag{15.7}$$

   where Lf_i is the MW line flow, Lf_i^{max} is the allowable maximum flow of line i (line capacity), and NL is the number of transmission lines subject to line capacity constraints.

4. System stability constraints:

$$\left|\partial_i - \partial_j\right| \le \partial_{ij}^{\max} \qquad i, j = 1, 2, \ldots ND \tag{15.8}$$

where ∂_i, ∂_j are voltage angle of bus i and j. $\partial_{ij}{}^{max}$ is the allowable maximum voltage angle

15.2.3 Line Flow Constraints

Let $X_i = (x_1, x_2, \ldots, xN_g)$ be a vector denoting the ith individual of the ant colony, where Ng is the number of units and X_i is the generated output of unit $_i$. At initialization phrase, X_i is selected randomly from the selected region S.

15.2.4 Objective Function and Feasible Region

In order to minimize the objective function of ED the constraints have to be obeyed. Penalty function was used to transform those constraints which have difficulties to deal with in the feasible region including power balance, etc. The feasible region S is determined by the unit operation constraints.

15.2.5 Parameter Setting

The parameter used here are $N = 5$, $\tau_0 = 1.0$, $\gamma_1 = \lambda_2 = 1$, $\rho = 0.6$, $q_0 = 0.6$.

15.3 Ant Colony Optimization Technique

Ant Colony Optimization (ACO) was first developed by Dorigo et al. [5]. It is proposed as a viable new approach to stochastic combinatorial optimization. The idea is based on the following observation. A colony of ants is able to succeed in a task to find the shortest path between the nest and the food source. It was found that ants deposit a chemical substance trail, called *pheromone* on the ground when they move. This pheromone can be observed by other ants and motivates them to follow the path with a high probability. The following example shows how over time, the shortest paths are found through this self-reinforcing process.

In Fig. 15.1a, suppose that ants move between food source *A* and nest *E* on a straight line. Then, as shown in Fig. 15.1b, suddenly an obstacle appears and the path was cut off. At position *B*, ants have to decide whether to turn right or left. Under this situation, they may choose to turn right *(BCD)* or left *(BHD)*. For all that, due to path *BCD* is shorter than *BHD*, the first ant following path *BCD* will reach *D* before the first ant following path *BHD*. So, a higher level of pheromone intensity on the right will give information to the followers as a higher probability to turn right, which is shown in Fig. 15.1c. For a while, the shorter path will collect larger amount of pheromone trail than the longer path. Therefore, more ants will be increasingly guided to move on the shorter path. The result is that, very soon all ants will choose the shortest path in their movement.

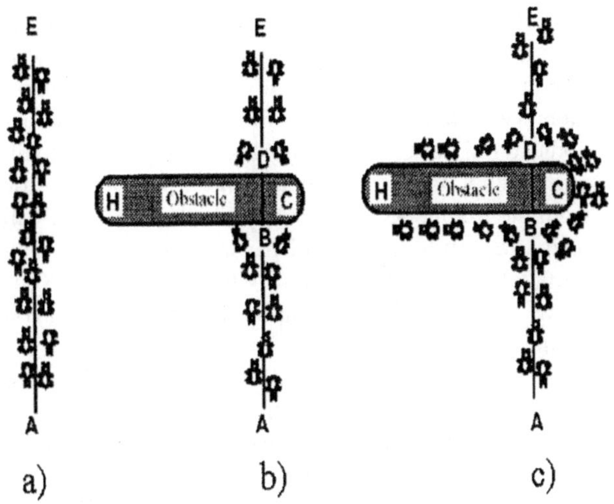

Fig. 15.1 Illustration of the real ant behavior

Like neural networks, ACO is based on biological research – in this case is the research into the foraging behavior of ant colonies. While individual ants essentially move at random, ant colonies can be seen as a system of collaborating agents pursuing a common goal: finding the quickest path to a food source. Ants communicate with the aid of a chemical called pheromone in a process referred to a "stigmergy". Pheromones are produced by ants and they deposit them on trails when walking in search of food [1]. ACO is based on the indirect communication of a colony of simple agents, called (artificial) ants, mediated by (artificial) pheromone trails. The pheromone trails in ACO serve as distributed, numerical information, which the ants use to probabilistically construct solutions to the problem being solved, and which the ants adapt during the algorithm's execution to reflect their search experience.

15.4 Ant Colony Optimization (ACO) Algorithm

ACO algorithm is inspired by the behaviour of real ant colonies used to solve combinatorial optimization problem [3]. The real ants lay down in some quantity an aromatic substance, known as pheromone, in their way to food. The pheromone quantity depends on the length of the path and the quality of the discovered food source. An ant chooses an exact path in connection with the intensity of the pheromone.

The pheromone trail evaporates over time if no more pheromone is laid down. Other ants are attracted to follow the pheromone trail. Therefore, the path will be marked again and it will attract more ants to use the same path. The pheromone trail on paths leading to rich food sources close to the nest be more frequented and will therefore grow faster. In this way, the best solution has more intensive pheromone and higher probability to be chosen. The described behaviour of real ant colonies can be used to solve combinatorial optimization problems by simulation: artificial ants searching the solution space by transiting from nodes to nodes. The artificial ants move usually associated with their previous action, stored in the memory with a specific data structure. The pheromone consistencies of all paths are updated only after the ant finished its tour from the first node to the last node. Every artificial ant has a constant amount of pheromone stored in it when the ant proceeds from the first node. The pheromone that stored in will be distributed average on the path after artificial ants finished its tour. The amount of pheromone will be high if artificial ants finished its tour with a good path. The pheromone of the routes progressively decreases by evaporation in order to avoid artificial ants stuck in local optima [8].

The overall steps of the ACO can be represented in the flow chart in Fig. 15.2. In this study, there is some modifications performed on the ACO algorithm in order to make it suitable for the application in power system. The following sections explain the ACO algorithms in detail.

Step 1: Initialization

In the beginning, the ACO parameters have to be specified during initialization process. The main difficulty with the ACO program is that; an appropriate choice of ACO control parameters is necessary before applying the ACO program. The choice of parameters has been obtained by trial and error. In order to get better result in the development of the ACO's program; the parameters must be selected carefully.

On the other hand, every parameter requires to be set for limiting the search range in order to avoid large computation time. During initialization; the following parameters need to be initialized for the purpose of ACO implementation.

n	: no. of nodes
M	: no. of ants
t_{max}	: maximum iteration
d_{max}	: maximum distance for every ants tour
	: parameter, which determines the relative importance of pheromone versus distance ($\beta > 0$)
	: heuristically defined coefficient ($0 < \rho < 1$)
	: pheromone decay parameter ($0 < \alpha < 1$)
Q_0	: parameter of the algorithm ($0 \leq q_0 \leq 1$)
τ_o	: initial pheromone level

The maximum distance for every ants tour (d_{max}) can be calculated by using this formula:

$$d_{\max} = \left[\sum_{i=1}^{n-1} di \right] \tag{15.9}$$

$$di = |r - \max(u)| \tag{15.10}$$

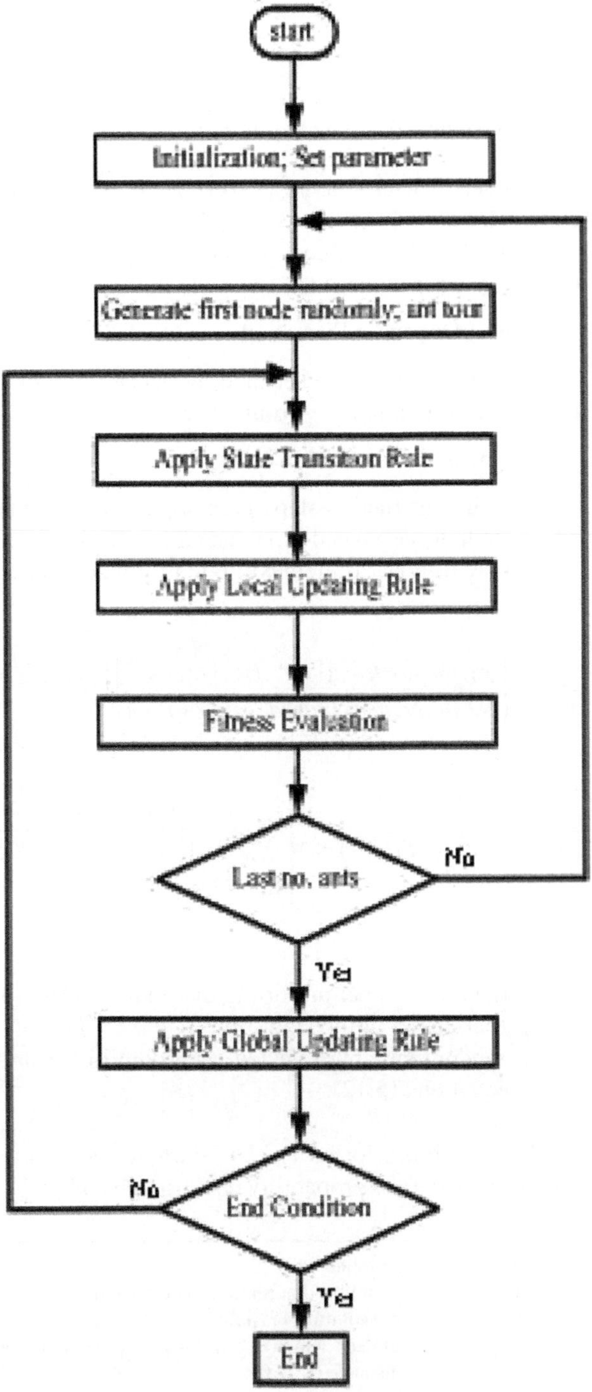

Fig. 15.2 Flow chart for ACO

where:

a: current node
u: unvisited node
d_i: distance between two nodes

In order to determine d_{max}, every node in the list is required to be set as the first node. The node in the list can only be selected once for every tour of ant. Each ant will select the next node which has higher distance from the current node. This process will repeat until the end node in the list.

Step 2: Generate First Node

The first node will be selected by generating random number based on the uniform distribution, ranging from 1 to n.

Step 3: State Transition Rule

At each construction step, ant k applies a state transition rule in order to decide which node to be visited next. The ant k, which is currently positioned at current node (r) will move to the next node (s) by applying the state transition rule given by:

$$s = \begin{cases} \arg\max_{u \varepsilon} J_{k(r)} \left\{ [\tau(r,u)] . \left[\eta(r,u)^{\beta} \right] \right\}, & if\ q \leq q_0\ (\exp\ loitation) \\ S, & otherwise\ (biased\ \exp loration) \end{cases} \tag{15.11}$$

$$P_k(r,s) = \begin{cases} \dfrac{[\tau(r,s)].[\eta(r,s)^{\beta}]}{\sum_{u \varepsilon J_{k(r)}} [\tau(r,u)].[\eta(r,u)^{\beta}]}, & if\ s \varepsilon J_{k(r)} \\ 0, & otherwise \end{cases} \tag{15.12}$$

where:

q: random number uniformly distributed in [0 to 1]
q_0: parameter of the algorithm ($0 \leq q_0 \leq 1$)
S: random variable selected according to the probability distribution given in equation (15.12)

The probability for an ant k at current node (r) to choose the next node (s) is calculated using the probability equation given by:

τ	:	Pheromone
$J_{k(r)}$	:	set of nodes that remain to be visited by ant k positioned on node (to make the solution feasible)
β	:	parameter, which determines the relative importance of pheromone versus distance ($\beta > 0$)
η	:	$1/d$, is the inverse of the distance $d(r,s)$

In equation (15.11) and (15.12) the pheromone on path $\tau(r,s)$ is multiplied by the heuristic value $\eta(r,s)$ in order to determine the selection of paths which are shorter and have a greater amount of pheromones. The parameter q_0 determines the relative importance of the exploitation versus exploration condition: an ant at node r (current node) has to choose a node s (next node) to move. This is determined by the value of q randomly where $(0 \leq q \leq 1)$. If $(q \leq q_0)$ the best path will be determined based on equation (15.11), (i.e. in the exploitation mode). On the other hand, if $(q \geq q_0)$ the best path will be determined based on equation (15.12), (i.e. in the exploration mode). The process to determine the next node (s) starts by calculating the probability of choosing the next node using equation (15.12). After the calculation of probability, the value of q is then generated randomly. If $(q \leq q_0)$, node 4 was selected as next node (s) which has the highest probability (i.e. in the exploitation mode). On the other hand, if $(q \geq q_0)$, the next node (s) will be selected randomly from the list of unvisited nodes (i.e. in the exploration mode).

Step 4: Local Updating Rule

Local updating rule is a process used to change the amount of pheromone to the visited paths during the construction of solution. The amount of pheromones can be updated based on the following equation:

$$\tau(r,s) \longleftarrow (1-\rho)\tau(r,s) + \rho \cdot \Delta\tau(r,s) \qquad (15.13)$$

where:

ρ = heuristically defined coefficient $(0 < \rho < 1)$
$\Delta\,\tau(r,s) = \tau_o$
τ = initial pheromone trail

This local updating rule will shuffle the tours, so that the early nodes in one ant's tour may be explored later in other ants' tours. The amount of pheromone on visited paths will be reduced so that the visited paths become less desirable and therefore will be chosen with lower probability by the other ants in the remaining steps of an iteration of the algorithm. When the new τ value is lower, the probabilities to select the same nodes also become lower.

Step 5: Fitness Evaluation

Fitness evaluation is performed after all ants have completed their tours. In this step, the control variable (x) is computed using the following equation:

$$x = \frac{d}{d\max} \times x\max \qquad (15.14)$$

where:

d : distance for every ants tour
$x_{\max}$: maximum x
d_{max} : maximum distance for every ants tour

The values of variable x will be assigned for the fitness in the ACO algorithm.

Step 6: Global Updating Rule

Global updating rule is a process used to update the amount of pheromones generated by the ant which has constructed the shortest tour from the beginning of the tour. There will be only one ant is allowed to update the amount of pheromone which determines the best fitness. The amount of pheromones is updated after all ants have completed their tours by applying equation (15.15).

$$\tau(r,s) \longleftarrow (1-\alpha)\tau(r,s) + \alpha.\Delta\tau(r,s) \tag{15.15}$$

where:

$$\Delta(r,s) = \begin{cases} (L_{gb}), & if\ (r,s)\ \varepsilon\ global - best - tour \\ 0, & otherwise \end{cases}$$

α: pheromone decay parameter ($0 < \alpha < 1$)
L_{gb}: length of the globally best tour from the beginning of the tour

In this case, the paths belonged to the globally best tour (i.e. the best fitness) of current iteration will receive reinforcement. For the next iteration; the first node of globally best tour in the first iteration will be selected as first node by the each ant.

Step 7: End Condition

The algorithm stops the iteration when a maximum number of iterations (t_{max}) have been performed. Every tour that was visited by ants should be evaluated. If a better path is discovered in the process, it will be kept for the next reference. The best path selected between all iterations engages the optimal scheduling solution. As a consequence, ants never converge to common path. This is observed experimentally, and it is a desirable property. If ants explore different paths then there is a higher probability that one of them will find an improving solution. There are cases where solutions converge to same tour which would make the use the no. of ants (m) pointless.

15.5 Results And Discussion

The results tabulated in Table 15.1 are obtained when the ACO parameter are set to the following values at the initialization process:

$n = 10, m = 5,\ t_{\max} = 5, d_{\max} = 49, \beta = 2, \alpha = 0.1, q_o = 0.6, \tau = 0.6$ and $\tau_o = 1$.

Table 15.1 Generating limit of generating units

Generating unit	Minimum MW	Maximum MW
1	100	500
2	50	200
3	80	300
4	50	150
5	50	200
26	50	120

The developed technique was employed to determine the economic dispatch for the IEEE 26-Bus RTS that has 6 generating units. The test was performed with bus 19 subjected to the load variation. The generator's operating costs in \$/h, with P_i in MW are as follows [5]:

$$C_1 = 240 + 7.0P_1 + 0.0070P_1^2$$
$$C_2 = 200 + 10.0P_2 + 0.0095P_2^2$$
$$C_3 = 220 + 8.5P_3 + 0.0090P_3^2$$
$$C_4 = 200 + 11.0P_4 + 0.0090P_4^2$$
$$C_5 = 220 + 10.5P_5 + 0.0080P_5^2$$
$$C_{26} = 190 + 12.0P_26 + 0.0075P_{26}^2$$

The total cost of generation is given by,

$$C_T = C_1 + C_2 + C_3 + C_4 + C_5 + C_{26}$$

In this study, C_T is taken as the fitness which was minimized during the optimization process. On the other hand, Table 15.2 tabulates the minimum generation cost obtained by ACO while varying the load (P_{load}) at a selected bus (i.e. bus 19) until it reaches the stability limit. The total loss increased accordingly as the load is increased. The constrains need to be considered are: $V_m \geq 0.95\ p.u$ and $V_m \leq 1.06\ p.u.$ and $T_{loss} \leq 20$. The developed ACO has successfully **identified** the generated power by each generating unit to minimize the total cost in the system. The minimization of total generation cost has also taken account the systems' limits and constraints. Table 15.2 shows the minimum generation cost obtained by ACO while varying the load (P_{load}) at a selected bus (i.e. bus 19) until it reaches the stability limit. The total loss increased accordingly as the load is increased.

Table 15.3 tabulates the output power by all the generating units in the system with the optimized cost of \$15448.76 as highlighted in the table. In order to achieve this minimal cost, the amount of generating unit for each generator is given in the table.

To realize the effectiveness of the proposed ACO technique, comparative studies were performed by reevaluating the process using AIS; with results given in Table 15.4. It is obvious that ACO outperformed AIS in terms of cost minimization, implemented with much faster execution time.

Table 15.2 Minimum cost of generation while varying the load at bus 19

Load at bus 19 (MW)	Load (MW)	Total losses (MW)	Generated power (MW)	Generation cost ($/h)
0	1199	10.9335	1209.934	14445.00
10	1209	11.1885	1220.189	15117.00
20	1219	11.4680	1230.468	14712.00
30	1228	11.6693	1239.669	14846.00
40	1240	11.9216	1251.922	14983.00
50	1250.2	12.3615	1262.562	15118.00
60	1260.5	12.5051	1273.005	15253.00
70	1270.8	12.7972	1283.597	15390.00
80	1270.8	13.0903	1283.89	15526.00
90	1291.4	13.4135	1304.814	15664.00
100	1301.8	13.7669	1315.567	15802.00
110	1312.1	14.1313	1326.231	15940.00
120	1322.5	14.5272	1337.027	16080.00

Table 15.3 Power output by generating units at minimum generating cost

P_1/MW	449.8322
P_2/MW	170.2403
P_3/MW	264.5261
P_4/MW	124.3220
P_5/MW	170.2403
P_{26}/MW	96.7710
P_D/MW	1275.9319
Fitness/$\$h^{-1}$	**15448.76**

Table 15.4 Comparison in minimum generating cost

Optimization technique	Fitness/$\$h^{-1}$	Execution time/sec.
Artificial Immune System (AIS)	15481.74	113.14000
ACO	**15448.76**	**8.5230**

15.6 Conclusion

Ant colony optimization has been proposed for solving economic dispatch problems with cost minimization as the objective function. In this study, the developed ACO engine is capable to implement the power to be generated by each generating unit at the minimum cost and meet the demand requirement. The results obtained are also compared with AIS to investigate its computation capability. The comparison shows that ACO is versatile, robust and efficient. Further work is required to develop techniques for searching the neighborhood, and present more efficacious sufficient conditions for convergence.

References

1. K.S. Swarup, Ant colony optimization for economic generator scheduling and load dispatch, Proceedings of the 6th WSEAS Int. Conf. on evolutioanary computing, Lisbon, Portugal, June 16–18, 2005, pp. 167–175.
2. Y.-H. Hou, Y.-W., L.-J. Lu and X.-Y. Xiong, Generalized ant colony optimization for economic dispatch of power systems, IEEE Trans. Power Apparatus and Systems, pp. 225–229.
3. M. Yoshimi, K.S. Swarup and Y. Izui, Optimal economic power dispatch using genetic algorithm, Proceedings of the Second International Forum on Applications of Neural Networks to Power Systems, ANNPS '93, 19–22 April 1993, pp. 157–162.
4. B.H. Chowdhury and S. Rahman, A review of recent advances in economic dispatch, IEEE Transactions on Power Systems, Vol. 5, No 4, Nov. 1990 pp. 1248–1259.
5. D. Nualhong, S. Chusanapiputt, S. Phomvuttisarn, T. Saengsuwan and S. Jantarang, Diversity control approach to ant colony optimization for unit commitment problem, 2004 IEEE Region 10 Conference, TENCON 2004, 21–24 Nov. 2004, Vol. C, pp. 488–491.
6. L. Slimani and T. Bouktir, Economic power dispatch of power system with pollution control using multiobjective ant colony optimization, *International Journal of Computational Intelligence Research (IJCIR)* 2007, Vol. 3, No. 2, pp. 145–153.
7. Y. Wen and T.-J. Wu , Dynamic window search of ant colony optimization for complex multistage decision problems, IEEE International Conference on Systems, Man and Cybernetics, 2003. 5–8 Oct. 2003 , Vol. 5, pp. 4091–4097.
8. H. Ying-Tung , C. Cheng-Long and C. Cheng-Chih, Ant colony optimization for best path planning, International Symposium on Communications and Information Technologies 2004 (ISCIT 2004), Sapporo, Japan, 26–29 October 2004, pp. 109–113.
9. D. Lukman and T.R. Blackburn, Loss minimization in load flow simulation in power system, 4th IEEE International Conference on Power Electronics and Drive Systems, 2001, 22–25 Oct. 2001, Vol. 1, pp. 84–88.
10. C.J. Bridenbaugh, D.A. DiMascio and R. D'Aquila, Voltage control improvement throught capacitor and transformer tap optimization, IEEE Transactions on Power Systems, Vol. 7, No. 1, February 1992, pp. 222–227.
11. L. Li, Z. Liao, S. Huang and G. Wang, A distributed model for power system restoration based on ant colony optimization algorithm, IEEE/PES Asia and Pacific Transmission and Distribution Conference and Exhibition, 2005, pp. 1–5.
12. K.Y. Lee and J.G. Vlachogiannis, Optimization of power systems based on ant colony system algorithms: an overview, 13th International Conference on Intelligent Systems Application to Power Systems, 2005. 6–10 Nov. 2005 pp. 22–35.

Chapter 16
Direction of Arrival Estimation Using a Phase Array Antenna

H.K. Hwang and Zekeriya Aliyazicioglu

Abstract Accurate estimation of a signal direction of arrival (DOA) has received considerable attention in communication and radar systems of commercial and military applications. Although an ideal array antenna has uniform spacing between the array elements, there will be always some small deviation of spacing from an array element to adjacent elements. This small deviation causes an error in the electrical angle of each element. Additionally, when the array system is not perfectly calibrated, there will be a random phase error. Statistical analysis is carried out in this chapter to investigate the effect of random element position variation and random phase variation on DOA estimation.

Keywords Direction of arrival · Array antenna · Electrical angle · Phase error · Statistical analysis

16.1 Introduction

Accurate estimation of a signal direction of arrival (DOA) has received considerable attention in communication and radar systems of commercial and military applications. Radar, sonar, seismology, and mobile communication are a few examples of the many possible applications. For example, in defense application, it is important to identify the direction of a possible threat. One example of commercial application is to identify the direction of an emergency cell phone call in order to dispatch a rescue team to the proper location.

DOA estimation using a fixed antenna has many limitations. Antenna mainlobe beamwidth is inversely proportional to its physical size. Improving the accuracy of angle measurement by increasing the physical aperture of the receiving antenna is not always a practical option. Certain systems such as a missile seeker or aircraft antenna have physical size limitations; therefore they have relatively wide main-

H.K. Hwang (✉)
California State Polytechnic University, Pomona, CA, USA

S.-I. Ao et al. (eds.), *Trends in Communication Technologies and Engineering Science*, Lecture Notes in Electrical Engineering 33, DOI 10.1007/978-1-4020-9532-0_16,

lobe beamwidth. Consequently, the resolution is quite poor and if there are multiple signals falling in the antenna mainlobe, it is difficult to distinguish between them.

Instead of using a single antenna, an array antenna system with innovative signal processing can enhance the resolution of a signal DOA. An array sensor system has multiple sensors distributed in space. This array configuration provides spatial samplings of the received waveform. A sensor array has better performance than the single sensor in signal reception and parameter estimation. A sensor array also has applications in interference rejection [1], electronic steering [2], multi-beam forming [3], etc.

There are many different super resolution algorithms including spectral estimation, model based, and eigen-analysis to name a few [4–6]. In this chapter, we discuss the application of estimating the DOA of multiple signals using uniform linear array (ULA) antenna with a class of Multiple Signal Classification (MUSIC) algorithms known as root-MUSIC. It does not require using a scan vector; resolution improvement does not necessarily require additional processing power. Detailed simulation results for the algorithm to demonstrate the performance are presented in this chapter.

Although an ideal array antenna has uniform spacing between the array elements, there will be always some small deviation of spacing from an array element to adjacent elements. This small deviation causes an error in the electrical angle of each element. Additionally, when the array system is not perfectly calibrated, there will be a random phase error. Statistical analysis is carried out in this chapter to investigate the effect of random element position variation and random phase variation on DOA estimation.

16.2 Sensor Array System

A sensor array system has multiple sensors distributed in space. This array configuration provides spatial sampling of the received waveform. We use an array antenna with an M element uniform linear array (ULA) in this chapter. Figure 16.1 shows the general configuration for a ULA antenna having M elements arranged along a straight line with the distance between sensor elements, be $d = \lambda/2$, where λ is the incoming signal wavelength. The angle of the incoming signal, θ, is measured relative to the antenna bore sight.

For a conventional antenna, the main lobe beam width (MLBW) is given by, in radians

$$MLBW = m\frac{\lambda}{D} \tag{16.1}$$

where D is the length of the antenna array and m is a proportionality constant, in most cases $m \approx 1$ [6].

Consider a general array of M sensors as shown in Fig. 16.1. The coordinate of the ith sensor is $i = 0, 1, \ldots, M{-}1$. Suppose a plane target signal waveform comes

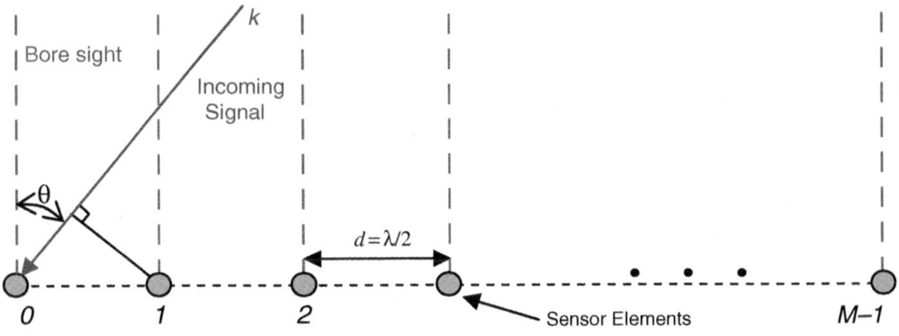

Fig. 16.1 One dimensional array antenna

from the direction of $k = \sin\theta$. The difference of the propagation path of this wave between the origin and the ith sensor Δd_i is $\Delta d_i = d_i \sin\theta$. The corresponding propagation time delay τ_i is

$$\tau_i = \frac{\Delta d_i}{c} = \frac{d_i \sin\theta_i}{c} \tag{16.2}$$

where $c$is the speed of light.

If the bandwidth of signals is sufficiently narrow, then the data picked up by different sensor elements are related by a pure phase factor. The relative phase shift of the ith sensor with respect to the reference sensor at the origin is

$$\beta_i = -\frac{2\pi}{\lambda} d_i \sin\theta_i. \tag{16.3}$$

To avoid the effect of grating lobes, the distance between the two neighboring sensors has to be no more than one half of the wavelength. For the sensor array with M elements, we can define the array input vector $\mathbf{x}(t)$ and the array weighting vector $\mathbf{w}$ as

$$\mathbf{x}(t) = \left[x_0(t), x_1(t), \ldots, x_{M-1}(t)\right]^T, \tag{16.4}$$

$$\mathbf{w} = \left[w_0, w_1, \ldots, w_{M-1}\right]^T. \tag{16.5}$$

where $x_i(t)$ is the data input to the ith sensor and w_i^* is the weight of the ith sensor. The sensor array output $\mathbf{y}(t)$ is

$$\mathbf{y}(t) = \mathbf{w}^H \mathbf{x}(t) \tag{16.6}$$

where the superscript H represents the complex conjugate transpose (Hermitian) of the matrix.

Suppose there are L independent signal sources impinging on the antenna and we want to use a sensor array system to identify their directions of arrival (DOA). The input signal to each individual sensor is the combination of L independent signals. Every sensor in the array also receives random environmental ambient noise. This noise is modeled as Additive White Gaussian Noise (AWGN). The input waveform of the ith sensor element $x_i(t)$ is given by

$$x_i(t) = \sum_{k=1}^{L} s_{k,i}(t) + n_i(t), \quad i = 0, 1, \ldots M - 1 \tag{16.7}$$

where $s_{k,i}(t) = s_{k,o}(t - \tau_{k,i})$, and $s_{k,o}(t)$ denote the kth signal picked up by the sensor at the origin, $n_i(t)$ is the noise at ith sensor, and $\tau_{k,i}$ is the relative delay of kth signal at the ith sensor.

For the narrowband input signals, signal $s_{k,i}(t)$ is related to the signal $s_{k,o}(t)$ by a phase shift factor of $\beta_{k,i}$. If the input signals have a wide bandwidth, the delay time of signal at ith sensor from reference signal at the origin may not be an integer multiple of the sampling time; additional interpolation filtering is required to emulate their delay. The weighted sum of samples of all sensors forms the array output. To estimate the DOA of wideband signals, each single weight is replaced by a tape delay lines filter. Such a processor is referred to as the Space Time Adaptive Processor (STAP) [7]. In this chapter, only narrowband signal is considered.

16.3 Root-Music Algorithm

The eigen-analysis method assumes that the received data can be decomposed into two mutually orthogonal subspaces. One is the signal plus noise subspace, the other is the noise only subspace. There are several important eigen-analysis methods; the Pisarenko Harmonic Decomposition [8], Multiple Signal Classification (MUSIC) [9] and Estimation of Signal Parameters via Rotational Invariance Techniques (ESPRIT) [10]. The MUSIC algorithm, using temporal averaging and spatial smoothing, is briefly presented for convenience.

Assume the number of signals impinging on the array antenna is L, and the array antenna has M elements, usually $M > L$. For finite number of data sequences, a sequence of received data can be considered as a vector in the sample space. If the noise is assumed to be white, it spans over the entire sample space. For signals with sufficient narrowband, they span over L dimensional subspace. For example, radar signal of a moving target is a sinusoid with a frequency equal to the Doppler frequency shift. If there are L targets, the received waveform at the reference sensor at the origin $x_0(n)$ can be expressed as

$$x_0(n) = \sum_{k=1}^{L} \alpha_k e^{j2\pi f_k n} + n_0(n) \tag{16.8}$$

where α_k, f_k, and $k = 1, 2, \ldots, L$ are the complex amplitude and frequency of the kth sinusoid, and $n_0(n)$ is the additive white noise with variance σ^2. Since the signal at the other sensors has a relative phase shift, the waveforms of the other sensors are

$$x_i(n) = \sum_{k=1}^{L} \alpha_k e^{j2\pi f_{k,i} n} e^{j\beta_{k,i}} + n_i(n), \quad i = 1, 2, \ldots, M - 1 \tag{16.9}$$

where $\beta_{k,i} = -\frac{2\pi}{\lambda} x_i \sin\theta_k, \quad k = 1, 2, \ldots, L$ for one dimensional array.

One of the eigen-analysis methods is the "Multiple Signal Classification" (MUSIC). MUSIC is especially practical in handling radar signals.

Define the signal matrix **S** as

$$\mathbf{S} = [\mathbf{s}_1, \mathbf{s}_2, \ \ldots \ , \mathbf{s}_L] = \begin{bmatrix} 1 & 1 & \cdots & 1 \\ e^{j\beta_{1,1}} & e^{j\beta_{2,1}} & \cdots & e^{j\beta_{L,1}} \\ \vdots & \vdots & \ddots & \vdots \\ e^{j\beta_{1,M-1}} & e^{j\beta_{2,M-1}} & \cdots & e^{j\beta_{L,M-1}} \end{bmatrix} \tag{16.10}$$

The correlation matrix of received data sequence **R** is

$$\mathbf{R} = \mathbf{SDS^H} - \boldsymbol{\sigma}^2\mathbf{I} \tag{16.11}$$

where $\mathbf{D} = diag[P_1, P_2, \ldots P_L]$ is the signal power matrix, **I** is the identity matrix, and $\boldsymbol{\sigma}^2$ is the variance of the Gaussian white noise,

The eigenvalues λ_i and eigenvectors $\mathbf{q}_i$ of the matrix **R** satisfy the following equation.

$$\mathbf{Rq}_i = \lambda_i \mathbf{q}_i \tag{16.12}$$

where

$$\lambda_i = \begin{cases} P_i + \sigma^2 & i = 1, 2, \ldots, L \\ \sigma^2 & i = L + 1, \ldots, M \end{cases} \tag{16.13}$$

and P_i, $i = 1, 2, \ldots, L$ are the eigenvalues of matrix $\mathbf{SDS^H}$.

Eigenvectors $\mathbf{q}_i$, $i = 1, \ldots, L$ span over the signal and noise subspace and eigenvectors $\mathbf{q}_i$, $i = L+1, \ldots, M$ span over the noise only subspace. Noise only subspace eigenvectors satisfy the following equation.

$$\mathbf{Rq}_i = \sigma^2\mathbf{q}_i, \quad i = L = 1, \ldots, M \tag{16.14}$$

The signal vectors $\mathbf{s}_i$ $i = 1, 2, \ldots, L$ belong to the subspace span by $\mathbf{q}_i$, $i = 1, \ldots, L$, thus they are orthogonal to noise only subspace eigenvector $\mathbf{q}_i$, $i = L+1, \ldots, M$. If the signal to noise ratio (SNR) is reasonably high, then there should be an obvious gap between the largest L eigenvalues to the rest. Thus the number of signals can be easily identified by using the L eigenvectors associated with the L

largest eigenvalues. We can compute the eigenvectors associated with the smallest *M-L* eigenvalues of the matrix **R** in the following form of matrix

$$\mathbf{V}_N = \left[\, \mathbf{q}_{L+1}\ \mathbf{q}_{L+2}\ \cdots\ \mathbf{q}_M \,\right]. \tag{16.15}$$

To estimate the DOA of the signal is to compute the roots of the polynomial *J*(*z*) defined by the following equation.

$$J(z) = \mathbf{a}^H \mathbf{V}_N \mathbf{V}_N^H \mathbf{a} \tag{16.16}$$

where

$$\mathbf{a} = \left[\, 1\ z\ z^2\ \ldots\ z^{M-1} \,\right]^T \tag{16.17}$$

and z for ULA can be given

$$z = e^{j\frac{2\pi d}{\lambda}\sin\theta}. \tag{16.18}$$

Then the roots of *J(z)* contain the directional information of the incoming signals. Ideally, the roots of *J(z)* would be on the unit circle at locations determined by the directions of the incoming signals; however, due to the presence of noise, the roots may not necessarily be on the unit circle. In this case, the *L* closest roots to the unit circle are the roots that correspond to the *L* incoming signals [11]. These selected roots, by themselves, do not directly represent the incoming angle. For each root, the incoming angle is found by solving Eq. (16.16).

$$\theta_k = \sin^{-1}\left[\frac{\lambda}{2\pi d}\arg(z_k)\right],\quad k = 1, 2, \ldots, L. \tag{16.19}$$

Obviously, when the root-MUSIC algorithm is implemented, there is no prior knowledge of the incoming signal directions or signal powers needed to construct the correlation matrix using Eq. (16.11). Therefore the correlation matrix must be estimated using only the information available from the sensor array. There are several methods commonly used to perform this estimation such as temporal averaging, spatial smoothing, or a hybrid combination of both temporal averaging and spatial smoothing [8]. In this chapter, we use only the temporal averaging method.

16.4 Computer Simulation Results

Consider a 16 element ULA with inter-element spacing equals half wavelength. In a conventional fixed antenna, the mainlobe beamwidth would be around 7°. This antenna array would not be able to resolve multiple signals if their angle separation is less than 7°. Using the root MUSIC algorithm, this ambiguity can be easily resolved.

Let $x_i(1),\ x_i(2), \ldots, x_i(N)$ represent the received sample data from *i*th element, where $i = 0, 1, \ldots, M\text{–}1$. The incoming data matrix **A** can be given

$$\mathbf{A}^H = \begin{bmatrix} x_0(1) & x_0(2) & \dots & x_0(N) \\ x_1(1) & x_1(2) & \dots & x_1(N) \\ \vdots & \vdots & \ddots & \vdots \\ x_{M-1}(1) & x_{M-1}(2) & \dots & x_{M-1}(N) \end{bmatrix} \tag{16.20}$$

The estimated correlation matrix $\mathbf{\Phi}$ is computed by

$$\mathbf{\Phi} = \mathbf{A}^{\mathrm{H}}\mathbf{A} \tag{16.21}$$

We can find the eigenvalues from the estimated correlation matrix $\mathbf{\Phi}$,. The columns of matrix $\mathbf{V}_{\mathrm{N}}$ are the eigenvectors associate with the *M-L* smallest eigenvalues of matrix $\mathbf{\Phi}$. Once this matrix is available, the signals' DOA can be derived from *L* roots of polynomial J(z) closest to the unit circle.

Figure 16.2 shows the roots computed from Eq. (16.16) when two signals impinging on a 16 element ULA from angles of 40° and 46°. In this simulation, the number of snapshot *N* is 32 and the signal to noise ratio (SNR) is 20 dB.

Figure 16.2 shows when the angle separation between two signals smaller than the mainlobe beamwidth, two distinct pair of roots closest to the unit circle can easily be identified. The zoom area shows one of the pair roots that they are very close to the theoretical root of the signal's DOA. The results are very similar to Fig. 16.2 when the angle separation is furher reduced to 3°, and 1.5°. Thus, the spatial resolution is improved by root MUSIC algorithm.

Equation (16.19) converts the roots of polynomial *J*(*z*) to the signals' DOA. Assume there are two signals impinging on a 16 element ULA with 20 dB SNR and taking 32 snapshots, the estimated signals DOA with different angle separations are shown in Fig. 16.3. The result shows that two incoming signals are clearly identified

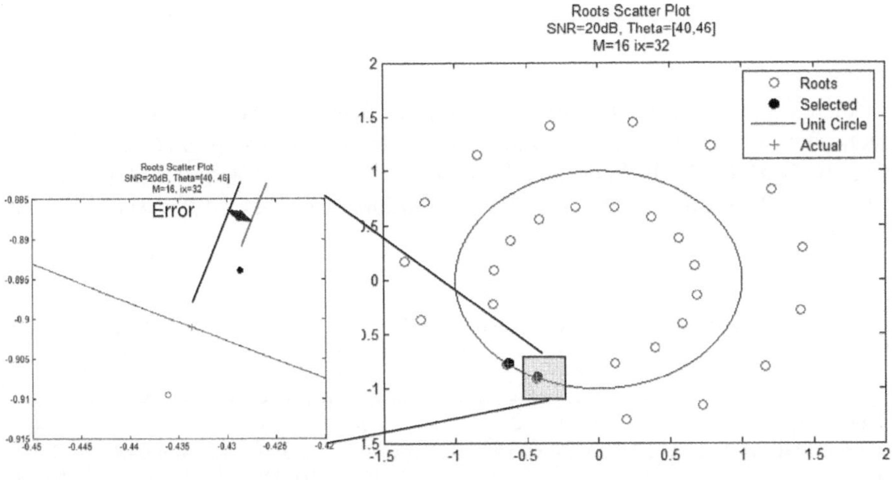

Fig. 16.2 Roots of polynomial J(z)

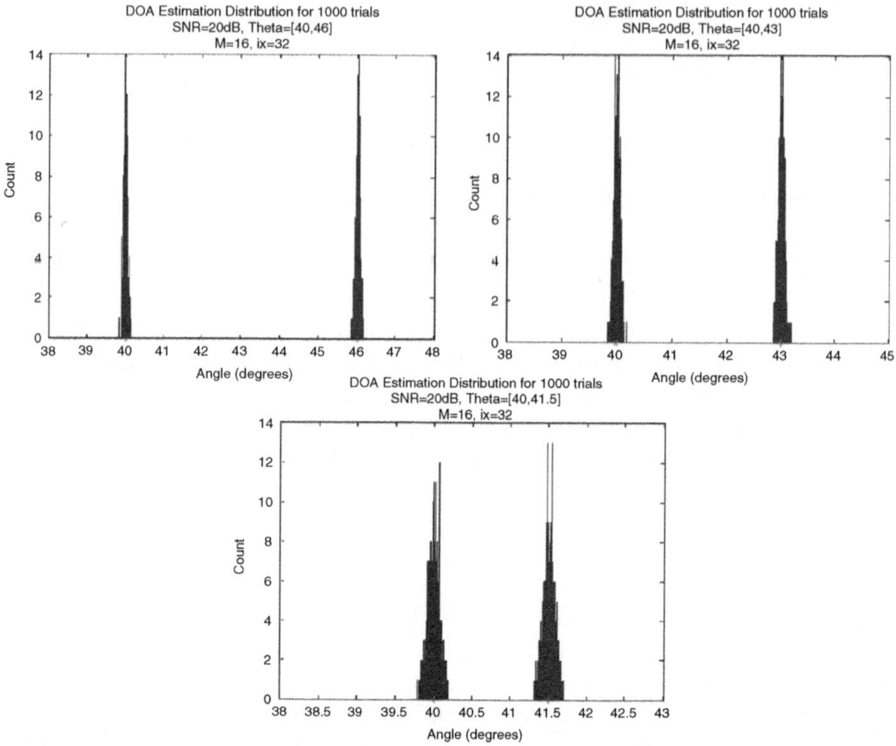

Fig. 16.3 Histogram of the estimated signals' DOA for angle separation equal 6°, 3°, and 1.5°

even as the separation between the two signals is well below the conventional main lobe beam width. The average estimation error is nearly zero in all cases.

The estimated means and variances based on 1000 trials are summarized in Table 16.1. Figure 16.3 and Table 16.1 show that the estimation variance increases as the angle separation becomes smaller.

Increasing the estimated correlation matrix from 32 snapshots to 96 snapshots reduces the estimated variance. Figure 16.4 compares the histogram of the estimated signals' DOA for 20 dB SNR, and the signals' DOA are 40° and 41.5° for 32 and 96 snapshots. The estimated mean values and variances based on 1000 trials are listed in Table 16.2.

Table 16.1 The estimated mean and variance of DOAs for SNR = 20 dB

Angle separation	6°		3°		1.5°	
True angles	40°	46°	40°	43°	40°	41.5°
Estimated mean	39.997°	46.004°	39.999°	42.999°	39.999°	41.498°
Variance	0.0009	0.0012	0.0023	0.0025	0.0047	0.0042

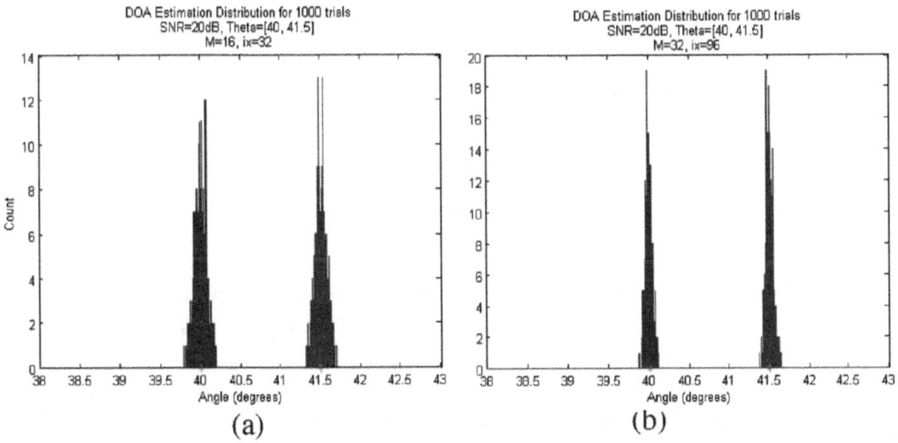

Fig. 16.4 Histogram of the estimated signals' DOA with SNR = 20 dB and angel separations 40° and 41.5° (**a**) 32 snapshots, (**b**) 96 snapshots

Table 16.2 The estimated mean and variance of DOAs for SNR = 20 dB

Number of snapshots	32		96	
True angles	40°	41.5°	40°	41.5°
Estimated mean	39.9999°	41.4987°	39.9992°	41.5008°
Variance	0.0047	0.0042	0.0015	0.0016

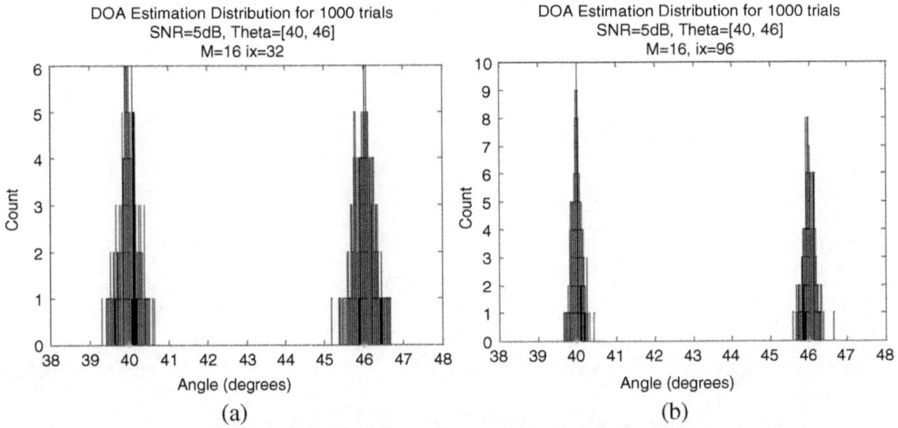

Fig. 16.5 Histogram of the estimated signals' DOA with SNR = 5 dB, (**a**) 32 Snapshots (**b**) 96 Snapshots, Signals' DOA are 40° and 46°

Table 16.3 The estimated mean and variance of DOAs for SNR = 5 dB

Number of snapshots	32		96	
True angles	40°	46°	40°	46°
Estimated mean	39.9708°	46.0279°	39.9893°	46.0094°
Variance	0.0358	0.0430	0.0112	0.0140

The above simulation results assume that the system operates in a high SNR environment such as SNR value at 20 dB. If the SNR is only 5 dB, the simulation result yields a larger estimation variance. Figure 16.5 shows the histogram of the estimated signals' DOA for 5 dB SNR, and the signals' DOA are 40° and 46°. This result is based on 1000 independent simulations where the number of snapshots of each simulation is 32 and 96.

The estimated mean values and variances based on 1000 trials for 5 dB SNR and the two different numbers of snapshots are listed in Table 16.3. This simulation result shows that as we increase the number of snapshots, the estimation variance decreases.

16.5 Sensor Spacing And Phase Sensitivity

The root-MUSIC algorithm in the previous section assumes that the positions of the elements ULA are exactly uniformly spaced and the weights of elements are perfect. However, the real system always has some variation in element position and phase. Figure 16.6 shows the sensor element position has a slight deviation from the ideal element position.

The spacing error of each sensor, e_i, is a Gaussian random variable added to the ideal spacing. Taking this error into account, Eq. (16.3) is used to create the phase

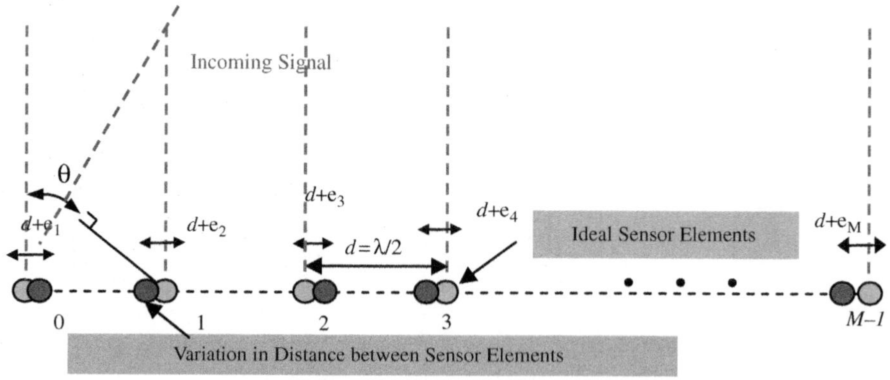

Fig. 16.6 Non-ideal ULA sensor spacing

shift between sensor elements for the incoming signals and becomes

$$\beta(\theta_i) = \frac{2\pi(d + e_i)}{\lambda}\sin(\theta_i)$$

where $d_k = (k\text{-}1)d$, and e_k represents the ideal positions and the position error of the kth element, and θ is the signal's DOA. Increasing the position error increases the estimation variance. Suppose the position error has a Gaussian distribution with standard deviation equal to 1% and 5% of the theoretical inter-element spacing d. Using 20 dB SNR and two signals impinging on the ULA at $\theta = 40^o$ and 46°, the histograms based on 1000 simulations are shown in Fig. 16.7. The number of snapshots is assumed to be 32.

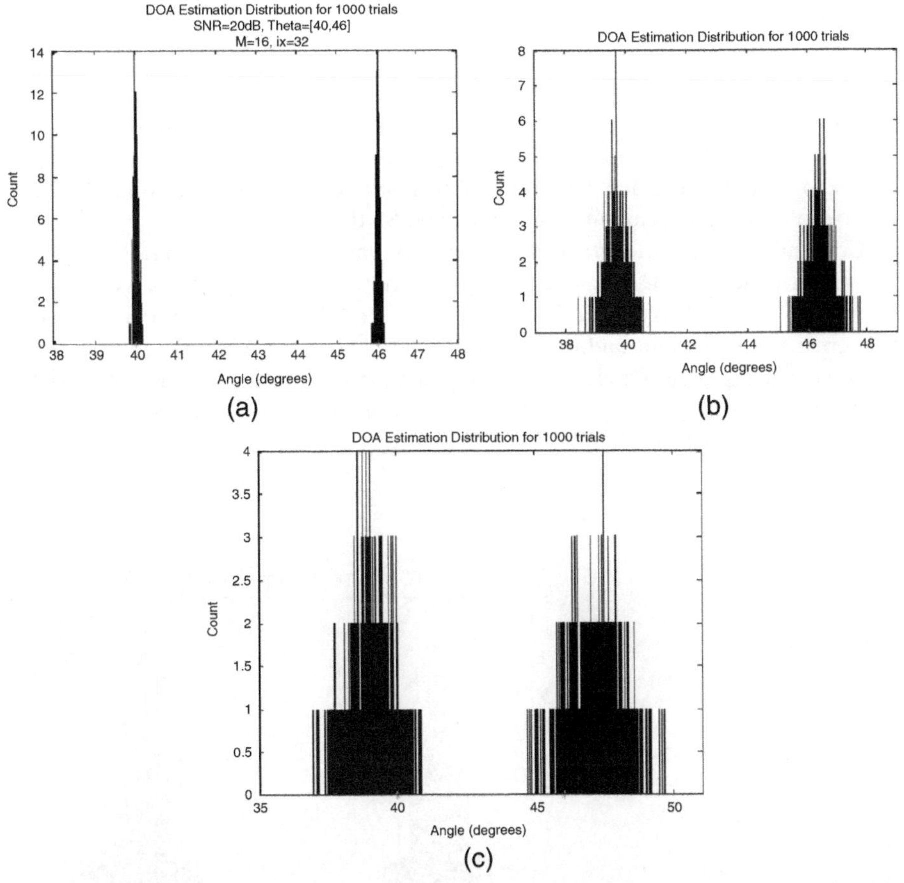

Fig. 16.7 Histograms of Two Signals with DOA = 40° and 46°, SNR = 20 dB with (**a**) Ideal Element Spacing d, (**b**) Gaussian Random Spacing with Standard Deviation equal .01d, (**c**) Gaussian Random Spacing with Standard Deviation equal .05d

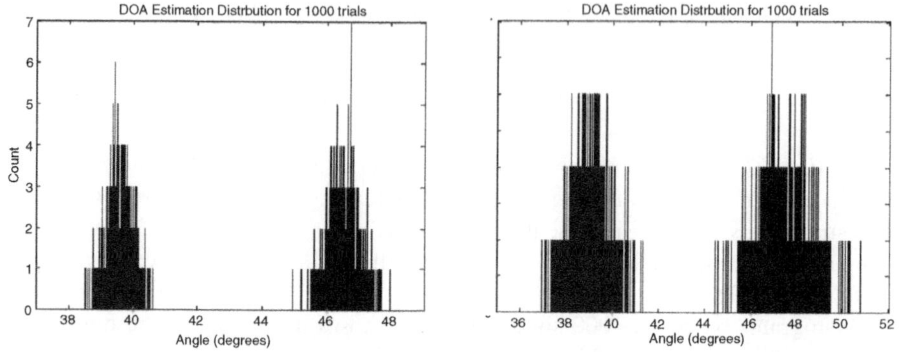

(a) (b)

Fig. 16.8 Histograms of two signals with DOA = 40° and 46°, SNR = 20 dB with (**a**) Gaussian random phase error with standard deviation = 1°, (**b**) Gaussian random phase error with standard deviation = 5°

If the phase error of each received signal increases, the estimation variance also increases. Example of a random Gaussian phase error with standard deviation equal to 1° and 5° and histograms based on 1000 simulations are shown in Fig. 16.8. The number of snapshots is assumed to be 32, and SNR = 20 dB.

Combining the effect of random phase error and position error further degrades the performance of the ULA. Figure 16.9 shows the histogram of two signals impinging on the ULA with DOA equal = 40° and 46°, SNR = 20 dB. The position error is a Gaussian random variable with a standard deviation equal to 1% of the ideal spacing between elements. The phase error is a Gaussian random variable with standard deviation equal to 1°. This result is based on 1000 simulations with 32 snapshots in each simulation.

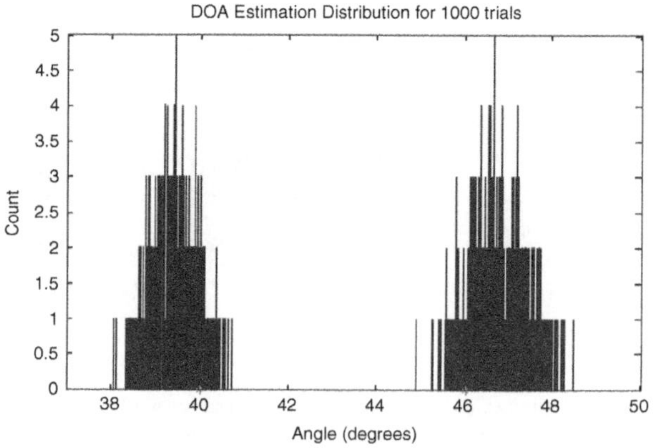

Fig. 16.9 Histogram with combine phase error and position error

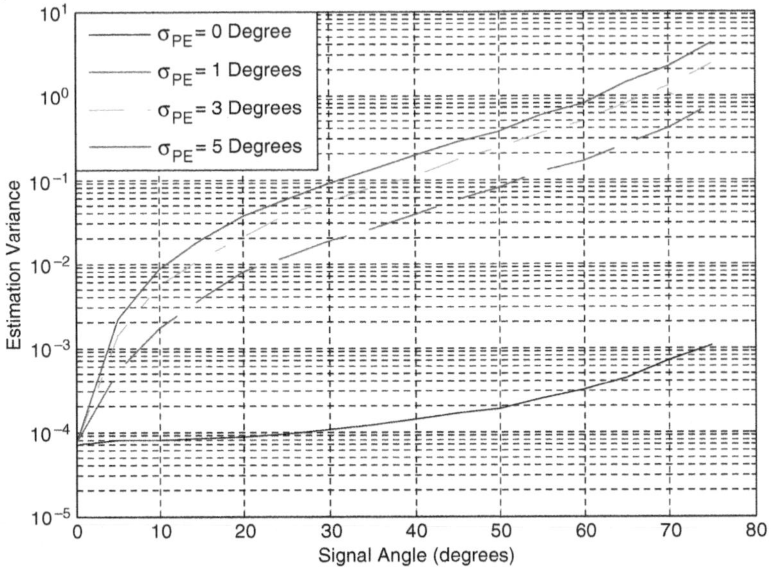

Fig. 16.10 Estimation variance of different gaussian random phase error versus signal's DOA for 20 dB SNR

Figure 16.10 shows the estimation variance of different Gaussian random phase error versus signal's DOA for 20 dB SNR. The estimated variance increases rapidly as the signal's DOA approaches the end fire of the ULA. If the signal's DOA is greater than 75°, the estimation variance become so large that reliable DOA estimation is difficult to achieve.

Figure 16.11 compares the variance of different Gaussian random phase error versus signal's DOA for 10 dB and 20 dB SNR. This simulation result shows that whenever the SNR reduce the estimation variance increase. Figures 16.10 and 16.11 are simulation results when the correlation matrices are computed based on 32 snapshots. With finite snapshots, the estimated correlation matrix is not very reliable, thus if the signal impinging the ULA with large angle, the estimated signal's DOA has very large variance. To reduce the estimation variance, the estimated correlation matrix should be based on averaging large number of snapshots.

16.6 Conclusion

This chapter investigates the possibility of combining the array antenna and advanced signal processing techniques to enhance the estimation of the direction of signal sources.

A conventional method to detect the direction of signal source is to use a fixed antenna to scan over a certain searching region. This primitive estimation technique has many limitations. First, its resolution is limited by the antenna mainlobe

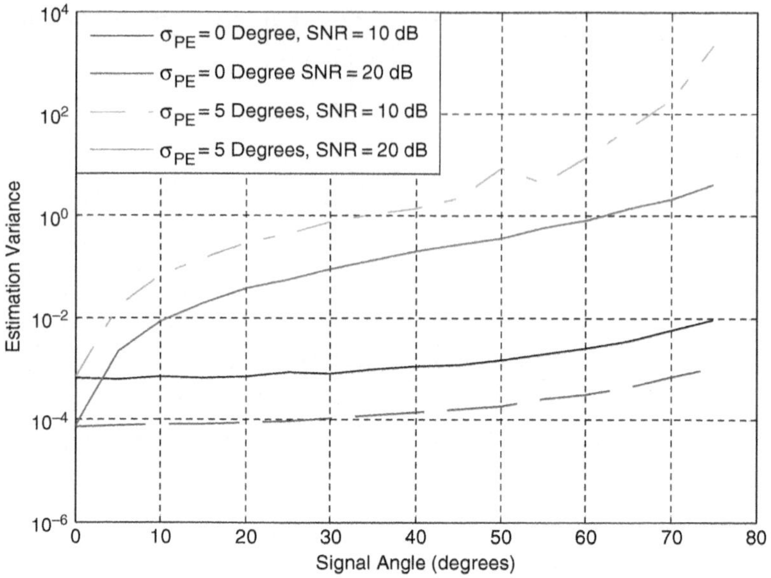

Fig. 16.11 Estimation variance of different Gaussian random phase error versus signal's DOA for 10 and 20 dB SNR

beamwidth. Also, if there are multiple signal sources, a conventional fixed antenna has difficulty in detecting them simultaneously.

Using the advanced signal processing techniques, the DOA estimation can be improved. The root-MUSIC method presented in this chapter is based on the eigenvector of the sensor array correlation matrix to estimate angle of incoming signals. Extensive computer simulation is used to demonstrate the performance of the algorithms, which enhance the DOA estimation.

The simulation results of the root-MUSIC algorithm suggests the following:

1. The capability to resolve multiple targets with separation angles smaller the main lobe beam width of the array antenna.
2. The estimation variance can be reduced by increasing the number of snapshots in correlation matrix estimation
3. The estimation variance increases as the angle separation between signals becomes smaller

The estimation variance depends on the direction of the signal. A signal coming from the bore sight has minimum estimation variance.

References

1. A. Lee, L. Chen, H.K. Hwang and et al. Simulation study of wideband interference rejection using adaptive array antenna, IEEE Aerospace Conference, March 5–12, 2005

2. M.G.M. Hussain, Performance analysis and advancement of self-steering arrays for nonsinusoidal waves, IEEE Trans. on Electromagnetic Compatibility, May 1988
3. X. Zhang and D. Su, Digital processing system for digital beam forming antenna, IEEE International Symposium on Microwave, Antenna Propagation and EMC Technologies, 2005
4. J. Zhu, M. Chan and H.K. Hwang, Simulation study on adaptive antenna array, IEEE International Signal Processing Conference, Dallas, 2003
5. M. Grice, J. Rodenkirch, A. Yakovlev, H. K. Hwang, Z. Aliyazicioglu, A. Lee, Direction of arrival estimation using advanced signal processing, RAST Conference , Istanbul-Turkey, 2007.
6. M. Skolnik, *Introduction to RADAR Systems*. 3rd ed., New York: Mc Graw Hill, 2001.
7. J.R. Guerci, *Space Time Adaptive Processing for Radar*, Norwood: Artech House Publisher, 2003.
8. V.F. Pisarenko, The retrieval of harmonics from a covariance function, Geophysical Journal of Royal Astronomical Society, Vol. 33, 1973
9. R. Schmidt, Multiple emitter location and signal parameter estimation, IEEE Transactions on Antenna and Propagation, Vol. 34, pp 276–280, 1986
10. R. Roy, A. Paulraj and T. Kailath, ESPRIT: A subspace rotation approach to estimation of parameters of cisoids in noise, IEEE ASSP -34, 1986
11. K. Forsythe, Utilizing waveform features for adaptive beamforming and direction finding with narrowband signals, Lincoln Laboratory Journal, Vol. 10, No. 2, 1997, 99–126.

Chapter 17
Grid Computing for Ubiquitous Computing Environment (GCUCE) Mobility Model

Dong-Bum Seo, Tae-Dong Lee and Chang-Sung Jeong

Abstract In this paper, we describe GCUCE (Grid Computing for Ubiquitous Computing Environment), which supports the unified efficient ubiquitous service interacting with Grid service modules using Access Grid computing. Also, we describe two mobility models: enterprise model and automata model. The former is focused on the service demand aspects, and the latter concentrates on the state transition. Based on those models, we shall show the performance of GCUCE by evaluating several experiments for DOWS (Distributed Object-oriented Wargame Simulation). The two models comprising enterprise model and automata model of software state suggest the formal and mathematical model about software mobility in GCUCE, and provide the overall views and direction of ubiquitous software development.

Keywords Ubiquitous computing · Ubiquitous mobility · Ubiquitous software mobility · Access grid · Grid computing

17.1 Introduction

Ubiquitous computing presents both an opportunity for users and a challenge for system engineers [1–4]. For users, a wealth have computing, informational, and communication resources available everywhere should allow them to work more effectively. For system engineers, however there is a challenge of using these resources without overburdening users with management of the underlying technology and infrastructure. One particularly important aspect of this problem is to support continuity in the face of time varying resources.

While ubiquitous computing promises to make many more resources available in any given location, a set of resources that can be used effectively is subject to frequent change because the resource pool itself can change dynamically, and a user

D.-B. Seo (✉)
Department of Information and Communication Engineering, Korea University, Seoul, South Korea

S.-I. Ao et al. (eds.), *Trends in Communication Technologies and Engineering Science*, Lecture Notes in Electrical Engineering 33, DOI 10.1007/978-1-4020-9532-0_17,

may move to a new environment, making some resources available and others not accessible. As we detail later, traditional solutions normally associated with mobile computing [5] are inadequate to solve this problem either because they are unable to exploit resources as they become available in a user's environment, or because users must pay too high price to manage those resources.

For the solution to this problem, we insert the concept of Grid computing [6, 5] into the existing concept of ubiquitous computing [7]. Many ubiquitous computing systems provide application developers with a powerful framework [8]; however, its design is not intended to support applications either requiring many resources that are needed to integrate instruments, displays, computational and information managed by diverse organizations, or requiring collaborative environment by audio/video conferencing.

Moreover, the existing ubiquitous systems do not consider coordinating and managing the resources to complete the task efficiently and effectively. The structure of the paper is as follows: In Section 17.2, we describes GCUCE(Grid Computing for Ubiquitous Computing Environment) architecture. In Section 17.3, we discuss enterprise model, and in Section 17.4 automata for software mobility in GCUCE. In Section 17.5, we describe two experiments on Grid computing, and in Section 17.6, conclusions are given.

17.2 Architecture

GCUCE is a context-aware ubiquitous computing [9–10] environment supported by grid computing [6]. It is composed of three layers as shown in Fig. 17.1: Grid Layer, Context-aware Layer, and Ubiquitous Main Layer.

17.2.1 Grid Layer

The Grid Layer [11–12] is divided into two elements: Computation Grid Framework (CGF), Access Grid Framework (AGF). The CGF provides the functions of grid computing using a Java cog kit, which supports the fault tolerance and high performance computation through resource sharing, monitoring, and allocation. The CGF is composed of four managers: Resource Manager (RM), Data Manager (DM), Job Manager (JM), and Runtime Information Manager (RIM). RM uses the resource management services offered by Grid, DM provides speed and reliability for files being transferred, JM acts as an agent for the tasks in a job, providing a single entity from which the tasks will request resources, and RIM provides the information to applications or middleware. AGF supplies collaboration with audio/video streaming and view of sharing through shared applications on collaborative environments [5]. It consists of four managers: Collaboration Manager (CM), Collaboration Session Manager (CSM), Shared Application Manager (SAM), and Multimedia Manager (MM). CM gathers the information about the shared stubs from the venue, CSM provides venue addresses that the user can access, SAM integrates the whole

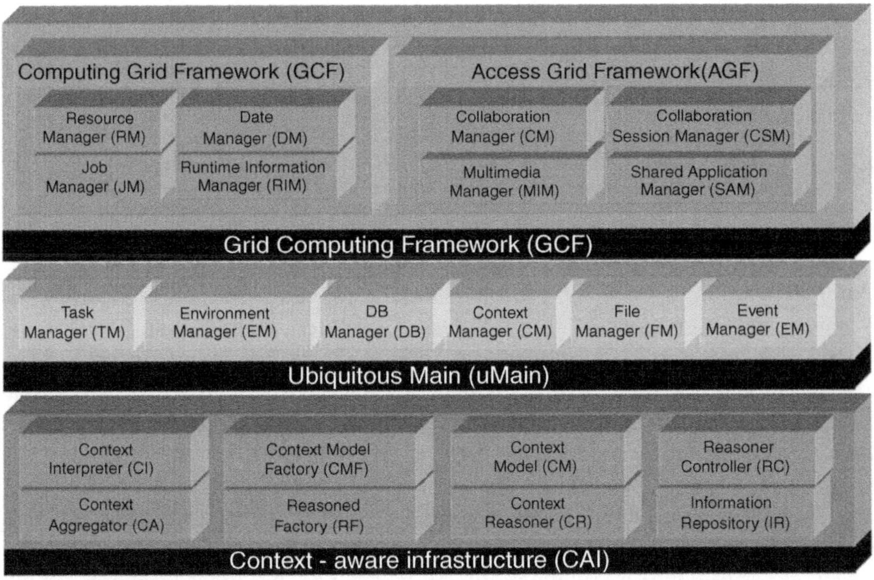

Fig. 17.1 GCUCE architecture

features of the shared applications used in the AccessGrid, and MM controls the base service of AccessGrid like Audio/Video streaming service.

17.2.2 Context-Aware Layer

The Context-aware Layer [9–10] provides a Context-Aware Infrastructure (CAI), which has a responsibility for functions, which support the gathering of context information from different sensors and the delivery of appropriate context information to applications. Also, it supports the context model by ontology methods. The development of formal context models satisfies the need to facilitate context representation, context sharing and semantic interoperability of heterogeneous systems. It provides an abstract context ontology that captures general concepts about basic context, and also provides extensibility for adding domain specific ontology in a hierarchical manner. In CAI, there are eight elements: Context Interpreter (CI), Context Aggregator (CA), Context Model Factory (CMF), Context Model (CM), Reasoner Factory (RF), Context Reasoned (CR), Reasoner Controller (RC), and Information Repository (IR).

CI gathers contextual information from sensors, manipulates the contextual information, and makes it uniformly available to the platform. CA processes and aggregates the data through sensor network after context extraction, which provides high-level contexts by interpreting low-level contexts. CMF defines a context based on concept of specific domain ontology through internal/external providers, and associates a data set with some reasoners to create a CM. RF creates the specified reasoners, and CR has the functionality of providing the deduced contexts based on direct contexts, resolving context conflicts and maintaining the consistency of IR, and RC starts and stops the specific CR.

17.2.3 Ubiquitous Main Layer

The Ubiquitous Main Layer [2–4, 13] (called uMain) is responsible for shielding the user from the underlying complexity and variability through self-tuning environment by mobility and adaptation, which are weak points in CGF and AGF. Whenever the user moves from one place to another, the tasks and devices such as grid authentication, environment variables, video/audio device or large display are automatically set up or executed, keeping their environment. The uMain has six managers: Task Manager (TM), Environment Manager (EM), DB Manager, Context Manager, File Manager, and Event Manager (EVM). TM has something concerned with user's task processes. EM supports services concerned with making same user's environment in everywhere. Database Manager DBM manages the recording about user information. CM has a role of detection about user's activities such as entering or leaving to/from the environment. File Manager (FM) has to take a charge for both remote and local file operations. EVM is to send and receive messages among them locally.

17.3 Enterprise Model in GCUCE

The system architecture in uMain has the components: ubiCore and ubiContainer as shown in Fig. 17.2. The ubiCore is responsible for the application mobility [9]. When a user moves from one place to another; ubiCore provides the automatic movement of computing environment through ubiContainer, which supplies the user information such like IP, user preference, etc.

The communication component in GCUCE uses socket, Java RMI (Remote Method Invocation). File Manager uses the socket for transferring the files. RMI

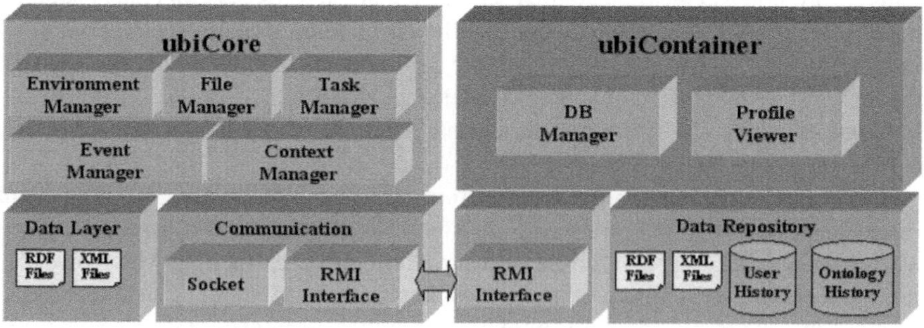

Fig. 17.2 Architecture of uMain

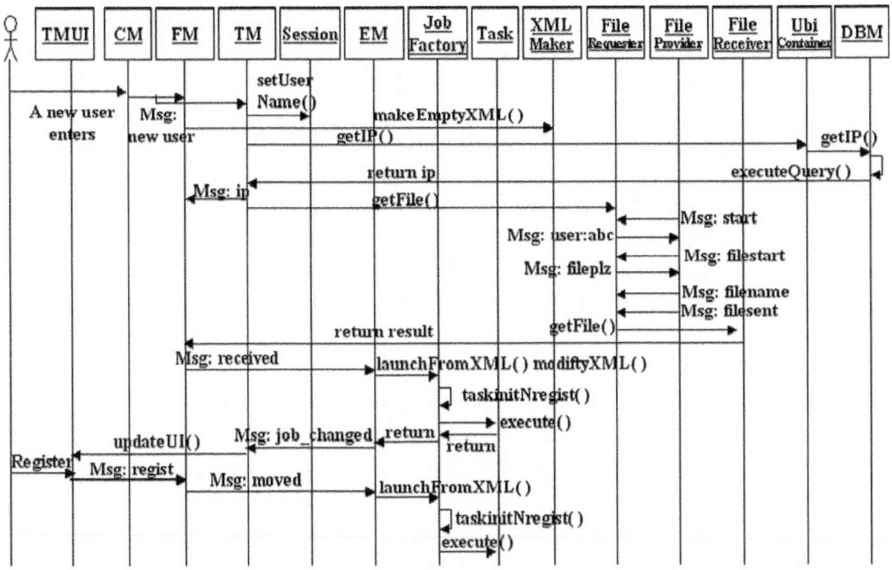

Fig. 17.3 Sequence diagram of uMain

is Java mechanism for supporting distributed object based computing, and it allows client/server based distributed applications to be developed easily because a client application running in a Java virtual machine at one node can invoke objects implemented by a remote Java virtual machine (e.g., a remote service) in the same way as local objects.

The RMI mechanism in Java allows distributed application components to communicate via remote object invocations, and enables applications to exploit distributed object technology rather than low-level message passing (e.g., sockets) to meet their communication needs.

Figure 17.3 shows the sequence diagram of uMain. When a new user enters into a new place or device, CM detects the entrance of the user, and TM brings the information related to the user, and FM makes the directory. TM copies the files associated with the user by File Requester. File Requester sends and receives messages to File Provider on remote host, and File Provider sends the files to File Receiver. After the end of file copies, EM executes the registered tasks through JobFactory, which commands the start of tasks to Task. If a user registers the task through the TMUI(TM user interface), the related files are moved to a user directory and the XML file is updated.

17.3.1 Enterprise Model

Let us think about the enterprise model of competitive ubiquitous grid service [13, 11, 14]. Assume a set G of N grid service providers. That is, $G = \{1, 2, \ldots n, \ldots N\}$. Each provider can be distinguished by three service parameters such as $\{r_n, l_n, m_n\}$, where r_n is the response time of nth service provider for unit resource demand from its subscribers and l_n is the loss probability experienced by that service provider, and m_n is the mobility probability accumulated by nth service provider. The selection of these parameters has a significant impact on completion of the job of ubiquitous users within a limited time frame. Due to the competitiveness in between the service providers, the ubiquitous user gets an option to shift from one service provider to another for job completion at the earliest possible time. So, the nth grid service provider experiences a demand d_n, which depends not only on his own response time, loss probability, mobility probability but also on the response time, loss probabilities, and mobility probabilities offered by its competitors. So, d_n depends upon entire response time vector $r = [r_1, r_2, ..., r_N]$, loss probabilities vector $l = [l_n, l_2, ..., l_N]$, and mobility probabilities vector $m = [m_1, m_2, ..., m_N]$. The strategy of each grid service provider will be always to provide a response time, loss probability, mobility probability which are in between the maximum and minimum values offered by all of its competitors. Then, the strategy space, S_n of nth grid service provider, as in

$$S_n = \left\{ \begin{array}{l} (r_n, l_n, m_n) : 0 \le r_{\min} \le r_n \le r_{\max}; \\ 0 \le l_{\min} \le l_n \le l_{\max}; 0 \le m_{\min} \le m_n \le m_{\max} \end{array} \right\}.$$

We assume $r_{\min}$ depends on l_n in the sense that if the value of the loss probability increases, then the service provider has to decrease its response time r_n. The upper bound on the response time and loss probability express that after a certain value, the demand will be zero.

17.3.2 Demand Model

For a particular grid service provider, the demand d_n for its services decreases as its response time r_n increases; on the other hand, it increases with the increase of its competitor response time r_m, for $m \neq n$. The analogous relationship holds for loss probabilities, but then, d_n increases with decrease of l_n and increase of l_m, for $m \neq n$. The reverse relationship holds for mobility probabilities, but then, d_n increases with increase of m_n and decrease of m_m, for $m \neq n$. We now consider the case where the demand function (d_n) is linear in all QoS parameters [11]. That is,

$$d_n(r, l, m) = \left\{ \begin{array}{l} \sum\limits_{m \subseteq G, m \neq n} \alpha_{nm} r_m - \theta_n r_n + \\ \sum\limits_{m \subseteq G, m \neq n} \beta_{nm} l_m - \delta_n l_n - \\ \sum\limits_{m \subseteq G, m \neq n} \chi_{nm} m_m + \eta_n m_n \end{array} \right\} \quad (17.1)$$

where $\alpha_{nm}, \theta_n, \beta_{nm}, \delta_n, \chi_{nm}, \eta_n$ are constants, and m_n is mobility value at nth grid service provider. Here we make some minimal assumptions regarding the demand function.

$$\begin{bmatrix} \frac{\partial d_n(r,l,m)}{\partial \gamma_n} \le 0, & \frac{\partial d_n(r,l,m)}{\partial l_n} \le 0, & \frac{\partial d_n(r,l,m)}{\partial m_n} \ge 0; \\ \frac{\partial d_n(r,l,m)}{\partial \gamma_m} \ge 0, & \frac{\partial d_n(r,l,m)}{\partial l_m} \ge 0, & \frac{\partial d_n(r,l,m)}{\partial m_m} \le 0; \\ \forall m \ne n & & \end{bmatrix}$$

This results in a decrease of own demand while increasing those of its competitors, if a service provider increases response time (loss probability), assumption the demand d_n is non-negative over the strategy space (i.e., response time and loss probability are decreasing) and negative over the non-strategy space (i.e. response time and loss probability are increasing). Now each service provider n will charge a cost, n_c, per unit of the demand provided to the ubiquitous user. Then the *gain* earned by the nth service provider is

$$G_n = c_n \times d_n(r,l,m) = c_n \times d_n \tag{17.2}$$

We define that ε is the ratio of proportionate change in quantity demanded to proportionate change in cost

$$\varepsilon = -\frac{(\partial d_n)/d_n}{(\partial c_n)/c_n} = -\left(\frac{\partial d_n}{\partial c_n}\right)\left(\frac{c_n}{d_n}\right) \tag{17.3}$$

Since the revenue of nth service provider is given by $G_n = c_n \times d_n$, then taking the partial derivative of both sides we get,

$$\begin{bmatrix} \frac{\partial G_n}{\partial c_n} = d_n + c_n \times \frac{\partial d_n}{\partial c_n} \\ \frac{\partial G_n/\partial c_n}{G_n} = \frac{(d_n + c_n \times \partial d_n/\partial c_n)}{G_n} \\ \frac{\partial G_n/\partial c_n}{G_n} = \frac{1-\varepsilon}{c_n} = -\left(\frac{\varepsilon - 1}{c_n}\right) \end{bmatrix} \tag{17.4}$$

Now integrating equation (17.2) and considering the initial value to demand as k, and then we get,

$$d_n = \frac{k}{c_n^{\varepsilon}} \Rightarrow c_n = \left(\frac{k}{d_n}\right)^{1/\varepsilon} \quad (k \text{ is initial value}) \tag{17.5}$$

Equation (17.5) represents a more generalized demand function by incorporating constant price elasticity model, which is more sensible and appropriate for our scenarios.

17.3.3 Gain Maximization

The revenue maximization problem is

$$\max_{(c_n, d_n)} G_n = c_n \times d_n\,(r, l, m) \tag{17.6}$$

Subject to the following constraints:

$$c_n \geq 0, \forall n \in N \tag{17.7}$$

$$\left\{ \begin{array}{l} \sum\limits_{m \subseteq G, m \neq n} \alpha_{nm} r_m - \theta_n r_n + \sum\limits_{m \subseteq G, m \neq n} \beta_{nm} l_m - \delta_n l_n - \\ \sum\limits_{m \subseteq G, m \neq n} \chi_{nm} m_m + \eta_n m_n \leq d_n(r, l, m), \forall n \in N \end{array} \right\} \tag{17.8}$$

In the above formulation, the cost and demand variable $d_n\,(r, l, m)$ are both present. We will find it convenient to replace the price variable by equation (17.6) and retain only the demand variables. The optimum price may be recovered from the demands in the solution. Transforming the objective function gives: max

$$\begin{aligned} \max_{(c_n, d_n)} G_n &= \left(\frac{k}{d_n\,(r, l, m)} \right)^{1/\varepsilon} \times d_n\,(r, l, m) \\ &= k^{1/\varepsilon} \times d_n\,(r, l, m)^{\frac{\varepsilon - 1}{\varepsilon}} \end{aligned} \tag{17.9}$$

In economics, the optimal profit to a player is calculated subject to a constrained space of actions, where a Lagrange multiplier is the value of relaxing a given constraint. Here we use a Lagrange multiplier λ, associated with the constraint implied by demand satisfaction in (17.9) to form the Lagrange expression

$$L\,(d_n\,(r, l, m)\,, \lambda) = k^{1/\varepsilon} \times d_n\,(r, l, m)^{\frac{\varepsilon - 1}{\varepsilon}}$$

$$+ \lambda \left(d_n\,(r, l, m) - \left(\begin{array}{l} \sum\limits_{m \subseteq G, m \neq n} \alpha_{nm} r_m - \theta_n r_n + \\ \sum\limits_{m \subseteq G, m \neq n} \beta_{nm} l_m - \delta_n l_n - \\ \sum\limits_{m \subseteq G, m \neq n} \chi_{nm} m_m + \eta_n m_n \end{array} \right) \right) \tag{17.10}$$

Now the first-order condition for maximization of $L\,(d_n\,(r, l)\,, \lambda)$ is found by equating the partial derivative of L to zero. Thus,

$$\frac{\partial L}{\partial d_n} = 0 \Rightarrow \left[\frac{\varepsilon - 1}{\varepsilon} \left(\frac{k}{d_n(r,l,m)} \right)^{1/\varepsilon} + \lambda = 0 \right]$$
$$\Rightarrow c_n = \left(\frac{\varepsilon - 1}{\varepsilon} \right) \lambda \tag{17.11}$$

So, the strategy of each grid service provider are will be always to provide a response time, loss probability, mobility probability which are in between the maximum and minimum values offered by all of its competitors.

17.4 Automata Model for Mobility in GCUCE

Mobile Grid service in grid computing is a new paradigm of Grid service. Grid Mobile Service [15] provides a series of standard interfaces and intelligent mobile code service to computation. It is extension software agent and Grid technologies. In GCUCE, the software mobility is supported. When sensor detects any signal such as entrance of person, related software is moved into new place, and then executed. The physical happenings can be made into formal automata.

Figure 17.4 illustrates the software mobility state transition. The initial state is software state beginning the system, running state is software execution state, migration state is the state of transfer from a host to another host, and stop (frozen) state is the stopped state of software, and finally end state is the termination state of software. Figure 17.4 describes automata of software mobility using Table. There is a response time and loss probability parameters are related to running and abnormal running state. The two parameters are measured on that state. The mobility parameter is related to migration state.

In the case, we describe the relationship of three parameters. In the running/abnormal running state, the response time and loss probability are main factors, and mobility is important factor when the software is transferred. The automata model is software state transition on ubiquitous computing. The combination of two models provides the system characteristics which software is best on ubiquitous computing environment.

17.5 Experiment

We use DOWS (Distributed Object-oriented Wargame Simulation) on RTI (Run-Time Infrastructure) on Grid [7, 16]. DOWS is an object-oriented simulation system based on a director-actor model, which can be mapped efficiently on object-oriented and distributed simulation. The existing RTIs for HLA (High Level Architecture) do not consider coordinating and managing the resource for distributed simulation to complete the simulation efficiently and effectively. The RTI on Grid is a grid-enabled implementation of RTI for solving the problems.

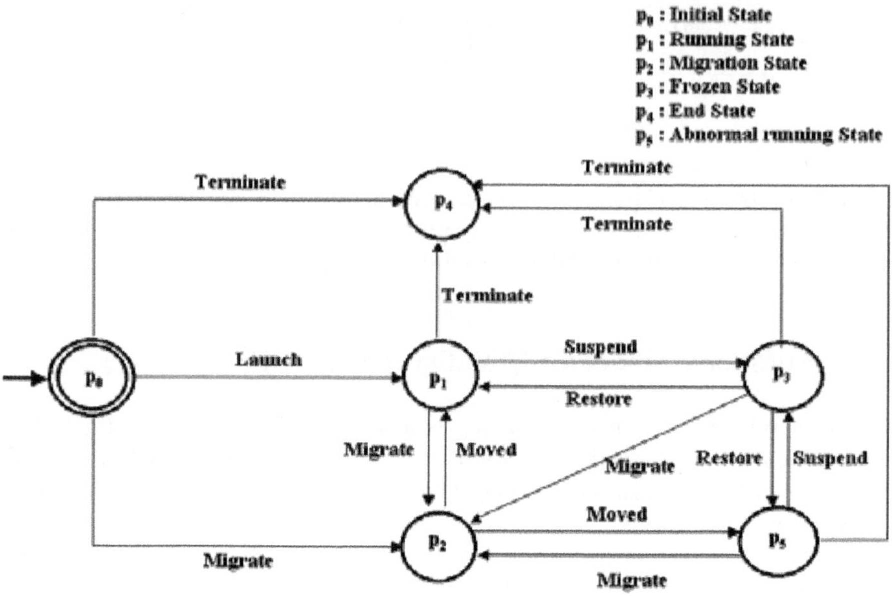

Input / Condition	R	E	F	M	A
00	E	No	E	R	E
01	F	No	R	A	F
10	A	No	A	No	R
11	M	No	M	No	M

Fig. 17.4 Automata of software mobility

In the scenario as shown in Fig. 17.5, John wants to know the result when DOWS simulation ends while he moves around the building, and continues to do other tasks. Currently, John is at lab B, and registers DOWS and RTI on computation Grid, and executes DOWS on RTI. After execution, he leaves the lab, and will go into the meeting room (Room C in Building C), which is 30 min away. The simulation will end 20 min after execution, and is transferred to GCUCE as text result. When John arrives at Room C in Building C, he will receive and analyze the result file. For test, the implementation is accomplished on 4 PCs as clients, and 10 clusters (5: Pentium IV 1.7 GHz, 5: Pentium III 1.0 GHz Dual) and one 486 computer as servers on a VO.

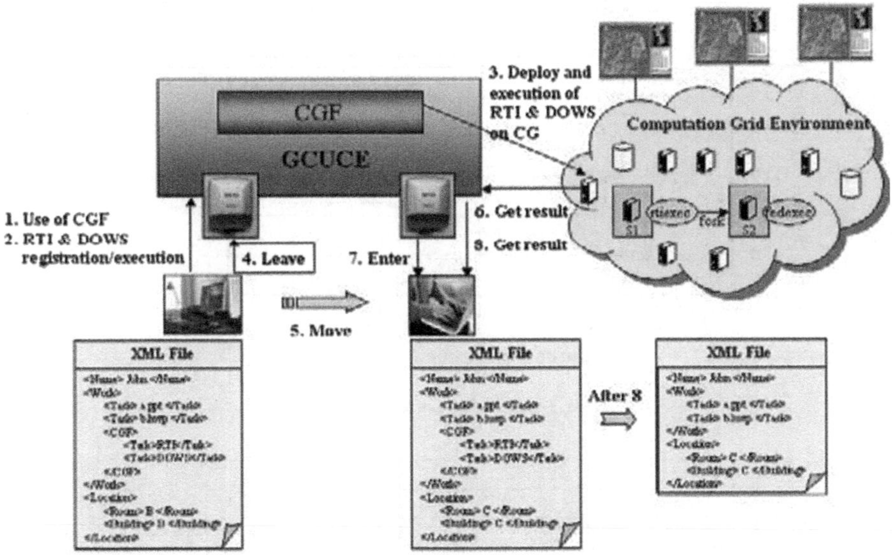

Fig. 17.5 DOWS on RTI using Grid for GCUCE

Our experiments are accomplished to confirm key services of CGF. The first experiment is for the automatic distribution service. We organize the system, which has five servers (we assume that 486 computer is included as server, because of the limitation in local condition), and the GCUCE that has a VO (Virtual Organization) of 11 servers (10 clusters and 486 PC). Then, we estimated the complete time of simulation as the number of forces increases. As we expected, the resource selection of the GCUCE did not choose the 486 computers.

As shown in the result of experiment, the demand in GCUCE is increasing because the response time in automatic distribution test is shorter, loss rate in dynamic migration test is smaller, and mobility rate in dynamic migration test is higher. In Fig. 17.6(a), GCUCE is superior as the scale of simulation is increasing, although the time consumption of initialization has an effect on the state of small forces. GCUCE can utilize abundant computing power and adapt for various environments, as well as provide convenient user interface. This brings a fast response time, and the demand becomes larger. To verify the dynamic migration service, we execute a second experiment. In this test, we measured accumulated received packets updated by 600 forces per 30 second. One packet has the size of 100 bytes. In 70 min, we intentionally made a failure on one server. The information related to execution like object information is stored periodically, and then the application resigns from the execution. The application sends the stop message to applications before resignation. RM gathers the information, and DM sends the application to a selected host. The application sends the restart message to all applications and receives the data stored before failure.

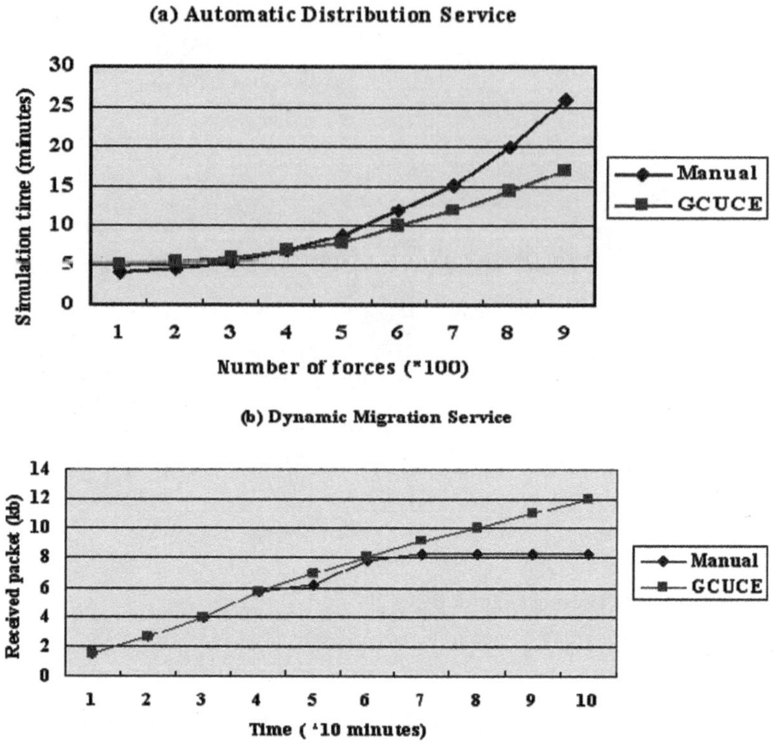

Fig. 17.6 Evaluation and result of GCUCE

As shown Fig. 17.6(b), GCUCE can fulfill its mission after the failure, while the original DOWS is halted. The enterprise model shows that responses time and loss probability are decreased, mobility is increased, and the demand of this service is increased.

17.6 Conclusion

We first have described the architecture of GCUCE that is the unified ubiquitous computing environment using grid computing. For collaborative computing, we develop the collaborative computing framework by supporting the real time video and audio stream using high-speed network, it can remote conference.

After explanation of architecture, we suggested the enterprise model of software and automata model of software state. Two models for GCUCE provide the system characteristics, which are designed with the philosophy that supports software mobility among a number of devices for ubiquitous computing makes the system vastly more complex.

The development for solution of the requirements as ubiquitous infrastructure is needed, and we have developed the GCUCE as the core module, which provides the automatic computing environment, and makes applications or services used everywhere.

The enterprise model compares the service aspects among software, and provides the direction of good service. The automata model describes the software state transition on ubiquitous computing. The combination of two models provides the system characteristics which software is best on ubiquitous computing environment.

For the traditional distributed computing model, we have developed two frameworks based on grid computing: Computation Grid Framework (CGF) and Access Grid Framework (AGF). For a large computation application, CGF provides the high performance-computing framework, which processes the real time data with high-speed computation through distributed resources using computation grid. For collaborative computing, AGF supplies the collaborative computing framework, which co-works the analysis, agreement and discussion among people with a lot of data using Access Grid. The framework gives the merits of collaboration with multimedia functions.

The heterogeneity and mobility among a number of devices for ubiquitous computing make the system vastly more complex. The development for solution of the requirements as ubiquitous infrastructure is needed, and we have developed the uMain (Ubiquitous Main) as the core modules, which provides the automatic computing environment, and provides applications or services used everywhere.

For GCUCE, we did two experiments for DOWS (Distributed Object-oriented Wargame Simulation) on CG, including completion time and packet bytes, after we explained the enterprise service model, an interactive simulation, and RTI-G (RunTime Infrastructure on Grid). The experiments showed the superior result of the performance of enterprise service model for GCUCE.

References

1. M. Weiser, "The computer for the twenty-first century," Scientific. Am., 1991, pp. 94–101.
2. G.D. Abowed, "Software engineering issues for ubiquitous computing," Software Engineering, 1999. Proceedings of the 1999 International Conference on, 16–22 May 1999 Page(s): 75–84.
3. M. Glesner, T. Hollstein and T. Murgan, "System design challenges in ubiquitous computing environments," Microelectronics, 2004. ICM 2004 Proceedings. The 16th International Conference on 6–8 Dec. 2004 Page(s): 11–14.
4. T. Kindberg and A. Fox, "System software for ubiquitous computing," Pervasive Computing, IEEE, 1(1), Jan.–March 2002 Page(s): 70–81.
5. R. Stevens, M. E. Papka, and T. Disz, "The Access GRID: Prototyping the workspaces of the future, Internet Computing, IEEE, 7(4), pp 51–58, July–Aug. 2003.
6. I. Foster, C. Kesselman and S. Tuecke, "The Anatomy of the grid: Enabling scalable virtual organizations," Int. J. Supercomput. Appl., 15(3), 2001.
7. IEEE Standard for Modeling and Simulation, "High level architecture (HLA) federate interface specification," IEEE Std 1516.1–2000.

8. R. Harrison, D. Obst and C.W. Chan, "Design of an ontology management framework," Cognitive Informatics, 2005. (ICCI 2005). Fourth IEEE Conference on 8–10 Aug. 2005 Page(s): 260–266.
9. T. Gu, H.K. Pung and D.Q. Zhang, "A middleware for building context-aware mobile services," Vehicular Technology Conference, 2004. VTC 2004 Spring, 2004 IEEE 59th Volume 5, 17–19 May 2004 Page(s): 2656–2660 Vol.5.
10. W.G Griswold, R. Boyer, S.W Brown and T.M. Truong, "A component architecture for an extensible, highly integrated context-aware computing infrastructure," Software Engineering, 2003. Proceedings 25th International Conference, 3–10 May 2003 Page(s): 363–372.
11. N. Roy, S.K Das, K. Basu, and M. Kumar, "Enhancing availability of grid computational services to ubiquitous computing applications," Parallel and Distributed Processing Symposium, 2005 Proceedings. 19th IEEE International 04–08, April 2005 Page(s): 92a–92a.
12. I. Foster and C. Kesselman, "Globus: A metacomputing infrastructure toolkit," Int. J. Supercomput. Appl., 11(2): 115–128, 1997.
13. N. Davies, A. Friday and O. Storz, "Exploring the grid's potential for ubiquitous computing," Pervasive Computing, IEEE, 3(2), April–June 2004 Page(s): 74–75.
14. N. Roy, "Providing better QoS assurance to next generation ubiquitous grid users", MS Thesis, University of Teaxs at Arlington, USA, Apr. 2004.
15. G. Shang-Fen, Z. Wei, M. Dan and Z. Wen-Li, "Grid mobile service: Using mobile software agent grid mobile service," Machine Learning and Cybernetics, Proceedings of 2004 International Conference on Publication Date: 26–29 Aug. 2004 Volume: 1, page(s): 178–182.
16. P. Ghosh, N. Roy, S. K. Das and K. Basu, "A game theory based pricing strategy for job allocation in mobile grids", Proceedings of 18th International Parallel and Distributed Processing Symposium, Snata Fe, New Mexico, Apr. 2004.

Chapter 18
A Novel Fault-Tolerant Multi-EPON System with Sharing Protection Scheme

I.-Shyan Hwang, Zen-Der Shyu and Liang-Yu Ke

Abstract In the EPON, many previous studies proposed dedicated protection architectures to protect the critical components which results in high cost for deployment. To achieve high reliability and low-cost for deployment, this article proposes a novel fault-tolerant Multi-EPON system with cost-effective shared protection scheme. Under the failures, the Bridge ONU controls the faulty EPON, plays the role of OLT and the transmission of faulty EPONs is restored by relaying to other interconnected adjacent EPONs. The minimum hop count relay algorithm and the relay window mechanism are also proposed in this article to help the data for relaying efficiently. Furthermore, the One-Wait DBA enables the controller of affected PONs to obtain more up-to-date buffer information from each ONU in order to enhance overall system performance. Simulation results show that the proposed Multi-EPON system can provide high system performance for different failed situations in terms of average delay, MAX delay and EF jitter, especially in high traffic loads.

Keywords Bridge ONU · DBA · EPON · Fault-tolerant · Protection

18.1 Introduction

In the EPON, all ONUs are served by the OLT using a discovery handshake in *Multi-Point Control Protocol* and share a common transmission channel towards the OLT by time-division multiple access schemes. Only a single ONU may transmit data in one timeslot to avoid signal collisions. After the ONU is registered by discovery process, the OLT controls PON and coordinates the transmission window of ONUs with granted GATE messages, which contain the transmission start time and transmission length of the corresponding ONU. To avoid signals collisions and allocate bandwidth fairly in the upstream direction, schedule algorithms which

I.-S. Hwang (✉)
Department of Computer Engineering and Science, Yuan-Ze University, Chung-Li, Taiwan 32026, ROC

S.-I. Ao et al. (eds.), *Trends in Communication Technologies and Engineering Science*, Lecture Notes in Electrical Engineering 33, DOI 10.1007/978-1-4020-9532-0_18,

have been extensively researched. There are two categories of bandwidth allocation schemes on EPONs: *fixed bandwidth allocation* (FBA) and *dynamic bandwidth allocation* (DBA). In the FBA scheme, each ONU is assigned fixed timeslots in data transmission from the ONUs to the OLT at full link capacity. In contrast to the FBA, the DBA further improves the system in a more efficient way; the OLT allocates a variable timeslot to each ONU dynamically based on the bandwidth request and ensures the Quality-of-Service (QoS) by guaranteed service level agreement (SLA). Furthermore, fault tolerance is also an important issue in PONs. There are two categories of network failures in EPON, one is link failure and the other is node failure. In link failure, the failure of feeder fiber will halt the whole PON system, but the failure of a branch will halt just one ONU. In the node failure, the failures of OLT or splitter will cause the whole PON system to fault. Therefore, the OLT, feeder fiber and splitter are the most critical components in the PON system. In order to protect PONs against these serious failures, many researchers proposed dedicated protection architectures, however, they are not cost-effective, as they require many redundant components. Sharing bandwidth to protect the neighboring PONs is an efficient way to reduce the cost of protection. To achieve high reliability and low-cost for deployment, a novel *fault-tolerant Multi-EPON*system with shared protection is proposed in this article to provide protection against OLT failure and feeder fiber failure.

The rest of this paper is organized as follows. Section 18.2 decries the related work. Section 18.3 proposes a novel *Fault-Tolerant Multi-EPON* System which is provided a robust fault-tolerant mechanism based on the concept of shared protection. Section 18.4 shows the simulation results in terms of the average packet delay, MAX packet delay and jitter performance. Finally, Section 18.5 draws conclusions and offers further suggestions.

18.2 Literature Review

Four protection architectures are discussed on PON in the ITU-T Recommendation G.983.1 [1]. However, those architectures have a lot of redundancy, and they are not economical solutions. Moreover, the protection scheme with one cold standby OLT is proposed in [2] and the standby OLT utilized to protect multiple PONs. The scheme still needs cold standby equipment and supports only one OLT failure. A resilient fast protection switching scheme [3], when a feeder fiber break or equipment failure occurs in central office (CO), and the switching is performed at the CO. Because of the switch, it is more complex and the scheme still needs a redundant feeder fiber to protect the feeder fiber. In [4], an automatic-protection-switching mechanism is proposed in the ONU to fight against distribution fiber breaks. The transmission of affected ONU is restored by other interconnected ONUs when branches are down. However, it cannot provide any protection for OLT and feeder fiber which are the most critical components of PON. In the ring topologies, the protection scheme in [5] has a large conventional ring and a cold backup transceiver and receiver to protect one point failure. The drawbacks of the ring topologies are more fiber usage, higher signal attenuation and serious near-far problems. To address those problems, the double feeder fibers with a hybrid small ring are introduced in [6]. The scheme minimizes the fiber usage and assures no packet loss by using hot standby components. However, the ONU is more complex in the scheme and 1+1 protection schemes have a low market penetration due to its high cost. Therefore, this paper proposed a shared protection scheme with interconnected adjacent PONs by Bridge ONUs to avoid redundancy.

To support differentiated service classes, a priority queuing scheme, such as DiffServ, and the queue management tasks are carried out by each ONU. The transmission of queued packets is decided by a specific scheduling scheme. The *strict priority scheduling* serves the buffered higher-priority packets first as defined in IEEE 802.1D. Lower-priority packets can only be transmitted when the higher-priority queues are empty. Therefore, the lower-priority packets suffer excessive delays and unfair increased packet loss. However, a scheduling scheme is also necessary to control the high-priority traffic if it exceeds the contract of SLA. In [7], priority-based scheduling is proposed to deal with the problems by employing *strict priority scheduling* within a specific time interval. This scheme provides a bounded delay for low-priority packets and ensures fairness by transmitting packets of all traffic classes.

The dynamic bandwidth allocation (DBA), such as limited bandwidth allocation (LBA), has been studied in [8]. In the LBA, the timeslots length of each ONU is upper bounded by the maximum timeslots length, $B_{\max}$, which could be specified by SLA. The drawback of LBA is the poor utilization for the upstream bandwidth and it restricts aggressive competition for the upstream bandwidth, especially under non-uniform traffic. In order to better utilize the leftover bandwidth from ONUs with some traffic backlogs, the authors in [7] proposed a DBA scheme called Excessive Bandwidth Reallocation (EBR) in which ONUs were divided into two categories, lightly-loaded and heavily-loaded, according to their guaranteed bandwidth $B_{\max}$. Total excessive bandwidth, B_{excess}, saved from lightly-loaded group is redistributed to heavily-loaded ONUs to improve efficiency. Unfortunately, the drawbacks of EBR are unfairness, *idle period problem* and allocating more than the requested bandwidth to ONUs [9], which was redefined as *redundant bandwidth problem*in our previous research [10]. To improve bandwidth utilization, an early allocation mechanism was proposed in [7], which grants the bandwidth of lightly-loaded ONUs immediately without any delay when it receives the REPORT message. Moreover, the efficient bandwidth allocation algorithm (EAA) with a time tracker is proposed to address the *idle period problem* [11]. It forces the OLT to grant the bandwidth to a heavily-loaded ONU without waiting reallocation when the upstream channel is going to idle. However, the drawback of [7, 11] is that the service order of ONUs changes in each service cycle, therefore, the estimation of the incoming high priority traffic is severely impaired because the waiting time in each ONU may change drastically. In [12], Xiaofeng et al. proposed another DBA scheme that maintains fairness mechanism of the excessive bandwidth reallocation operation well for heavily-loaded ONUs, but ignores the fairness of lightly-loaded

ONUs. The reason is that the request by the lightly-loaded ONU does not consider the possible packets arriving during the waiting time. Jitter performance studied in [13] is another important concern in EPON, especially for the high priority service which is delay-variation sensitive traffic. In this scheme, high priority service is protected in a separate sub-cycle, therefore, the jitter performance is considerably improved.

18.3 Proposed Novel Fault-Tolerant Multi-EPON System

18.3.1 Fault-Tolerant Architecture

To achieve high reliability and low-cost for deployment, each EPON system is connected the nearest EPON by a Bridge ONU in order to minimize fiber usage, shown in Fig. 18.1. When the feeder fiber is cut or OLT is down, the Bridge ONU controls the PON, plays the role of OLT and receives the data coming from the ONUs. Then, the Bridge ONU relays the data to the OLT of the adjacent PON. In normal situations, the Bridge ONU ignores the upstream signal of ONUs and monitors the signal of downstream channel to detect the failures. In [14], the passive optical splitter/combiner (PSC) broadcasting the upstream signal to all ONUs may cause potential security problems and higher signal power attenuation. To alleviate those problems, the PSC, constructed by a $2 \times N$ PSC and a 2×2 PSC, is considered in the proposed architecture, shown in Fig. 18.2. In the PSC, the upstream optical signal power is only transmitted to OLT and Bridge ONUs and the downstream optical signal transmitted by the Bridge ONU is broadcasted to all ONUs as the downstream signal of OLT. In the Bridge ONU, both interfaces connect with two adjacent EPON. Note that in addition to the conventional transceiver (a 1310 nm transmitter and a 1490 nm receiver) of normal ONU maintained at the Bridge ONU, the Bridge ONU

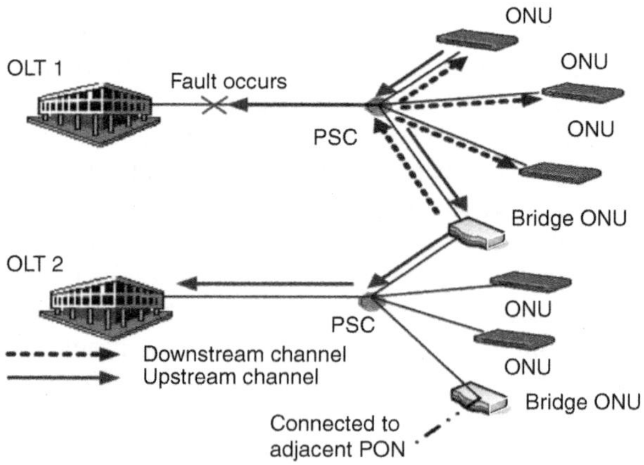

Fig. 18.1 The bridge ONU connects two adjacent PONs

Fig. 18.2 Architecture of the PSC

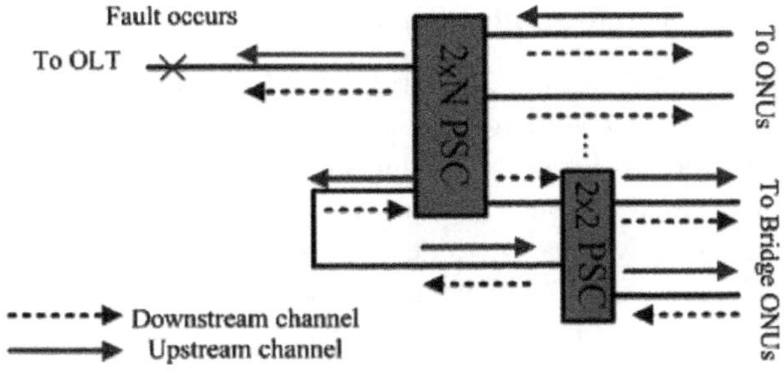

requires an extra 1310 nm receiver, an extra 1490 nm transmitter and a switch. The advantages of the proposed architecture are that no backup feeder fibers and backup OLT modules are needed and the fiber usage is close to standard EPON which is without any protection mechanism.

18.3.2 Minimum Hop Count Relay Path Main Algorithm

When faults occur, in order to let the affected PONs determine the system situation and decide the relay path with minimum hop count, the Bridge ONUs will execute four phases in the *Minimum Hop Count Relay Path Main Algorithm*, and they are described in the following subsection. Note that there are two definitions of the Bridge ONU in the boundary of fault area, one is that the Bridge ONU has a failed PON on only one side (i.e. Bridge ONU BC in Fig. 18.3), the other one is the Bridge ONU has two failed PONs on both sides and a failed PON of the two has only one Bridge ONU (i.e. Bridge ONU EF in Fig. 18.3).

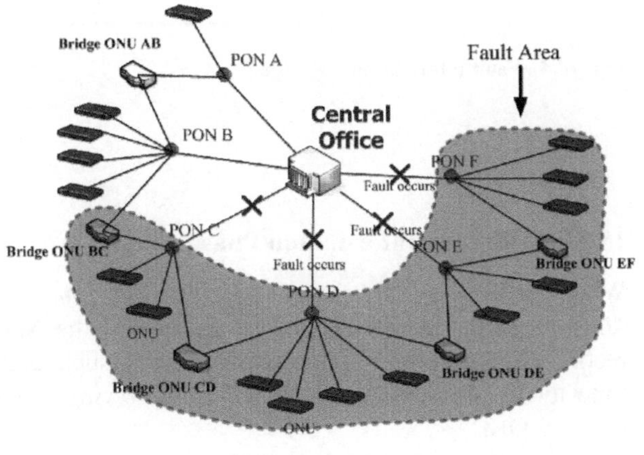

Fig. 18.3 Cost-effective shared protection system

18.3.2.1 Information Sending Phase

When fault occurs, the Bridge ONUs in the boundary of fault area send the *Fault Information Message*, shown in Fig. 18.4, to the Bridge ONU inside the fault area in order to let other Bridge ONUs and affected OLTs get the Multi-EPON system situation, such as the number of hop count and guaranteed timeslots in the left and right paths, to decide the relay path. After sending the *Fault Information Message*, the sender waits for the *ACK Message* reply by the receiver. If the waiting reaches timeout, the *Fault Information Message* will be resent after random delay.

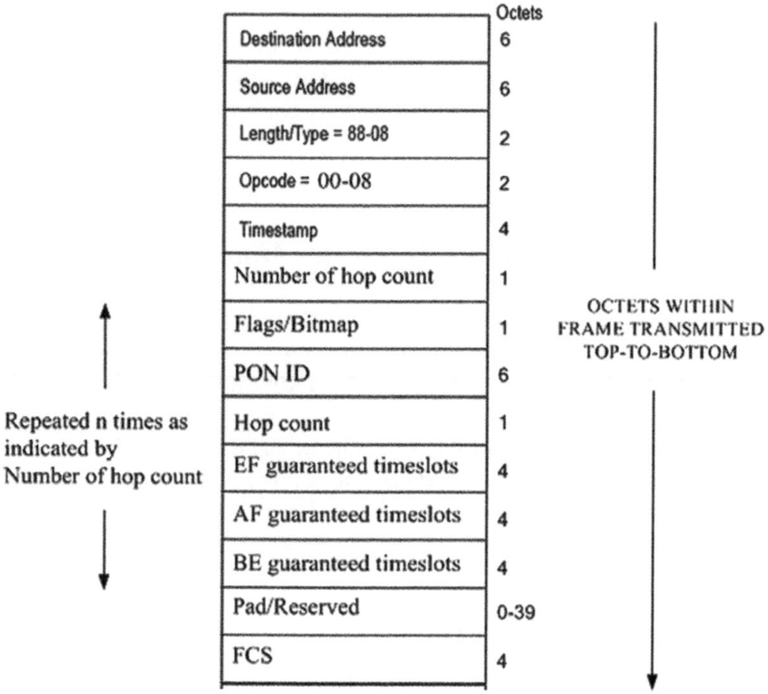

Fig. 18.4 Fault information message

18.3.2.2 Waiting Information Phase

When the Bridge ONU received the *Fault Information Message*, it replied an *ACK Message* to the sender. Then, it adds one to the *Number of hop count* field and records the information of the current PON, such as the hop count and guaranteed timeslots fields in the *Fault Information Message*, and sends it to the next Bridge ONU or OLT.

18.3.2.3 Deciding Relay Path Phase

After getting all information of the affected PONs, the Bridge ONU decides the relay paths. In order to reduce the packet delay during the failure time, each PON chooses the relay path with minimal hop count.

18.3.2.4 Calculating Available Bandwidth Phase

The controller of affected PONs calculates the available bandwidth of each PON as $PON\_BW_i = (T_{cycle}^{MAX} - Ng) \times (\sum_{k \in PON_i} S_k / \sum_{j \in ALL\_PON} S_j)$, where $\sum_{k \in PON_i} S_k / \sum_{j \in ALL\_PON} S_j$ is the proportion of PON_i can get minimum guaranteed bandwidth from the available bandwidth, T_{cycle}^{MAX} is the maximum cycle time, g is the guard time, N is the number of ONUs in the PON providing the bandwidth to Faulty PON, S_k is the sum of S_k^c, the minimum guaranteed timeslots for the *EF*, *AF* and *BE* traffic determined by SLA of ONU_k which is in the PON_i. And S_j is the sum of S_j^c, the minimum guaranteed timeslots for the *EF*, *AF* and *BE* traffic determined by SLA of ONU_j which is in the relay path that including PON_i. The S_j and S_k can be expressed as $S_k = \sum_c S_k^c$ and $S_j = \sum_c S_j^c$, $c \in \{EF, AF, BE\}$.

18.3.3 Relay Mechanism and Local DBA

The PON with fault is referred to as *Faulty PON* and the PON, which provides protection by sharing bandwidth to the Faulty PONs, is referred to as *Protection PON*. After calculating the available timeslots of the PONs, the Bridge ONUs start to control the adjacent PON of the downstream relay path. The Bridge ONUs, in the Faulty PON, execute the One-Wait DBA for local ONUs and the OLT in the Protection PON executes the Protection DBA for local ONUs. Then both DBA schemes cooperate with the Relay Window Mechanism to help relay the data hop-by-hop in the Faulty PONs. For example, the Bridge ONU EF controls the PON F and the Bridge ONU DE controls the PON E, shown in Fig. 18.3. The Bridge ONU EF relays the data queuing in its buffer to the Bridge ONU DE. Then, the Bridge ONU DE has two kinds of data in its buffer; one is PON E coming from local ONU; the other is PON F coming from Bridge ONU EF. The Bridge ONU DE relays the data queuing in its buffer to the controller of PON D that is the Bridge ONU CD and the Bridge ONU CD relays queuing data to the Bridge ONU BC. Finally, the Bridge ONU BC relays all queuing data to the controller of PON B. The proposed Relay Window Mechanism and two DBA schemes are described in the following subsection.

18.3.3.1 Relay Window Mechanism

There are two transmission windows in each PON of the relay path. One is the local DBA window transmitted by local ONUs and the other one is the relay window

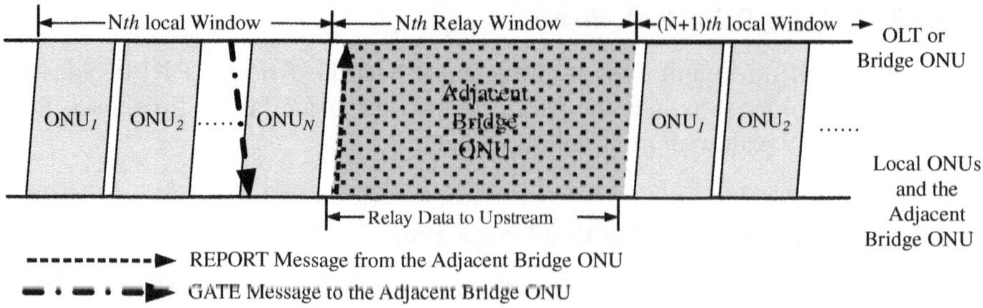

Fig. 18.5 Relay window

transmitted by the adjacent Bridge ONU in the downstream relay path, shown in Fig. 18.5. The OLT or the Bridge ONU specifies the start time of the relay window in the GATE message and transmits it to the adjacent Bridge ONU in the downstream relay path. When the relay window begins, the Bridge ONU specifies transmission length in a REPORT message and sends it to the controller of upstream PON. After the controller gets the REPORT message, it can know when the end of relay window is and then start to execute the local DBA. When a Bridge ONU transmits an upstream packet in the relay window, the packets from local ONUs or the adjacent Bridge ONU in the downstream relay path are still arriving. The EF traffic class with the highest priority for strictly delay sensitive services is typically a constant bit rate (CBR). To further reduce the EF packet delay time, the Bridge ONU predicts the EF bandwidth requirement during the relay window by the EF traffic rate with the rate-based prediction scheme presented in [15]. Then, the Bridge ONU reports the total length transmitting timeslots of this relay cycle with upper bounded by guaranteed bandwidth in the downside relay path, and it can be expressed as

$$G_{i,n+1}^{Total} = \min(R_{i,n}^{Total}, \sum_{k \in downstream\ PON\ in\ the\ relay\ path} PON\_BW_k), \tag{18.1}$$

where $R_{i,n}^{Total}$ is the sum of request timeslots for differentiated traffics in the queues of the Bridge ONU$_i$ in the nth cycle, i.e. $R_{i,n}^{Total} = R_{i,n}^{EF} + R_{i,n}^{AF} + R_{i,n}^{BE} + P_{i,n}^{EF}$, the $P_{i,n}^{EF}$ is the predicted timeslots of EF traffic coming during the relay window, and $\sum_{k \in downstream\ PON\ in\ the\ relay\ path} PON\_BW_k$ is the sum of available bandwidth of Faulty PONs in the downside of the relay path.

18.3.3.2 Two DBA Schemes for Local ONUs

In the Faulty PONs, the link utilization is impossible to use fully. The reason is that the available bandwidth is provided by the OLT in Protection PON and is shared with the affected PONs. The packet delay is more important than link utilization in the Faulty PONs. However, the link utilization and the packet delay are the same importance in Protection PON. Therefore, the two DBA are proposed to deal with the different network characteristics for improving the whole Multi-EPON system performance. First, the One-Wait DBA is proposed to reduce the packet delay in the Faulty PON. Second, the Protection DBA is proposed to improve the channel utilization and reduce the packet delay in the Protection PON.

One-Wait DBA Scheme in the Faulty PON

The operation of traditional DBA schemes are that the REPORT messages piggyback in data timeslots and report the queue length of the ONU, but without considering the packet arriving during the waiting time. It is observed that the packet delay in those DBA schemes is close to 1.5 transmission cycle time. Therefore, the packets arriving in the waiting time cannot be transmitted in the current cycle even if the ONU is lightly-loaded. This will result in longer packet relayed delay and is unfair to the lightly-loaded ONU. To improve the drawbacks and offer better QoS in the Multi-EPON system, the proposed One-Wait DBA shifts the report time of ONUs purposely in order to enable the Bridge ONU to obtain more up-to-date buffer occupancy information from each ONU. This point is further illustrated in Fig. 18.6. The ONU_i uploads its REPORT message between the (i–2)th and (i–1)th ONU. When the Bridge ONU receives the REPORT message of ONU_i, it starts to calculate available bandwidth of ONU_i and grant the available bandwidth in the GATE message to ONU_i immediately. After the end of transmission window of ONU_{i-1}, the REPORT message is uploaded by (i+1)th ONU, and then, the ONU_i starts to upload the queuing data in granted timeslots. Although the amount of cost guard time in the One-Wait DBA is twice as much as the number of the DBA schemes piggybacking their REPORT messages in data timeslots. However, the average packets delay is smaller than one cycle time and close to half of a cycle time. Therefore, the One-Wait DBA can reduce to almost one cycle time in the average packet delay.

When it receives a REPORT message from a local ONU, the One-Wait DBA assigns the timeslots to the ONU based on the guaranteed timeslots immediately and can be expressed as $G_i^{Total} = \min(R_i^{Total}, PON\_BW_i \times (S_i / \sum_{k \in PON_i} S_k))$,

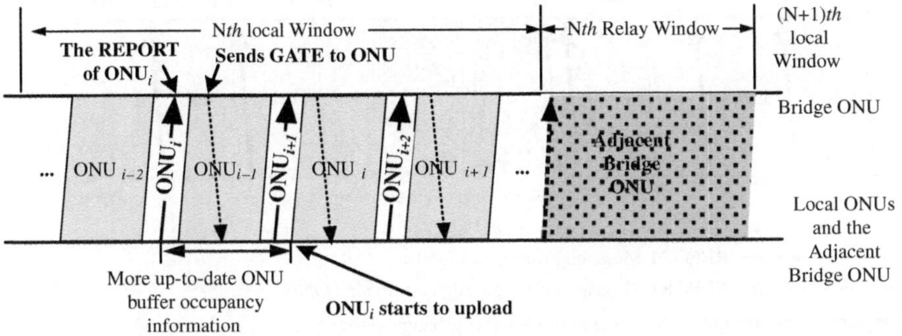

Fig. 18.6 One-wait DBA

where R_i^{Total} is the sum of requested BW for differentiated traffics of ONU_i, and $S_i / \sum_{k \in PON_i} S_k$ is the proportion of ONU_i can get granted bandwidth from the available bandwidth of this PON denoted as PON_BW_i. After finishing granting bandwidth to the ONU_i by sending GATE message, the One-Wait DBA specifies the REPORT sending time of ONU_{i+2} at the end of uploaded timeslots in ONU_i. Moreover, after granting the last ONU in local cycle, the controller of the PON invokes the *Relay Mechanism.* When the controller received the REPORT message of the adjacent Bridge ONU of the PON, it executes the One-Wait DBA again.

Protection DBA Scheme in the Protection PON

The Protection PON is the final hop in the relay path, the performance of packet delay in relay data and link utilization are the key factors on the system performance. To further improve bandwidth utilization and packet delay of relay data, there are two transmission windows in the *Protection DBA*. One is the relay window for Bridge ONUs to relay data; the other is local transmission window for local ONUs, shown in Fig. 18.7. Furthermore, to address the jitter performance of EF traffic, the proposed Protection DBA fixed the service order of all local ONUs and the Bridge ONU in the PON.

Firstly, at the beginning of *N*th Relay Window to transmit the data, the Bridge ONU limited the transmitting timeslots by equation (18.1) and broadcasts the REPORT message to the OLT in order to specify its transmission length. Then, the Bridge ONU starts to relay the data queue in its buffer to OLT, shown in Fig. 18.7. Secondly, after receiving the REPORT message of the Bridge ONU, the OLT executes the Protection DBA scheme to allocate bandwidth to local ONUs for (*N*+1)th transmission cycle and sends all GATE messages to local ONUs. Thirdly, the OLT

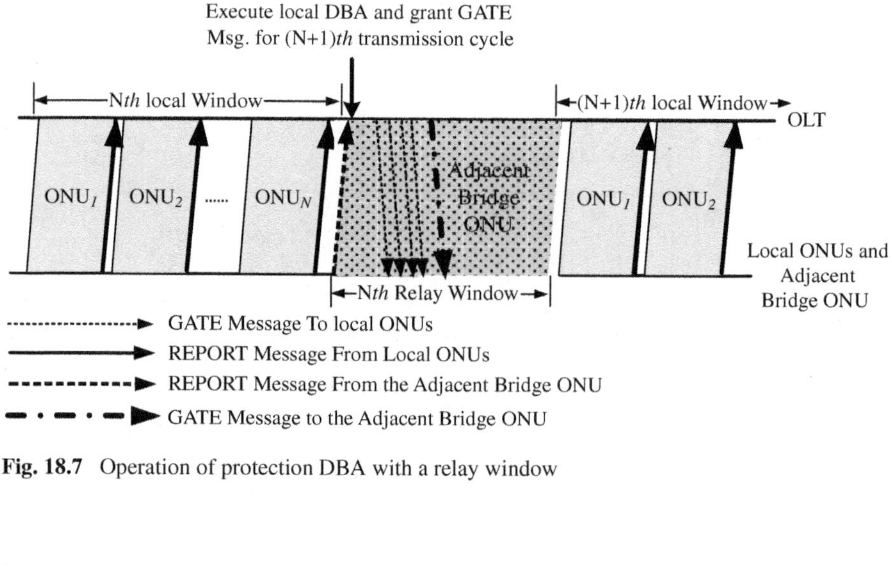

Fig. 18.7 Operation of protection DBA with a relay window

specifies the beginning timeslots of (N+1)th Relay Window by sending a GATE message to the Bridge ONU. Note that the transmissions of different nodes are separated by a guard time.

The detail of allocating timeslots for local ONUs in the proposed Protection DBA is described as follows. First, calculate $R_{i,n}^{Total}$ of all ONUs and initialize the available bandwidth, $B_{available}$, which is expressed as $B_{available} = PON\_BW_i$. Then, the proposed Protection DBA will select the ONU$_i$ with the maximal residue bandwidth, i.e. $\max\left(S_i - R_{i,n}^{Total}\right)$, from unassigned ONUs. The granted bandwidth for ONU$_i$, $G_{i,n+1}^{Total}$, in the next cycle is given as follows

$$G_{i,n+1}^{Total} = \min\left(B_{available} \times (S_i / \sum_{k \in unassigned} S_k) \ , \ R_{i,n}^{Total} \right),$$

where R_i^{Total} is the sum of requested BW for differentiated traffics of ONU$_i$ in the nth cycle, i.e. $R_{i,n}^{Total} = R_{i,n}^{EF} + R_{i,n}^{AF} + R_{i,n}^{BE} + P_{i,n}^{EF}$, where $P_{i,n}^{EF}$ is the predicted timeslots of *EF* traffic with the rate-based prediction scheme presented in [15] and $S_i / \sum_{k \in unassigned} S_k$ is the proportion of ONU$_i$ can get granted bandwidth from the available bandwidth, $B_{available}$. Furthermore, the granted bandwidth for *EF*, *A*F and *BE* classes are as follows:

$$\begin{cases} G_{i,n+1}^{EF} = \min(G_{i,n+1}^{Total}, R_{i,n}^{EF} + P_{i,n}^{EF}) \\ G_{i,n+1}^{AF} = \min(G_{i,n+1}^{Total} - G_{i,n+1}^{EF}, R_{i,n}^{AF}). \\ G_{i,n+1}^{BE} = G_{i,n+1}^{Total} - G_{i,n+1}^{EF} - G_{i,n+1}^{AF} \end{cases}$$

In the final, the available bandwidth becomes $B_{available} = B_{available} - G_{i,n+1}^{Total}$. The whole process continues until all local ONUs have been assigned, and the Protection DBA fixes the transmission order of ONUs to improve the jitter performance. Then, the OLT sends the GATE message to all local ONUs and specifies the start time of (N+1)th Relay Window by sending a GATE message to the Bridge ONU.

18.4 Performance Evaluation

The system performance is analyzed by simulating the four significant scenarios, shown in Fig. 18.8, and compared with EAA [11] in terms of average packet delay, MAX packet delay and jitter. The performance evaluation is studied using the OPNET simulator with one OLT and 32 ONUs in a PON. The downstream and upstream capacities are both 1 Gb/s. The distance from an ONU to the OLT is assumed 20 km and the distance between each PON is assumed 1 km. Each ONU has infinite buffer and the service policy is in first-in first-out discipline. The traffic model is characterized by self-similarity and long-range dependence (LRD) [16]. This model is used to generate highly bursty AF and BE traffic classes with a Hurst parameter of 0.7, and packet sizes are uniformly distributed between 64 and 1518

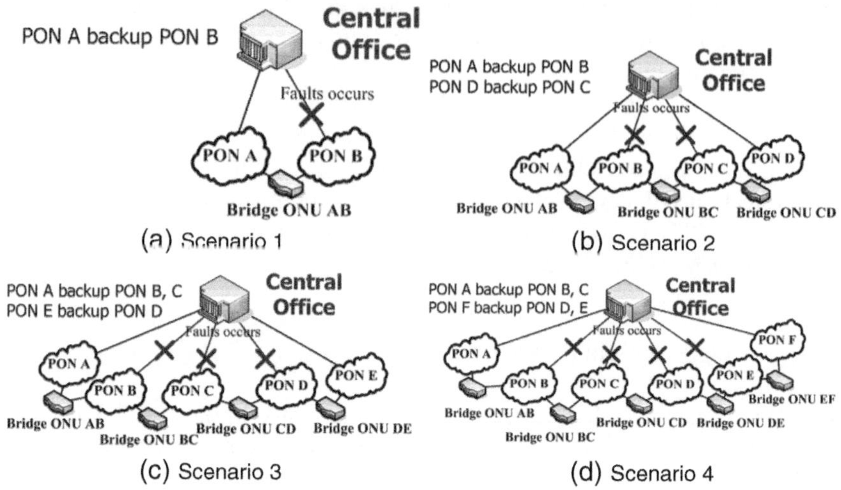

Fig. 18.8 Four significant scenarios

bytes. Furthermore, high-priority traffic is modeled using a Poisson distribution and packet size is fixed to 70 bytes. The traffic profile is as follows: 20% of the total generated traffic is considered high priority, and the remaining 80% equally distributed between low- and medium-priority traffic [7]. For simplicity, the total network load is evenly distributed amongst all ONUs in the same relay path and the ONUs are equally guaranteed bandwidth weighted [7, 8].

18.4.1 Average End-to-End Delay

Figure 18.9 compares the average packet delay from ONUs to central office among the four scenarios and EAA for EF, AF, BE, and total traffic, respectively. Figure 18.9(a) shows the proposed Multi-EPON system under failures has better performance than EAA when the traffic load is greater than 40%. The EF delay time of Multi-EPON reaches 1 *ms* when the traffic load is heavy, but still is less than 1.5 ms, which is specified by ITU-T Recommendation G.114. Figure 18.9(b) shows the delay time of Multi-EPON scenarios for AF traffic are still shorter than the EAA when traffic load exceeds 70%. Figure 18.9(c) shows that Multi-EPON yields notable improvements of EF and AF services without degrading the BE services.

18.4.2 MAX End-to-End Delay

Figure 18.10 compares the maximum packet delay among the four scenarios and EAA for all traffic classes. Figure 18.10(a) shows the proposed Multi-EPON system has better performance than the EAA when traffic load exceeds 60% in the EF

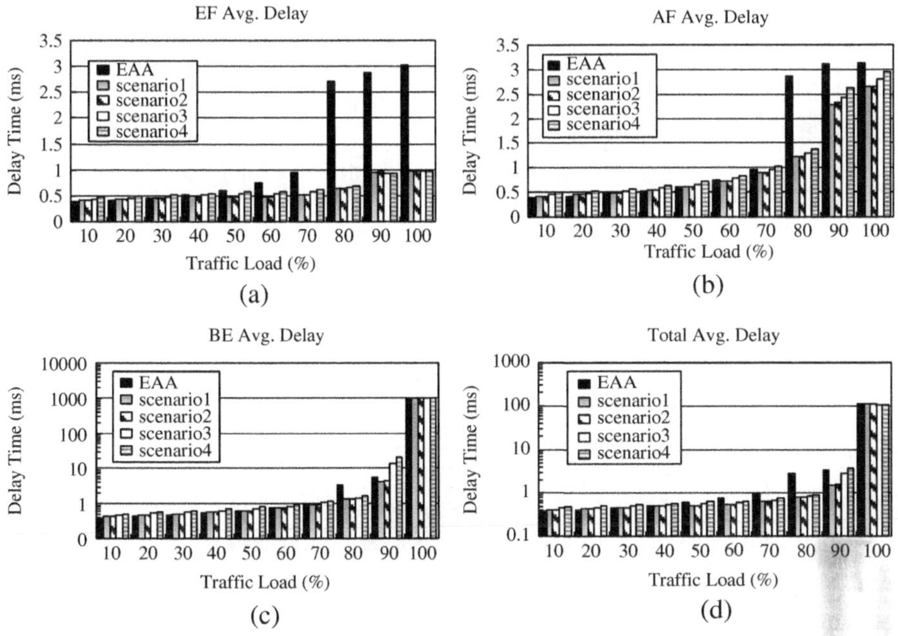

Fig. 18.9 Average end-to-end delay

traffic. In Multi-EPON scenarios, the MAX delay for EF traffic increases smoothly with traffic loads and implies the control ability in the Multi-EPON. The MAX delay for EF traffic at full traffic load is less than 90%. The reason is that the transmission cycle reaches T_{cycle}^{MAX} results in the rate-based prediction of EF traffic more accurately. In Fig. 18.10(b), the MAX delay performance of EAA for AF traffic is better than the Multi-EPON scenarios, because the AF packet encounters long relay time in the relay path. However, while the traffic is between 60 and 80%, the EAA has the longest MAX delay because of the transmission order of ONUs changing drastically. In Fig. 18.10(c), the MAX delay for BE traffic increases rapidly when the traffic load exceeds 90% because the system is in full load.

18.4.3 EF Jitter Performance

The delay variance is calculated as $\sigma^2 = \left(\sum_{i=1}^{N} (d_i^{EF} - \overline{d})^2\right) \cdot N^{-1}$, where the d_i^{EF} is the delay time of EF packet i and N is the total number of received EF packets. As Fig. 18.11 shows, the delay variances of Multi-EPON are higher than the EAA when traffic load is below 70%. The reason is that the EF packets coming from PONs with failures may be relayed by other Bridge ONUs. However, the delay variance of EAA scheme increases drastically when traffic load exceeds 80%. The reason is that the

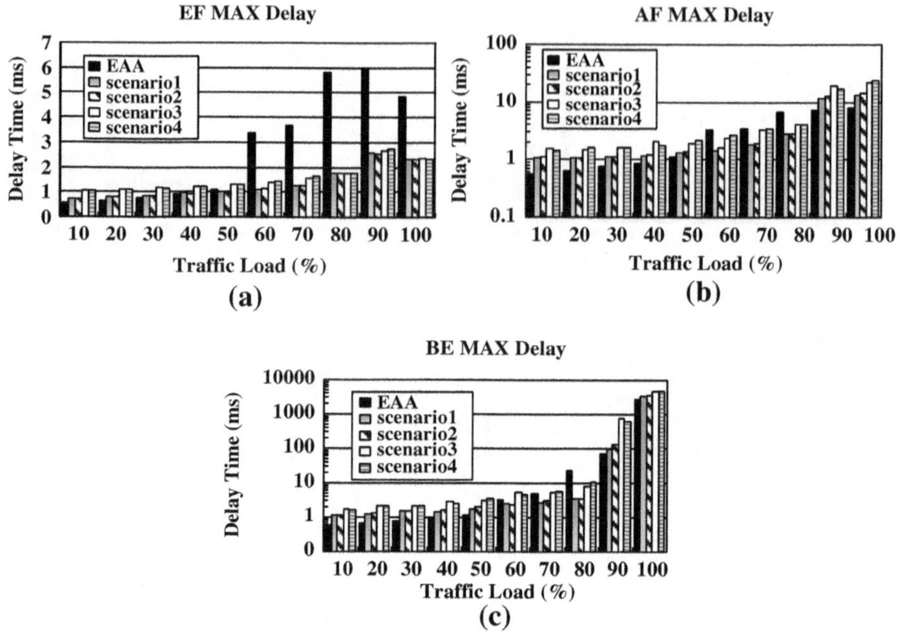

Fig. 18.10 (**a**) EF MAX delay; (**b**) AF MAX Delay; (**c**) BE MAX delay

Fig. 18.11 Jitter for EF traffic

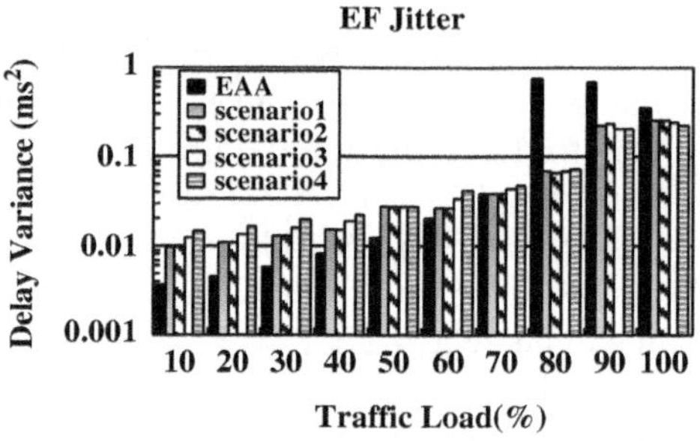

service cycle order in every cycle of the EAA is changed drastically and the order of Multi-EPON is always fixed.

18.5 Conclusion

In this article, a novel *fault-tolerant Multi-EPON* system is proposed to protect the critical PON element failure, such as OLT failure or feeder fiber cut. Because of no limited number of interconnected EPONs and without any redundant components, the proposed system is very flexible, cost-effective and uses simple shared protection architecture on EPON. Overall, the simulation results confirm that the *fault-tolerant Multi-EPON* system yields notable improvements in average packet delay, MAX packet delay and delay variation for EF traffic under different failures.

References

1. ITU-T, Study Group 15, G.983: High Speed Optical Access Systems based on Passive Optical Network (PON) techniques, 2001.
2. Y.M. Kim, J.Y. Choi, J.H. Ryou, H.M. Baek, O.S. Lee, H.S. Park, M. Kang, and J.H. Yoo, Cost effective protection architecture to provide diverse protection demands in Ethernet passive optical network, International Conference on Communication Technology, **1**, 721–724 (2003).
3. M.K. Abdullah, W.T. P'ng, P.W. Lau, and E.R. Tee, FTTH access network protection using a switch, The 9th Asia-Pacific Conference on Communications, **3**, 1219–1222 (2003).
4. N. Nadarajah, E. Wong, M. Attygalle, and A. Nirmalathas, Protection switching and local area network emulation in passive optical networks, *Journal of Lightwave Technology*, **24**(5), 1955–1967, (2006).
5. X. Zhao, X. Chen, and X. Fu, A novel protection switching scheme for PONs with ring plus tree topology, Proceedings of the SPIE, **6022**, 949–956 (2005).
6. D. Hossain, H. Erkan, R. Dorsinville, M. Ali, S. Shami, and C. Assi, Protection for a Ring-Based EPON Architecture, 2nd International Conference on Broadband Networks, **2**, 1548–1553 (2005).
7. C. Assi, Y. Ye, S. Dixit, and M.A. Ali, Dynamic bandwidth allocation for Quality-of-Service over Ethernet PONs, *IEEE Journal on Selected Areas in Communications*, **21**(9), 1467–1477 (2003).
8. G. Kramer, B. Mukherjee, and G. Pesavento, Interleaved polling with adaptive cycle time (IPACT): A dynamic bandwidth distribution scheme in an optical access network, *Photonic Network Communications*, **4**(1), 89–107 (2002).
9. B. Chen, J. Chen, and S. He, Efficient and fine scheduling algorithm for bandwidth allocation in Ethernet passive optical networks, *IEEE Journal of Selected Topics in Quantum Electronics*, **12**(4), 653–660 (2006).
10. I.S. Hwang, Z.D. Shyu, L.Y. Ke, and C.C. Chang, A novel early DBA mechanism with prediction-based fair excessive bandwidth reallocation scheme in EPON, *Computer Communications*, **31**(9), 1814–1823 (2008).
11. J. Zheng, Efficient bandwidth allocation algorithm for Ethernet passive optical networks, *IEE Proceedings Communications*, **153**(3), 464–468 (2006).
12. X. Bai, A. Shami, and C. Assi, On the fairness of dynamic bandwidth allocation schemes in Ethernet passive optical networks, *Computer Communications*, **29**(11), 2123–2135 (2006) .
13. A. Shami, B. Xiaofeng, C.M. Assi, and N. Ghani, Jitter performance in Ethernet passive optical networks, *Journal of Lightwave Technology*, **23**(4), 1745–1753 (2005).
14. S.R. Sherif, A. Hadjiantonis, G. Ellinas, C. Assi, and M.A. Ali, A novel decentralized Ethernet-based PON access architecture for provisioning differentiated QoS, *Journal of Lightwave Technology*, **22**(11), 2483–2497 (2004).
15. G. Kramer, B. Mukherjee, S. Dixit, Y. Ye, and R. Hirth, On supporting differentiated classes of service in EPON-based access networks, *Journal of Optical Networking*, **1**(8/9), 1–20 (2002).
16. W. Willinger, M.S. Taqqu, and A. Erramilli, A bibliographical guide to self-similar traffic and performance modeling for modern high-speed networks, in Stochastic Networks, 1996. Available: http://citeseer.ist.psu.edu/willinger96bibliographical.html.

Chapter 19
Design and Performance Evaluation of Knuth-Bendix Multi-Completion System Using Boolean Constrained Reduction Orders

Haruhiko Sato and Masahito Kurihara

Abstract A multi-completion system MKB developed by Kurihara and Kondo accepts as input a set of reduction orders in addition to equations and efficiently simulates parallel processes each of which executes the standard completion procedure with one of the given orderings. In this paper, we describe a new system MKBOOL, which is a refinement of MKB in the sense that it restricts the reduction orders to those classes which can be represented by boolean constraints on some domains. Such classes include the recursive path orders (with status) for a finite signature. Our implementation represents a conjunction of the constraints by a binary decision diagram. The comprehensive experiments with the implementation on a set of well-known test problems show that in exchange for that restriction, MKBOOL runs more efficiently than MKB for most of the problems.

Keywords Term rewriting system · Knuth-Bendix completion · Multi-completion · Recursive path ordering

19.1 Introduction

Equational reasoning, which is rather restricted class of first-order languages, is an important component in various areas, such as automated theorem proving, artificial intelligence, functional and logic programming languages, and algebraic specification of abstract data types. Therefore, developing specialized methods and tools for this type of reasoning is also important. Term rewriting [1,10] is a very powerful method for dealing with equations. Oriented equations, called *rewrite rules*, are used to replace equals by equals, but only in one direction. This constitutes a Turing-complete computational model. In general, term rewriting applies in any context where efficient methods for reasoning with equations are required.

H. Sato (✉)
Graduate School of Information Science and Technology, Hokkaido University, Hokkaido, Japan

S.-I. Ao et al. (eds.), *Trends in Communication Technologies and Engineering Science*, Lecture Notes in Electrical Engineering 33, DOI 10.1007/978-1-4020-9532-0_19,

A set of rewrite rules is called a *term rewriting system* (TRS). It represents a computational system based on rewriting. Given an initial expression (term), we apply the rewrite rules successively in order to transform it to other terms until we get a *normal form*, to which no more rules apply. The normal form represents the result of the computation. Since there may be different ways of applying rules, we may have distinct normal forms of a term. This indicates that term rewriting systems also provide a natural model of parallel, nondeterministic computation. Term rewriting systems are attractive computational model because of their simple syntax and semantics which do not involve bound variables.

In many applications, termination and confluence are crucially important properties of term rewriting systems. A term rewriting system is *terminating* if there are no infinite rewrite sequences. In other words, after finitely many rule applications we always reach a normal form. A term rewriting system is *confluent* if every term has at most one normal form. A term rewriting system which is terminating and confluent is said to be *convergent*. If a term rewriting system is convergent, every term has a unique normal form.

In order to compute a convergent term rewriting system, the standard completion procedure (KB) was proposed by Knuth and Bendix [5] and has been widely improved since then. Given a term rewriting system R (or a set of equations E) and a reduction order which is used to check termination, the procedure verifies the convergence of R analytically. If R is not convergent, the procedure tries to construct a convergent term rewriting system which is equationally equivalent to R (or E) by adding or modifying rewrite rules. The result of the computation is either success, failure, or divergence [1, 2, 5, 10]. The existence of divergence implies that the procedure can be non-terminating, thus the procedure is a semi-algorithm.

The success of the procedure heavily depends on the choice of the appropriate reduction order to be supplied. Such a choise is often difficult for general users unless they have good insight in termination proof techniques. Unfortunately, one cannot try out potentially-appropriate reduction orders one by one (*sequentially*), because one of those runs may lead to infinite, divergent computation and inhibit the exploration of the remaining possibilities.

This means that we have to consider, more or less, *parallel* execution of Knuth-Bendix completion procedures each running with one of the reduction orders. When the number of those orderings is very large, however, it is clear that the direct implementation of this idea on a fixed number of workstations would end up with serious inefficiently because of the large number of processes run on each machine.

Kurihara and Kondo [6] partially solved this problem by developing a completion procedure called MKB, which, accepting as input a *set* of reduction orders as well as equations, efficiently simulates (in a single process) parallel execution of KB procedures each working with one of those orders. The key idea is the development of the data structure for storing a pair $s : t$ of terms associated with the information to show which processes contain the rule $s \to t$ (or $t \to s$) and which processes contain the equation $s \leftrightarrow t$. This structure makes it possible to define a meta-inference system for MKB that effectively simulates a lot of closely-related inferences made in different processes all in a simgle operation. We call this type of procedure a *multi-completion* procedure.

In order to orient an equation $s \leftrightarrow t$, MKB needs to check $s \succ t$ for all reduction orders $\succ$ one by one. It causes unignorable degradation in performance when the number of orders is very large. To settle this problem, the new system MKBOOL presented in this paper represents a set of reduction orders by a boolean constraint on the basic objects (such as a precedence and a status) underlying the orders. Using this representation, the system can compute directly the set of all reduction orders $\succ$ such that $s \succ t$ for a given equation $s \leftrightarrow t$.

In this paper, we describe a general framework for MKBOOL, followed by a particular boolean encoding for recursive path orders (with status). The comprehensive experiments on a set of well-known test problems show that MKBOOL with this encoding runs more efficiently than MKB for most of the problems.

The paper is organized as follows. In Section 19.2 we briefly review the multi-completion procedure MKB. In Section 19.3 we propose a general framework for MKBOOL. In Section 19.4 we present a boolean encoding for recursive path orders, based on the encodings for the quasi-precedence and the status. In Section 19.5 we evaluate the performance of MKBOOL by comparing the experimental results with MKB. Finally, we conclude with future work in Section 19.6.

19.2 Multi-Completion

Given a set E of equations and a reduction order $\succ$, the standard completion procedure KB tries to compute a convergent set R of rewrite rules that is contained in $\succ$ and that induces the same equational theory as E.

Starting from the initial state $(E_0; R_0) = (E; \emptyset)$, the procedure obeys the inference system defined in Fig. 19.1 to generate a 'fair' sequence $(E_0; R_0) \vdash (E_1; R_1) \vdash \ldots$ of deduction, where $\rhd$ in the COLLAPSE rule is a well-founded order on terms such as the *encompassment* order. The result of the *successful* sequence is the empty set E_∞ of persisting equations and the convergent set R_∞ of rewrite rules.

DELETE:	$(E \cup \{t \leftrightarrow t\}; R) \vdash (E; R)$
ORIENT:	$(E \cup \{t \leftrightarrow u\}; R) \vdash (E; R \cup \{t \to u\})$ if $t \succ u$
SIMPLIFY:	$(E \cup \{t \leftrightarrow u\}; R) \vdash (E \cup \{t \leftrightarrow v\}; R)$ if $u \to_R v$
COMPOSE:	$(E; R \cup \{t \to u\}) \vdash (E; R \cup \{t \to v\})$ if $u \to_R v$
COLLAPSE:	$(E; R \cup \{u \to t\}) \vdash (E \cup \{v \to t\}; R)$ if $l \to r \in R, u \to_{\{l \to r\}} v$, and $u \rhd l$
DEDUCE:	$(E; R) \vdash (E \cup \{t \leftrightarrow u\}; R)$ if $t \leftarrow_R v \to_R u$

Fig. 19.1 Standard completion

A *multi-completion* procedure accepts as input a finite set $O = \{\succ_1, \ldots, \succ_m\}$ of reduction orders as well as a set E of equations. The mission of the procedure is basically the same as KB: it tries to compute a convergent set R of rewrite rules that is contained in *some* $\succ_i$ and that induces the same equational theory as E.

The multi-completion procedure MKB developed in [6] exploits the data structure called nodes. Let $I = \{1, 2, \ldots, m\}$ be the set of indexes for orders $O = \{\succ_1, \ldots, \succ_m\}$. A *node* is a tuple $\langle t : u, L_1, L_2, L_3 \rangle$, where $t : u$ (called a *datum*) is an ordered pair of terms, and L_1, L_2 and L_3 (called *labels*) are subsets of I such that

- $L_1 \cap L_2 = L_2 \cap L_3 = L_3 \cap L_1 = \emptyset$ and
- $i \in L_1$ implies $t \succ_i u$, and $i \in L_2$ implies $u \succ_i t$.

The node $\langle t : u, L_1, L_2, L_3 \rangle$ is considered to be identical with $\langle u : t, L_2, L_1, L_3 \rangle$.

The MKB procedure is defined by the inference system working on a set N of nodes, as given in Fig. 19.2. Starting from the initial set of nodes,

DELETE: $N \cup \{\langle t : t, \emptyset, \emptyset, L \rangle\} \vdash N$ if $L \neq \emptyset$

ORIENT: $N \cup \{\langle t : u, L_1, L_2, L_3 \cup L \rangle\} \vdash N \cup \{\langle t : u, L_1 \cup L, L_2, L_3 \rangle\}$ if $L \neq \emptyset, L_3 \cap L = \emptyset$, and $t \succ_i u$ for all $i \in L$

REWRITE-1: $N \cup \{\langle t : u, L_1, L_2, L_3 \rangle\} \vdash N \cup \left\{ \begin{array}{l} \langle t : u, L_1 \setminus L, L_2, L_3 \setminus L \rangle \\ \langle t : v, L_1 \cap F, \emptyset, L_3 \cap L \rangle \end{array} \right\}$ if $\langle l : r, L, \ldots, \ldots \rangle \in N, u \rightarrow_{\{l \rightarrow r\}} v$, $u \doteq l$, and $(L_1 \cup L_3) \cap L \neq \emptyset$

REWRITE-2: $N \cup \{\langle t : u, L_1, L_2, L_3 \rangle\} \vdash N \cup \left\{ \begin{array}{l} \langle t : u, L_1 \setminus L, L_2 \setminus L, L_3 \setminus L \rangle \\ \langle t : v, L_1 \cap L, \emptyset, (L_2 \cup L_3) \cap L \rangle \end{array} \right\}$ if $\langle l : r, L, \ldots, \ldots \rangle \in N, u \rightarrow_{\{l \rightarrow r\}} v$, $u \rhd l$, and $(L_1 \cup L_2 \cup L_3) \cap L \neq \emptyset$

DEDUCE: $N \vdash N \cup \{\langle t : u, \emptyset, \emptyset, L \cap L' \rangle\}$ if $\langle l : r, L, \ldots, \ldots \rangle \in N$, $\langle l' : r', L', \ldots, \ldots \rangle \in N, L \cap L' \neq \emptyset$, $v \rightarrow_{\{l \rightarrow r\}} t$, and $v \rightarrow_{\{l' \rightarrow r'\}} u$

GC: $N \cup \{\langle t : u, \emptyset, \emptyset, \emptyset \rangle\} \vdash N$

SUBSUME: $N \cup \left\{ \begin{array}{l} \langle t : u, L_1, L_2, L_3 \rangle, \\ \langle t' : u', L'_1, L'_2, L'_3 \rangle \end{array} \right\} \vdash N \cup \{\langle t : u, L_1 \cup L'_1, L_2 \cup L'_2, (L_3 \cup L'_3) \setminus (L_1 \cup L'_1 \cup L_2 \cup L'_2) \rangle\}$ if $t : u$ and $t' : u'$ are the same (up to renaming of variables).

Fig. 19.2 MKB inference rules

$$N_0 = \{\langle t : u, \emptyset, \emptyset, I\rangle | t \leftrightarrow u \in E\}$$

the procedure generates a fair sequence $N_0 \vdash N_1 \vdash \ldots$. From a successful sequence, the convergent set of rewrite rules can be extracted by projecting N_∞ onto a successful index i.

The semantics of MKB is based on the interpretation in which MKB simulates the parallel processes $P_1, \ldots, P_m$ with P_i executing the KB with the ordering $\succ_i$ and the *common* input E. The soundness, completeness, and fairness of MKB are discussed in [6].

19.3 Multi-Completion for Boolean Constrained Reduction Orders

In this section, we construct a general framework for a new multi-completion system MKBOOL on top of the basic framework of MKB by introducing boolean encoding for a set of reduction orders. In Section 19.3.1 we establish an abstract setting for boolean encoding for arbitrary sets. Then in Section 19.3.2 we use this setting to define MKBOOL.

19.3.1 Abstract Boolean Encoding

Let X be a set of *logical variables* and A_X be the set of all *assignments* (or interpretations) for X. A *constraint* F on X is a restriction on the combinations of the values taken by the logical variables in X. Let C_X be the set of all constraints on X. The constraint F may be represented by a propositional formula (using the connectives such as $\neg, \wedge, \vee$) or a boolean function $F(X)$ (in an expression using the operators such as the negation ($\bar{\ }$), product ($\cdot$), and sum ($+$)). We use these representations interchangeably in this paper.

Definition 19.1 Let A be a set and X be a set of logical variables. An encoding function for A in X is a mapping *enc*: $A \to C_X$ such that

- for all $a \in A$, $enc(a)$ is satisfiable and
- for all $a, a' \in A$, if $a \neq a'$ then $enc(a) \wedge enc(a')$ is unsatisfiable.

The decoding function *dec*: $A_X \to A$ is derived as follows:

$$dec(\alpha) = \begin{cases} a & if \alpha \models enc(a), a \in A \\ \bot & otherwise \end{cases}$$

Definition 19.2 Let B be a subset of A. We extend the definition of the encoding function as follows:

$$enc(B) = \bigvee_{b \in B} enc(b)$$

Let F be a constraint on X. We extend the definition of the decoding function as follows:

$$dec(F) = \{dec(\alpha) | \alpha \in A_X, \alpha \models F\}$$

Definition 19.3 An *encoding* for a set A is a pair (X, enc) consisting of a set of logical variables X and an encoding function enc for A in X.

Definition 19.4 The *base constraint* of an encoding (X, enc) for a set A is defined by $enc(A)$.

Lemma 19.5 *If (X_1, enc_1) is an encoding for A_1, (X_2, enc_2) is an encoding for A_2, and $(X_1 \cap X_2) = \emptyset$ then (X, enc) is an encoding for $A_1 \times A_2$ where*

$$X = X_1 \cup X_2$$
$$enc((a_1, a_2)) = enc_1(a_1) \wedge enc_2(a_2)$$
$$enc(A_1 \times A_2) = enc_1(A_1) \wedge enc_2(A_2).$$

19.3.2 MKBOOL

In the following, we consider encodings for a set of reduction orders. Let O be a set of reduction orders, $O_{t,u} = \{\succ \mid \succ \in O, t \succ u\}$, and (X, enc) be an encoding for O. We put some constraint $F_i \in C_X$ in each label field of the node structure of MKB. In this way, we define a new node structure

$$\langle t : u, F_1, F_2, F_3 \rangle$$

with constraint labels F_1, F_2 and F_3, in place of the MKB node structure $\langle t : u, L_1, L_2, L_3 \rangle$ with set labels, where $F_i = enc(L_i)$, $(i = 1, 2, 3)$. The constraint labels of a node n is denoted by $F_1[n]$, $F_2[n]$ and $F_3[n]$, respectively. The labels must satisfy the following conditions.

- $F_1 \wedge F_2$, $F_2 \wedge F_3$, $F_3 \wedge F_1$ are unsatisfiable.
- F_1 implies $enc(O_{t,u})$ and F_2 implies $enc(O_{u,t})$.

Clearly, the union and intersection of set labels are represented by the sum and product of the corresponding constraint labels, respectively; and the set difference $L_1 \setminus L_2$ is represented by $F_1 \cdot \overline{F_2}$. With these transformation in mind, we can rewrite MKB to the new system MKBOOL given in Fig. 19.3. Note that $enc(O_{t,u})$ is used in ORIENT to encode the set of reduction orders $\succ$ in which $t \succ u$.

Given a set of equations E, we start from the initial set of nodes

$$N_0 = \{\langle t : u, 0, 0, enc(O) \rangle | t \leftrightarrow u \in E\}.$$

$$
\begin{array}{ll}
\textsc{Delete}: & N \cup \{\langle t : t, 0, 0, F\rangle\} \vdash N \text{ if } F \neq 0 \\
\textsc{Orient}: & N \cup \{\langle t : u, F_1, F_2, F_3\rangle\} \vdash \\
& N \cup \{\langle t : u, F_1 + F_3 \cdot F, F_2, F_3 \cdot \bar{F}\rangle\} \\
& \text{if } F = enc(O_{t,u}) \neq 0 \\
\textsc{Rewrite-1}: & N \cup \{\langle t : u, F_1, F_2, F_3\rangle\} \vdash \\
& N \cup \left\{ \begin{array}{l} \langle t : u, F_1 \cdot \bar{F}, F_2, F_3 \cdot \bar{F}\rangle \\ \langle t : v, F_1 \cdot F, 0, F_3 \cdot F\rangle \end{array} \right\} \\
& \text{if } \langle l : r, F, \ldots, \ldots\rangle \in N, u \rightarrow_{\{l \rightarrow r\}} v, \\
& u \doteq l, \text{ and } (F_1 + F_3) \cdot F \neq 0 \\
\textsc{Rewrite-2}: & N \cup \{\langle t : u, F_1, F_2, F_3\rangle\} \vdash N \cup \\
& \left\{ \begin{array}{l} \langle t : u, F_1 \cdot \bar{F}, F_2 \cdot \bar{F}, F_3 \cdot \bar{F}\rangle \\ \langle t : v, F_1 \cdot F, 0, (F_2 + F_3) \cdot F\rangle \end{array} \right\} \\
& \text{if } \langle l : r, F, \ldots, \ldots\rangle \in N, u \rightarrow_{\{l \rightarrow r\}} v, \\
& u \rhd l, \text{ and } (F_1 + F_2 + F_3) \cdot F \neq 0 \\
\textsc{Deduce}: & N \vdash N \cup \{\langle t : u, 0, 0, F \cdot F'\rangle\} \\
& \text{if } \langle l : r, F, \ldots, \ldots\rangle \in N, \\
& \langle l' : r', F', \ldots, \ldots\rangle \in N, F \cdot F' \neq 0, \\
& v \rightarrow_{\{l \rightarrow r\}} t, \text{ and } v \rightarrow_{\{l' \rightarrow r'\}} u \\
\textsc{Gc}: & N \cup \{\langle t : u, 0, 0, 0\rangle\} \vdash N \\
\textsc{Subsume}: & N \cup \left\{ \begin{array}{l} \langle t : u, F_1, F_2, F_3\rangle, \\ \langle t' : u', F_1', F_2', F_3'\rangle \end{array} \right\} \vdash \\
& N \cup \{\langle t : u, F_1 + F_1', F_2 + F_2', \\
& (F_3 + F_3') \cdot \bar{F_1} \cdot \bar{F_1'} \cdot \bar{F_2} \cdot \bar{F_2'}\rangle\} \\
& \text{if } t : u \text{ and } t' : u' \text{ are the same} \\
& \text{(up to renaming of variables).}
\end{array}
$$

Fig. 19.3 MKBOOL inference rules

Given $\alpha \in A_X$, we define E- and R-projection from N to E and R as follows.

$$E[N, \alpha] = \bigcup_{n \in N} E[n, \alpha]. \quad R[N, \alpha] = \bigcup_{n \in N} R[n, \alpha].$$

$$E[n, \alpha] = \begin{cases} \{t \leftrightarrow u\}, & \text{if } \alpha \models F_3 \\ \emptyset, & \text{otherwise.} \end{cases}$$

$$R[n, \alpha] = \begin{cases} \{t \rightarrow u\}, & \text{if } \alpha \models F_1 \\ \{u \rightarrow t\}, & \text{if } \alpha \models F_2 \\ \emptyset, & \text{otherwise.} \end{cases}$$

where $n = \langle t : u, F_1, F_2, F_3\rangle$.

In order to check the success of the multi-completion, we introduce the following boolean constraint.

$$NoE[N] = enc(O) \cdot \prod_{n \in N} \overline{F_3[n]}.$$

Suppose that the deduction is fair in the sence defined in [6]. If $NoE[N] \not\equiv 0$, there exists $\alpha \in A_X$ such that $\alpha \models NoE[N]$. Since this assignment satisfies $E[N, \alpha] = \emptyset$, the KB process working with the reduction order given by $dec(\alpha)$ has no equations. In this case, MKBOOL terminates and returns the convergent set of rewrite rules $R[N, \alpha]$.

MKBOOL can be more efficient than MKB for two reasons. First, in ORIENT rule MKB needs to repeat orientation check for each reduction order. Instead, MKBOOL calculates $enc(O_{t,u})$. If the set of reduction orders O is defined inductively on the structure of terms (like the recursive path orders), $enc(O_{t,u})$ can be also calculated inductively without enumerating the entire set O. Therefore, the computation time for ORIENT rule of MKBOOL only depends on the structure of the datum $t : u$, independent of the number of reduction orders. Second, the binary operations on two labels (such as $L_1 \cap L$ or $F_1 \cdot F$) can be more efficient for a similar reason. The time required for the operation depends on the structure of the constraint labels in MKBOOL. In practice, the amount of time for MKBOOL is far smaller than that for MKB, when the sizes of terms are small and the number of reduction orders to be considered is very large.

19.4 Encoding for Recursive Path Orders

The abstract framework for Boolean encoding for reduction orders given in the previous section is a generalization of Boolean encodings known in the literature, such as the atom-based encoding [7] and the symbol-based encoding [4] for the lexicographic path orders (LPO). The encoding for the Knuth-Bendix orders given in [11] is also seen as an instance of this framework.

In this section, we present an encoding for recursive path orders (RPO) [8] with quasi-precedence and status. First, we recall the definition of RPO.

Throughout the section, we let $\mathcal{F} \equiv \{f_1, \ldots, f_p\}, \mathcal{V}, \mathcal{T}(\mathcal{F}, \mathcal{V})$ be a finite set of function symbols, a set of variables, and the set of terms constructed from $\mathcal{F}$ and $\mathcal{V}$, respectively.

19.4.1 Basic Definitions

Definition 19.6 The set of permutations of $\{1, \ldots, n\}$ is denoted by $Per(n) = \{\pi_1^n, \ldots, \pi_{n!}^n\}$. (The superscript n is omitted if n is clear from the context.) The set of lexicographic statuses for function symbols with arity n is denoted by $Lex(n) = \{lex_\pi | \pi \in Per(n)\}$. The multiset status is denoted by mul. The set of all statuses for function symbols with arity n is denoted by $Status(n)$:

$$Status(n) = \begin{cases} Lex(n) & if\ n \le 1 \\ Lex(n) \cup \{mul\} & if\ n \ge 2 \end{cases}$$

Definition 19.7 For a quasi-order $\succsim$ on terms, the quasi-order $\succsim^{s_1,s_2}$ is defined on sequences of terms as follows:

$$\langle t_1, \ldots, t_m \rangle \succsim^{mul,mul} \langle u_1, \ldots, u_n \rangle$$
$$\iff \{t_1, \ldots, t_m\} \succsim_{mul} \{u_1, \ldots, u_n\}$$
$$\langle t_1, \ldots, t_m \rangle \succsim^{lex_{\pi_1}, lex_{\pi_2}} \langle u_1, \ldots, u_n \rangle$$
$$\iff \langle t_{\pi_1(1)}, \ldots, t_{\pi_1(m)} \rangle \succsim_{lex} \langle u_{\pi_2(1)}, \ldots, u_{\pi_2(n)} \rangle$$

Definition 19.8 A quasi-precedence $\succsim$ is a quasi-order on $\mathcal{F}$. A status function τ maps every $f \in \mathcal{F}$ to $Status(n)$ where n is the arity of f. Let t, u be terms. We write $t = f(t_1, \ldots, t_m)$, $u = g(u_1, \ldots, u_n)$ when they are not variables. We define the quasi-recursive path order $\succsim^{\tau}_{rpo}$ and the recursive path order $\succ^{\tau}_{rpo}$ as follows:
$t \succsim^{\tau}_{rpo} u$ if and only if at least one of the following four conditions are satisfied:

- $t = u$
- $t_i \succsim^{\tau}_{rpo} u\, for\, 1 \le \exists i \le m$
- $f \succ g\ and\ t \succ^{\tau}_{rpo} u_j\, for\, 1 \le \forall j \le n$
- $f \sim g\ and\ t \succ^{\tau}_{rpo} u_j\ for\ 1 \le \forall j \le n, and((\tau(f) = mul\ and\ \tau(g) = mul)\ or(\tau(f) \in Lex(m)\ and\ \tau(g) \in Lex(n)))\ and\ \langle t_1, \ldots, t_m \rangle \succsim^{\tau(f),\tau(g)}_{rpo} \langle u_1, \ldots, u_n \rangle$

$t \succ^{\tau}_{rpo} u$ *if and only if* $(t \succsim^{\tau}_{rpo} u) \wedge \neg(u \succsim^{\tau}_{rpo} t)$.

We will leave out the superscript τ, if it is understood. Let $Prec^{\mathcal{F}}$ be the set of all *total* quasi-precedences on $\mathcal{F}$. Let $Stat^{\mathcal{F}} \equiv Status(arity(f_1)) \times \cdots \times Status(arity(f_p))$ and $RPO^{\mathcal{F}} \equiv Prec^{\mathcal{F}} \times Stat^{\mathcal{F}}$. An element $(\succsim^{\mathcal{F}}, (s_1, \ldots, s_p)) \in RPO^{\mathcal{F}}$ corresponds to an RPO such that its quasi-precedence is $\succsim^{\mathcal{F}}$ and its status function maps f_i to s_i. In the following, we present encodings for $Prec^{\mathcal{F}}$ and $Stat^{\mathcal{F}}$.

19.4.2 Encoding for Precendence and Status

We present an encoding for $Prec^{\mathcal{F}}$ by following the approach of Codish et al. [4] who have proposed to assign an integer value to function symbols for representing the priority. A quasi-precedence $\succsim$ is defined in terms of the $priority$ by $f \succsim g$ if and only if $priority(f) \ge priority(g)$.

Definition 19.9 Let $p = |\mathcal{F}|$ and $k = \lceil log_2(p) \rceil$. Then k is the number of bits required for encoding the priority per function symbol. Let $X^f_{prec} = \{b^f_i | 1 \leq i \leq k\}$ be the set of logical variables for encoding the priority of a function symbol f. The variable b^f_k represents the most significant bit. The set of variables for encoding the precedence is denoted by $X_{prec} = \bigcup_{f \in \mathcal{F}} X^f_{prec}$. We define the encoding function for the precedence as follows.

$$enc_{prec}(\succsim) = \prod_{f,g \in \mathcal{F}, f \neq g, f \succsim g} P^k_{f,g},$$

$$P^k_{f,g} = \begin{cases} 1 & if\ k = 0 \\ b^f_k \bar{b}^g_k + (b^f_k b^g_k + \bar{b}^f_k \bar{b}^g_k) \cdot P^{k-1}_{f,g} & if\ k > 0 \end{cases}$$

Note that $P^k_{f,g} = 1$ if and only if $priority(f) \geq priority(g)$.

Lemma 19.10 (X_{prec}, enc_{prec}) *is an encoding for* $Prec^{\mathcal{F}}$ *and* $enc_{prec}(Prec^{\mathcal{F}}) = 1$.

Let us proceed to an encoding of $Status(n)$. If $n \leq 1$ then the encoding is trivial ($X = \emptyset$ and $enc(s) = 1$). From here, we assume $n \geq 2$.

Definition 19.11 We introduce the set of variables X^f_{stat}.

$$X^f_{stat} = \left\{x^f_{mul}\right\} \cup \left\{x^f_{ij} | 1 \leq i < j \leq n\right\}$$

Intuitively, a variable x^f_{mul} means that we use multiset comparison for arguments of f and $x^f_{ij} (i < j)$ or $\bar{x}^f_{ji} (j < i)$ means that the i-th argument is compared before the j-th argument in the lexicographical comparison of the arguments of f. We define the encoding function for the set $Status(arity(f))$ of the statuses of f as follows:

$$enc^f_{stat}(s) = \begin{cases} x^f_{mul} & \text{if } s = mul \\ \bar{x}^f_{mul} P^f_\pi & \text{if } s = lex_\pi \end{cases}$$

where

$$P^f_\pi = \prod_{k=1}^{n-1} \prod_{l=k+1}^{n} L^f_{\pi(k),\pi(l)}, \qquad L^f_{ij} = \begin{cases} x^f_{ij} & \text{if } i < j \\ \bar{x}^f_{ji} & \text{if } j < i \end{cases}$$

By Proposition 1, the encoding for $Stat^{\mathcal{F}}$ is derived from $(X^{f_1}_{stat}, enc^{f_1}_{stat}), \ldots, (X^{f_p}_{stat}, enc^{f_p}_{stat})$. We denote it by (X_{stat}, enc_{stat}), where $X_{stat} = \bigcup_{f \in \mathcal{F}} X^f_{stat}$, $enc_{stat} = \prod_{f \in \mathcal{F}} enc^f_{stat}$.

Lemma 19.12 (X_{stat}, enc_{stat}) *is an encoding for* $Stat^{\mathcal{F}}$.

19.4.3 Encoding for RPO

Finally, we present the inductive definition of $enc(RPO^{\mathcal{F}}_{s,t})$. Let (X, enc) be the encoding for $RPO^{\mathcal{F}}$ derived from (X_{prec}, enc_{prec}) and (X_{stat}, enc_{stat}) defined in the previous section, i.e., $X = X_{prec} \cup X_{stat}$, $enc = enc_{prec} \cdot enc_{stat}$. In the following definition, the constraint $RPO_{t,u}$ is true iff $t \succ_{rpo} u$ for some RPO $\succ_{rpo}$, and $QRPO_{t,u}$ is true iff $t \succsim_{rpo} u$ for some quasi-RPO $\succsim_{rpo}$.

Definition 19.13 Let t, u be terms such that $t = f(t_1, \ldots, t_m), u = g(u_1, \ldots, u_n)$ when t and u are not variables and let $k = \lceil log_2(|\mathcal{F}|) \rceil$. We define the following constraints.

$$RPO_{t,u} = QRPO_{t,u} \cdot \overline{QRPO}_{u,t}$$

$$QRPO_{t,u} = \begin{cases} 0\, if\ (t \in \mathcal{V}, t \neq u)\ or\ \mathcal{V} \ni u \in \mathcal{V}ar(t) \\ 1\, if\ t = u\ or\ \mathcal{V} \ni u \in \mathcal{V}ar(t) \\ \sum_{i=1}^{m} QRPO_{t_i,u} + P^k_{f,g} \cdot \prod_{j=1}^{n} RPO_{t,u_j} \cdot (\bar{P}^k_{g,f} \\ \quad + \sum_{s_1 \in Status(arity(f))} \sum_{s_2 \in Status(arity(g))} EX^{s_1,s_2}_{t,u}) \\ \quad ift \neq u\ and\ t, u \in \mathcal{V} \end{cases}$$

where

$$LEX_{\langle t_1,\ldots,t_m\rangle,\langle u_1,\ldots,u_n\rangle} = \begin{cases} 1\, if\ n = 0 \\ 0\, if\ m = 0\ and\ n > 0 \\ RPO_{t_1,u_1} + QRPO_{t_1,u_1} \cdot LEX_{\langle t_2,\ldots,t_m\rangle,\langle u_2,\ldots,u_n\rangle} \\ \quad ifn > 0\ and\ m > 0 \end{cases}$$

$$\begin{aligned} EX^{mul,lex_\pi}_{t,u} &= EX^{lex_\pi,mul}_{t,u} = 0 \\ EX^{lex_{\pi_t},lex_{\pi_u}}_{t,u} &= enc^f_{stat}(lex_{\pi_t}) \cdot enc^g_{stat}(lex_{\pi_u}) \\ &\quad \cdot LEX_{\langle t_{\pi_t(1)},\ldots,t_{\pi_t(m)}\rangle,\langle u_{\pi_u(1)},\ldots,u_{\pi_u(n)}\rangle} \\ EX^{mul,mul}_{t,u} &= enc^f_{stat}(mul) \cdot enc^g_{stat}(mul) \\ &\quad \cdot \prod_{y \in N-M} \sum_{x \in M-N} QRPO_{x,y} \end{aligned}$$

where M and N are multiset: $M = \{t_1, \ldots, t_m\}$, $N = \{u_1, \ldots, u_n\}$.

Based on Lemmas 1, 2, and 3, we can define an encoding function for the recursive path order by $enc(\succ^\tau_{rpo}) = enc_{prec}(\succsim) \wedge enc_{stat}(\tau)$ where $\succsim$ and τ are the quasi-precedence and status function of $\succ^\tau_{rpo}$, respectively. In order to execute the ORIENT rule of MKBOOL, however, we need to compute $RPO^{\mathcal{F}}_{t,u}$, which is the set of all RPOs $\succ^\tau_{rpo}$ such that $t \succ^\tau_{rpo} u$. We also need to compute $enc(RPO^{\mathcal{F}})$, for starting MKBOOL. The following theorem gives the results. The proof is omitted.

Theorem 19.14

$$enc(RPO^{\mathcal{F}}_{t,u}) = RPO_{t,u} \text{ and } enc(RPO^{\mathcal{F}}) = enc_{stat}(Stat^{\mathcal{F}})$$

19.5 Implementation and Experiments

We have implemented MKBOOL and experimented on a set of the standard benchmark problems [9]. For example, the problem 3-1 is the most well-known problem from the group theory. For efficiency, we have used binary decision diagrams (BDDs) [3] as a representation for boolean constraints. The results are summarized in Table 19.1. The problems selected are all the problems that could be solved by the systems. We have considered all the RPOs with total quasi-precedence and status. The table gives the numbers of such orders and CPU times (in seconds) run on a PC with Pentium 4 CPU and 512 MB main memory. Clearly, MKBOOL is more efficient than MKB in all the problems examined.

Table 19.1 Computation time of MKB and MKBOOL

Problem	3–1	3–3	3–4	3–5	3–6	3–7	3–20	3–28	3–29
# of orders	675	675	225	39	14607	34083	1722204423	4683	545835
MKB(sec)	20.7	16.7	435.0	3.9	88.0	188.6	>3600	47.9	359.3
MKBOOL(sec)	8.1	6.7	331.3	3.6	3.7	3.2	2.8	6.8	1.4

19.6 Conclusion

We have presented a new multi-completion system MKBOOL, based on the abstract framework for boolean constraints on reduction orders. In particular, we have presented an encoding for the recursive path orders with quasi-precedence and status. The experiments show that MKBOOL is more efficient than MKB in all the problems examined. As future work, we plan to incorporate the dependency pair method into multi-completion. We also plan to support more classes of reduction orders in addition to RPO.

Acknowledgments This work was partially supported by JSPS Grant-in-Aid for Scientific Research (C), No. 19500020.

References

1. F. Baader and T. Nipkow. *Term Rewriting and All That*. Cambridge University Press, Cambridge, 1998.
2. L. Bachmair. *Canonical Equational Proofs*. Birkhäuser, Basel, 1991.
3. R. E. Bryant. Graph-based algorithms for boolean function manipulation. *IEEE Trans. Comput.*, Vol. C-35, No. 8, pages 677–691, 1986.

4. M. Codish, V. Lagoon and P. Stuckey. Solving partial order constraints for LPO termination. In *Proc. 17th RTA*, volume 4098 of *LNCS*, pages 4–18, 2006.
5. D. E. Knuth and P. B. Bendix. Simple word problems in universal algebras. in J. Leech (ed.), *Computational Problems in Abstract Algebra*, pages 263–297, Pergamon Press, New York, 1970.
6. M. Kurihara and H. Kondo. Completion for multiple reduction orderings. *Journal of Automated Reasoning*, Vol. 23, No.1, pages 25–42, 1999.
7. M. Kurihara and H. Kondo. BDD encoding for partial order constraints and its application to expert systems in software verification domains. In *Proc. IEEE International Conference on Systems, Man and Cybernetics*, pages 2062–2067, 2000.
8. J. Steinbach. Extensions and comparison of simplification orderings. in N. Dershowitz (ed.), *Proc. 3rd RTA*, volume 355 of *LNCS*, pages 434–448, 1989.
9. J. Steinbach and U. Kühler. Check your ordering - termination proofs and problems. Technical Report SR-90-25, Universität Kaiserslautern, 1990.
10. Terese. *Term rewriting systems*. Cambridge University Press, Cambridge 2003.
11. H. Zankl and A. Middeldorp. KBO as a satisaction problem. Technical Report, University of Innsbruck, 2006.

Chapter 20
Generating Distributed Code from Cola Models

Wolfgang Haberl, Michael Tautschnig and Uwe Baumgarten

Abstract Model driven development (MDD) is well established as a means of tackling the complexity involved in designing such distributed embedded systems. The complexity of the modeled system not only necessitates proper abstractions, but also calls for automation to take a model to an executable object, and later to a functional integrated system. An automated translation reduces errors, guarantees reproducible results, and thus improves overall quality. We focus on automation of the translation steps. Using COLA, we need not handle a multitude of heterogeneous models, but rather benefit from the consistent modeling formalism and start with a single, behavioral, model. It provides a clustering of application components which define the tasks that must be allocated to hardware nodes contained in the platform model. The obtained executables are tailored towards the specific platform and require no further manual intervention before effective installation (flashing) onto the target hardware system.

Keywords Distributed code · Component language · Automated translation · Hardware system · Code generation

20.1 Introduction

Today, more than 90% of all processors are part of embedded systems [3,15]. These are integrated in laundry machines, medical systems, cars, and aircrafts, just to name a few. In the work presented here, we focus on distributed embedded systems, built up from dozens or even hundreds of computing nodes, interconnected by various bus systems. Such systems contain or constitute life-critical electronic resources. Faults, of any kind, thus may be fatal. Even if not fatal, they bare large warranty costs for the designers and integrators of the product.

W. Haberl (✉)
Institut für Informatik, Technische Universität München, Boltzmannstr. 3,
85748 Garching b. München, Germany
e-mail: haberl@in.tum.de

S.-I. Ao et al. (eds.), *Trends in Communication Technologies and Engineering Science*, Lecture Notes in Electrical Engineering 33, DOI 10.1007/978-1-4020-9532-0_20,

Model driven development (MDD) is well established as a means of tackling the complexity involved in designing such distributed embedded systems. As the large scale prohibits engineers from grasping the entire system at once, a hierarchy of abstractions is applied to attain manageability at each level. The *Co*mponent *La*nguage (COLA) [9] was built to tackle the complexity of such large scale systems in a framework consistent across all layers of abstraction. COLA has a slender syntax, which caters for easy understanding, and it is defined by a rigorous formal semantics. At higher levels of abstraction then formal verification, e.g., using model checking, can be applied to guarantee conformance with requirements. These characteristics make COLA well-suited for development of mission-critical real-time systems.

The complexity of the modeled system not only necessitates proper abstractions, but also calls for automation to take a model to an executable object, and later to a functional integrated system. An automated translation reduces errors, guarantees reproducible results, and thus improves overall quality. The process of translation is best compared to that of a software compiler. Given a functional model, or a piece of source code in case of a software compiler a runnable entity is produced. A simple model of the system will comprise three layers of abstraction: the application, an operating system, and the hardware. Compiler and linker will be given all constraints imposed by the operating system and the target hardware platform to obtain appropriate, and often also optimized, application code.

In embedded systems, we will call this process *systems compilation* as the compilation must be accomplished for the overall system model, where the involved software and hardware components may be of various types. The target platform to operate on likely invalidates several assumptions made at model level, or exposes properties that are not captured by functional/behavioral modeling. Furthermore, the distributed nature of the target hardware must be accounted for.

Here, both translation and optimization are by no means straightforward. As per translation, often a heterogeneous heap of models and requirements must be considered to obtain a valid runnable entity [5]. Further, in general a certain level of black magic performed by engineers is required to fit the software and hardware components onto the target platform.

We focus on automation of the translation steps. Additional system-wide optimizations based on this framework are discussed in Kugele et al. [8]. Using COLA, we need not handle a multitude of heterogeneous models, but rather benefit from the consistent modeling formalism and start with a single, behavioral, model. It provides a *clustering* of application components which define the tasks that must be allocated to hardware nodes contained in the platform model. Therefrom, and based on the formal semantics of COLA, we generate C code that may be compiled (using an ordinary software compiler) for a target platform. As part of systems compilation, task allocation and scheduling are computed. Based on this information, the configuration of the middleware layer, which handles all communication, is generated automatically. At this point, the obtained executables are tailored towards the specific platform and require no further manual intervention before effective installation (flashing) onto the target hardware system.

20.2 Automated Code Generation

Whenever a platform description is available, executable code may be generated from the behavioral (graphical) models. Examples of such compilers include TargetLink, Real-Time Workshop/Embedded Coder, ASCET-SC, and SCADE Drive. An overview and comparisons of these tools have been presented in Reuter [12] and Wybo and Putti [17].

These modeling approaches generally lack a formal semantics, which is one of the requirements expressed in Whalen and Heimdahl [16] to ensure correctness of the translation: both the source and the target language shall have a rigorous formal semantics. While this is hardly viable for the target language C, our source language COLA provides these. Consequently, we only rely on a subset of C and a template-like set of transformation rules such that preservation of behavior of the generator remains easy to check. At this point it should be noted that COLA has a slender syntax and thus requires only a small number of transformation rules. The elements of COLA are introduced in Section 20.2.1.

Correctness of a distributed system not only relies on the semantically correct transformation of parts of the model into C code, but also on the interaction of tasks. A COLA model is intended to cover complete systems, and thus defines data dependencies between groups of tasks. A *cluster* is the COLA model representation of a task from the operating systems point of view. As such, a COLA model is partitioned into clusters after a first iteration of functional design. Code generation then must take care of inter-cluster communication (see Section 20.3.1) in a way independent of the allocation, which is performed as a subsequent step in systems compilation. This transparency is provided by our middleware layer, which has been described in Haberl et al. [4].

20.2.1 Overview of COLA

The key concept of COLA is that of *units*. These can be composed hierarchically, or occur in terms of *blocks* that define the basic (arithmetic) operations of an application.

Each unit has a set of typed *ports* describing the interface. These ports form the *signature of the unit*, and are categorized into input and output ports. Units can be used to build more complex components by building a *network* of units and by defining an interface to such a network. The individual connections of sub-units in a network are called *channels* and connect an output port with one or more suitably typed input ports [10].

In addition to the hierarchy of networks, COLA provides a decomposition into *automata*, i.e., finite state machines, similar to Statecharts [2]. If a unit is decomposed into an automaton, each state of the automaton is associated with a corresponding sub-unit, which determines the behavior in that particular state. This definition of an automaton is therefore well-suited to partition complex networks of units into disjoint *operating modes* [1], the activation of which depends on the input signals of the automaton.

The collection of all units forms a COLA *system*, which models the application, possibly including its environment. Such a system does not have any unconnected input or output ports as there would be no way to provide input to systems. For effective communication with the environment not describable within the functional COLA model, *sources* and *sinks* embody connectors to the underlying hardware. Sources represent sensors and sinks correspond to actuators of the used hardware platform.

Distributable software modules are specified by *clusters*, which correspond to tasks at the operating system level. Each cluster has a distinguished root unit. Whenever this is a network or an automaton, the respective sub-units become part of the same cluster. Consequently, a cluster cannot have any sub-clusters. Code generation then yields a C source file for each cluster.

20.2.1.1 Semantics

COLA is a synchronous data flow language, i.e., it is assumed that operations start at the same instant of time and are performed simultaneously with respect to data dependencies. The computation of the system over time can be subdivided into discrete steps, called *ticks*, and the execution is performed in a stepwise manner over the discrete uniform time-base. Data dependencies are implied by the employed channels. At each tick a unit emits new values to the channels connected to its output ports. These values become available immediately for ports connected to the reading side of the channel.

To retain data for a series of ticks, the concept of *delays* is introduced. These blocks model memory by saving the actual input value and providing the input of the previous tick at the output port. At the first tick, where no prior input is available, a default value specified in the model is used.

20.2.2 Coding Basic Model Elements

In the following we describe the transformation of COLA elements into C code. The mapping is exemplified presenting COLA diagrams and the according code snippets from our case study (see Section 20.4).

20.2.2.1 Units and Signatures

For each unit found in the given system the code generator creates a C function with an appropriate signature. As COLA units may carry more than one output port in their signature we decided to include a variable of appropriate type for each input and output port of the unit in the function's signature. The C types corresponding to the type names used in the model are predefined in the code generator. All variables are passed as pointers. In case of input ports this saves memory when complex data types are used as the pointer requires only a fixed amount of memory independent

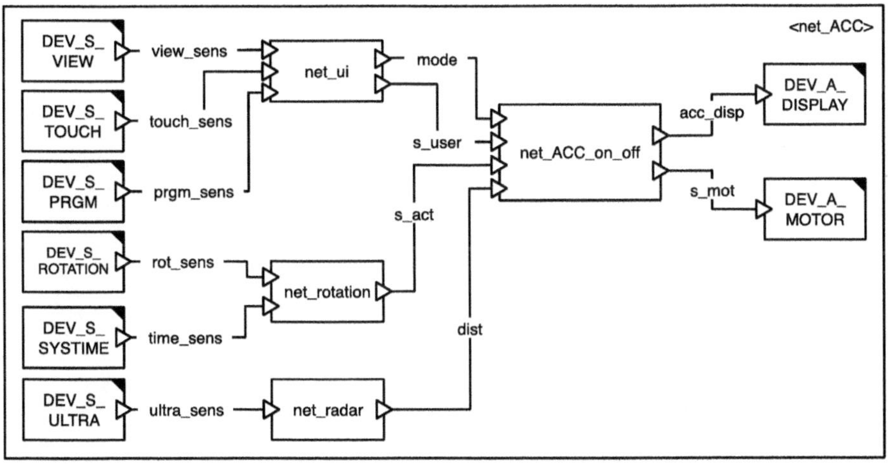

Fig. 20.1 The `net_ACC` diagram

of the dimension of the data type pointed to. Plus, for output ports, this is the only way to allow for multiple return values in C. In addition to these variables each signature of a stateful unit, i.e., automata and delays or parent units of one of these elements, includes a struct named `unit_state`. In this struct the actual state values are retained. If neither the unit itself, nor any of its sub-units are stateful, the state struct is superfluous and thus omitted.

In Fig. 20.1 the main network, i.e., the system diagram of the case study used throughout the article is shown. The system consists of twelve top-level units. These are connected using channels, which forward data from the output ports to the designated input ports of the succeeding unit. Listing 20.1 shows the generated code for Fig. 20.1. The shown parent unit, named `net_ACC`, carries no ports, hence the signature of the according C function is empty. In the following we explain the construction of the function body for this network.

20.2.2.2 Networks

The body of a generated C function implements the unit's behavior. For a network this means that the generated function for each sub-unit included in the network is called. Of course the sequence of calls has to preserve the order induced by semantics of data flow. To do so, the set of sub-units provided for the network is searched for units not dependent on other units in the network. Each such unit can instantly be inserted in the resulting C code and removed from the set. This is true for all units which are connected to an input port of the parent unit, to an output port of a unit already coded, or to a unit which represents a constant value, a source or sink, or a delay. This iteration over the set of sub-units is repeated until the set is empty. Then all sub-units are coded in a sequential order preserving the data flow semantics.

```
void net_acc()
{
    net_acc_state unit_state;
    int s_act_0;
    int dist_1;
    int mode_2;
    int s_user_3;
    int acc_disp_4;
    int s_mot_5;
    int rot_sens_6;
    int time_sens_7;
    int ultra_sens_8;
    int view_sens_9;
    int touch_sens_10;
    int prgm_sens_11;
    mw_restore_task_state(14, &unit_state);
    mw_receive(2, &rot_sens_6);
    mw_receive(1, &time_sens_7);
    mw_receive(3, &ultra_sens_8);
    mw_receive(5, &view_sens_9);
    mw_receive(4, &touch_sens_10);
    mw_receive(6, &prgm_sens_11);
    net_rotation200399(&(unit_state.state_rotation200399_num0),
     &rot_sens_6, &time_sens_7, &s_act_0);
    net_radar200400(&(unit_state.state_radar200400_num1),
     &ultra_sens_8, &dist_1);
    net_ui200398(&(unit_state.state_userinterface200398_num2),
     &view_sens_9, &touch_sens_10, &prgm_sens_11, &mode_2,
     &s_user_3);
    net_acc_on_off200401(&(unit_state.state_acc200401_num3),
     &s_user_3, &s_act_0, &dist_1, &mode_2, &acc_disp_4,
     &s_mot_5);
    mw_send(13, &s_mot_5);
    mw_send(12, &acc_disp_4);
    mw_save_task_state(14, &unit_state);
}
```

Listing 20.1 The `net_ACC` code

In Listing 20.1 the preservation of causality according to data flow semantics becomes apparent. The formal semantics for the evaluation order of networks is given in Kugele et al. [9]. Looking at the example in Figure 20.1, six sensors, indicated by the `DEV_S_` prefix in their names, can be seen. The input data of these devices, sources in terms of COLA, are read from the middleware in lines 17–22. Further there are three sub-units `net_ui`, `net_rotation`, and `net_radar`, which can be evaluated independently. Thus they are called first in the resulting code, cf. lines 23–28 in Listing 20.1. The order of the three calls is arbitrary. Using their results, `net_ACC_on_off` can be executed and then finally the actuators `DEV_A_DISPLAY` and `DEV_A_MOTOR` are written to. The resulting code is shown in lines 29–32 of Listing 20.1. The last two units are sinks, again indicating interaction with the hardware platform. Sources and sinks are detailed in Section 20.3.2.

In addition to the evaluation order, the listing shows how each channel connecting two sub-units is realized. A variable is used to pass data from one function call to the next. It is being written to by the ancestor unit and read from by the descendant one.

20.2.2.3 Automata

In COLA, automata provide the means of control flow. As described in Kugele et al. [9], an automaton's behavior is implemented by the currently active state, which, again, is a unit.

COLA automata describe general, possibly non-deterministic, Moore-type finite automata [11], but the transformation to C code must yield deterministic behavior. In our current prototypical tool-chain it is the responsibility of the modeler to employ only deterministic automata. Further efforts are put in an automated test for non-determinism while assuming data types with finite domains. Note that the problem is undecidable in case of unbounded data types (undecidability of equivalence).

For an implementation to determine the active state, first the current state must be known and then possible outgoing transitions must be evaluated. The former is stored in the `unit_state` struct, and transitions are implemented by a sequence of `switch ... case ...` and `if ... else ...` statements.

In Fig. 20.2 a COLA unit implemented by an automaton with two states is given. The states are named `atm_dist_35_check` and `net_emergency`. The actual state is changed either if the value of `dist` equals zero, or is greater than zero, depending on the actual state. Listing 20.2 shows the code for the automaton given in Fig. 20.2.

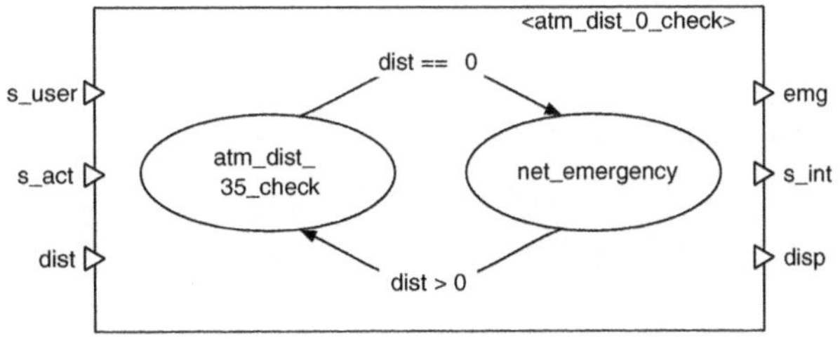

Fig. 20.2 The `atm_dist_0_check` diagram

```
void dist_0_check200695(state_dist_0_check200695 *unit_state,
 int *s_user_in_0, int *s_act_in_1, int *dist_in_2,
 int *emg_out_0, int *s_int_out_1, int *disp_out_2)
{
    int guard_result;
    switch(unit_state->atm_state)
    {
        case 0:
            emergency_guard200747(s_user_in_0, s_act_in_1,
            dist_in_2, &guard_result); if(guard_result)
            {
                unit_state->atm_state = 1;
                atm_dist_35_check200787(&(unit_state->
                 state_dist_gt_0_behavior200787_num1),
                 s_user_in_0, s_act_in_1, dist_in_2,
                 emg_out_0, s_int_out_1, disp_out_2);
                break;
            }
            net_emergency200764(s_user_in_0, s_act_in_1,
             dist_in_2, emg_out_0, s_int_out_1, disp_out_2);
            break;
        case 1:
            dist_35_guard200730(s_user_in_0, s_act_in_1,
            dist_in_2, &guard_result); if(guard_result)
            {
                unit_state->atm_state = 0;
                net_emergency200764(s_user_in_0, s_act_in_1,
                 dist_in_2, emg_out_0, s_int_out_1,
                 disp_out_2);
                break;
            }
            atm_dist_35_check200787(&(unit_state->
            state_dist_gt_0_behavior200787_num1), s_user_in_0,
              s_act_in_1, dist_in_2, emg_out_0, s_int_out_1,
              disp_out_2);
            break;
    }
}
```

Listing 20.2 Code for `atm_dist_0_check`

During code generation each state of an automaton is given a numeric identifier. In line 6 the automaton's state is used to decide on the transitions to check and the behavior to process subsequently. Here two states are coded, as can be seen in lines 8 and 20, represented by a case-switch based on the stored automaton state. In either case the guards for the outgoing transitions are evaluated. A guard is coded as a separate function returning a Boolean result. If one of the guards, called in lines 9 and 21, respectively, in our example evaluates to *true*, the transition is taken. Thus the automaton's state is changed and the behavior of the target state is executed as exemplified in lines 10–16 and 22–28, respectively. Conversely, if the guards evaluate to *false*, the behavior of the actual state is processed. In our example this is shown in lines 17 and 29. As mentioned in Section 20.2.2, the state struct is omitted for stateless units. An example of this approach can be seen in the calls in line 17 and 25.

In Listing 20.2 the equality of the automaton's signature and the signature of its states is apparent. The variables passed to the automaton in line 1 are forwarded to the function calls in lines 13, 17, 25, and 29. The only variable differing is the `unit_state` as it is distinct for each unit.

20.2.2.4 Functional Blocks and Delays

Having only dealt with COLA elements allowing for hierarchical composition so far, we will now describe the coding of blocks. They form the base of the model representing elementary operations and marking the endpoint of the hierarchy. We distinguish functional blocks and timing blocks, i.e., delays. Strictly speaking, sources and sinks are blocks as well, but we defer their discussion to Section 20.3.2. The functional blocks allow for constants and arithmetic and Boolean operations, while timing blocks provide a means of retaining data over time.

To avoid the function call overhead, all blocks are expanded inline instead of calling a function. An example of coding blocks and delays is given in Listing 20.3. It shows the code generated for the diagram in Fig. 20.3. All units in this diagram are

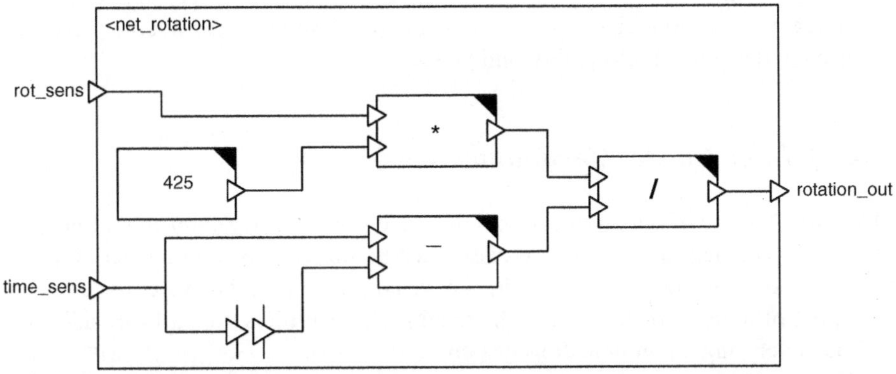

Fig. 20.3 The `net_rotation` diagram

```
void rotation200399(state_rotation200399 *unit_state,
 int *rotation_in_0, int *time_in_1, int *rotation_out_2)
{
    *rotation_out_2 = ((*rotation_in_0 * 425) / (*time_in_1 -
     (unit_state->delay200513)));
    (unit_state->delay200513) = *time_in_1;
}
```

Listing 20.3 Code for `net_rotation`

marked with a black triangle in the upper right corner, indicating a functional block. A constant block with value 425 is given. It is multiplied by the value delivered by the port `rotation`, and the result divided by the value calculated in the lower part of the diagram. The according code is given in lines 4 and 5 of Listing 20.3. The delay included in Fig. 20.3 is depicted by two parallel vertical lines, each of them carrying a port. This timing block is a one-step FIFO. It outputs the value given to it during the previous invocation of the network. Thus, in our example, the previous value of port `time` is delivered. Subsequently the data pending at its input port is stored for the next invocation. The two described working steps of the delay can be seen at the end of line 5 and in line 6 of the listing.

20.3 Running the Generated Code

The presented code generator is part of an implementation of a model driven system development workflow as demanded for future embedded real-time systems [14]. Assuming the use of a standardized middleware, as described below, enables for the use of our generator in combination with various platforms. Thus the effort for connecting the application to the underlying hardware is shifted to the implementation of a single middleware. As this layer's interface is consistent every application, the complexity of the application is decreased and the development resources saved can be used to develop a high-quality middleware.

20.3.1 Inter-Cluster Communication

To enable the use of our approach in large scale systems, communication links, i.e., channels specified in the COLA model, must be mapped to communication primitives between the nodes. For this purpose we employ our middleware. Its task is the mapping of logical addresses, which are numeric identifiers, to hardware addresses of the underlying communication systems.

Figure 20.4 depicts the inter-cluster communication via our middleware. Task 1 allocated on node 1 sends a message with identifier 42. All other tasks in the

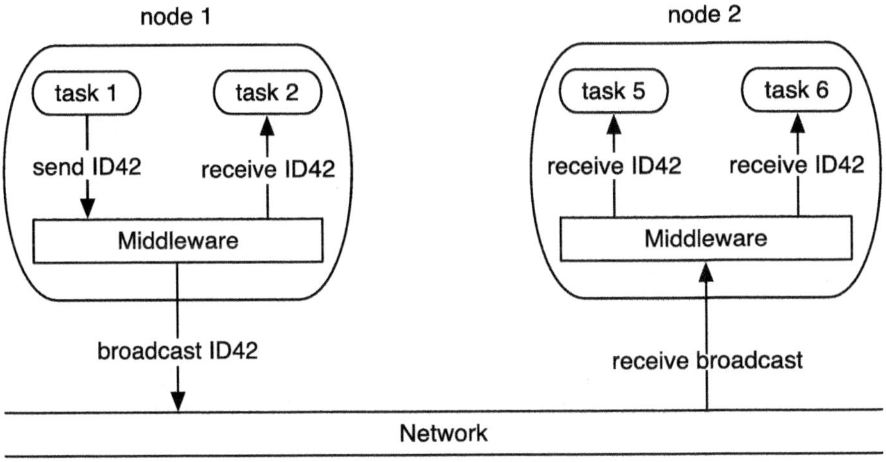

Fig. 20.4 Inter-cluster communication via middleware

system can subsequently receive the datum using this identifier. The middleware distributes the datum locally as well as remote. The bus communication is realized using broadcasts, thus allowing an arbitrary number of listening nodes.

Besides retaining port connections, the synchronous semantics of the model have to be preserved. To guarantee the timely delivery of data, we chose a time division multiple access (TDMA) scheme for communication similar to the TTA, as presented in Kopetz [6]. Because of the avoidance of collisions in such a system, arrival times of data can be anticipated and the model's causality demands, as well as timing demands, are fulfilled.

Logical addresses used by the middleware are generated using a dependency graph covering all clusters of the modeled system [7]. Each node of the graph representing a data store, which corresponds to a channel connecting clusters, is given a numerical address. These addresses are later referred to in the middleware API calls, inserted by the code generator.

In the case study presented here, only one computing node was available. Thus the COLA model contains only a single cluster. In general, inter-cluster communication calls are analogous to those used in lines 17–22 and 32–33 of Listing 20.1 for interfacing hardware devices.

20.3.2 Interfacing Sensors and Actuators

Embedded systems in general require a vast amount of software/hardware interaction. Thus, COLA as a domain-specific language, and consequently code generation, must cater for proper interfaces to sensors and actuators. The synchronous semantics of COLA require that all operations occur instantly, which shall thus also be true for hardware interaction. Hence, if the model contains sources representing, e.g., wheel

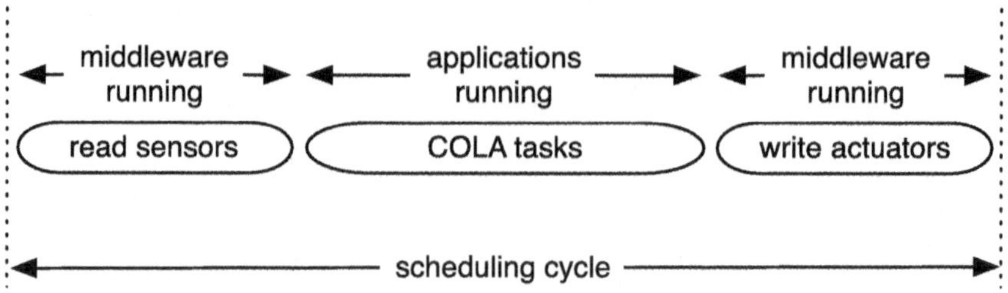

Fig. 20.5 Hardware interaction within a scheduling cycle

sensors for all four wheels of a car, these sensors are read at the same instant, from the model's point of view, which of course is infeasible for any effective implementation.

To approximate the synchrony assumption in a best effort manner, hardware interaction is performed in a batch style. That is, all sensors referenced in the model are read from before executing any application tasks. Consequently, all actuators are written to after execution of tasks has finished. This scheduling cycle is depicted in Fig. 20.5.

Similar to inter-cluster communication, the generated code uses the middleware functions `mw_receive()` and `mw_send()` for any hardware interaction. As said before, the calls take the logical address corresponding to the device in question. The middleware decides whether the address references a cluster or a piece of hardware according to its configuration, as detailed in Haberl et al. [4]. Examples of interfacing the devices shown in Fig. 20.1 are found in lines 17–22 and 30 plus 31, respectively, of Listing 20.1.

In contrast to regular COLA blocks, sources and sinks must always have exactly one port. An output port for sensors, delivering the sensor's values and an input port for actuators accepting the calculated value of the control loop algorithm. To avoid the possibility of race conditions, each sensor or actuator may only be inserted once into a COLA model. Since every actuator features just one port, this demands the developer to explicitly model the control of every concurrent hardware interaction.

20.4 Case Study

To prove the practical viability of our approach, we performed a case study using LEGO® Mindstorms™ controllers as hardware platform, equipped with the BrickOS[1] operating system. The demonstrator should realize the functionality of an adaptive cruise control (ACC) [13]. This is a control device for cars providing the functionality of keeping the car's speed at a value set by the driver, while maintain-

[1] http://brickos.sourceforge.net

Fig. 20.6 The Mindstorms demonstrator

ing a minimum distance to the car driving ahead. A picture of the demonstrator is shown in Fig. 20.6.

The presented example is an imitation of the concerns and requirements of automotive design, and does not represent a real set of control algorithms for an actual product or prototype.

20.4.1 Functionality of the Demonstrator

The intended functionality of the demonstrator includes the possibility to turn the ACC on and off. If the device is turned off, the motor speed set by the user is forwarded to the engine control without any modification. The display indicates the current ACC state. By engaging the ACC, the speed and distance regulation are activated. This includes the measurement and comparison of the pace set by the user and the actual measured car velocity. If the desired user speed s_{user} differs from the actual speed s_{act}, the target speed for the motor control is corrected by $(s_{user} - s_{act})/20$. This results in a speed correction of 5% of the difference between actual and desired speed. This regulation is used as long as no object is detected within 35 cm ahead of the car. If the distance drops below this threshold, the actual speed is continuously decreased by 5%. The minimum distance allowed constitutes 15 cm. If the actual distance is lower, the car performs an emergency stop. After either reducing speed or coming to a halt, the ACC should speed up the car smoothly again, if the obstacle is out of the critical region.

20.5 Conclusions

We presented a translation scheme of models given in the Component Language (COLA) to C code, and a prototypical implementation thereof. The formal semantics of COLA and its small number of syntactic elements allow for a transformation that fully retains the behavior of the model. A performance evaluation showed that, using an optimizing compiler, the execution time of the generated code only had an overhead of 20% when compared to a hand-crafted fully optimized version. We assume that such a low overhead immediately pays off as the generated code is guaranteed to realize the behavior of the model, which has undergone functional validation.

Tests of the generated code showed the Mindstorms car to behave as desired, which serves as the first practical validation of our prototypical implementation.

Future enhancements of the code generator will focus on generating more efficient code, i.e., using less memory and CPU time, and integrating more syntactic elements as the vocabulary of COLA is still growing. Recent additions to COLA include the introduction of multidimensional data types and additional timing operators.

References

1. Bauer, A., Broy, M., Romberg, Jan, S., Bernhard, B., Peter, F., Ulrich, M., Núria, S., Robert, and Ziegenbein, D. (2005). AutoMoDe —Notations, Methods, and Tools for Model-Based Development of Automotive Software. In *Proceedings of the SAE 2005 World Congress*. Society of Automotive Engineers.
2. Booch, G., Rumbaugh, J., and Jacobson, I. (1998). *The Unified Modeling Language User Guide*. Addison-Wesley, New York.
3. Broy, M. (2005). Automotive software and systems engineering (panel). In *Proceedings of the 3rd ACM & IEEE International Conference on Formal Methods and Models for Co-Design (MEMOCODE 2005)*, pages 143–149. IEEE.
4. Haberl, W., Birke, J., and Baumgarten, U. (2008). A Middleware for Model-Based Embedded Systems. In *Proceedings of the 2008 International Conference on Embedded Systems and Applications, ESA 2008*, Las Vegas, Nevada, USA.
5. Henzinger, T. A. and Sifakis, J. (2007). The discipline of embedded systems design. *IEEE Computer*, 40(10):32–40.
6. Kopetz, H. (1998). The time-triggered architecture. In *ISORC '98: Proceedings of the The 1st IEEE International Symposium on Object-Oriented Real-Time Distributed Computing*, page 22, Washington, DC, USA. IEEE Computer Society.
7. Kugele, S. and Haberl, W. (2008). Mapping Data-Flow Dependencies onto Distributed Embedded Systems. In *Proceedings of the 2008 International Conference on Software Engineering Research & Practice, SERP 2008*, Las Vegas, Nevada, USA.
8. Kugele, S., Haberl, W., Tautschnig, M., and Wechs, M. (2008). Optimizing automatic deployment using non-functional requirement annotations. In *Proceedings of International Symposium On Leveraging Applications of Formal Methods, Verification and Validation (ISoLA)*. Springer, New York To appear.
9. Kugele, S., Tautschnig, M., Bauer, A., Schallhart, C., Merenda, S., Haberl, W., Kühnel, C., Müller, F., Wang, Z., Wild, D., Rittmann, S., and Wechs, M. (2007). COLA – The component language. Technical Report TUM-I0714, Institut für Informatik, Technische Universität München.

10. Kühnel, C., Bauer, A., and Tautschnig, M. (2007). Compatibility and reuse in component-based systems via type and unit inference. In *Proceedings of the 33rd EUROMICRO Conference on Software Engineering and Advanced Applications (SEAA)*. IEEE Computer Society Press.
11. Moore, E. F. (1956). Gedanken-experiments on sequential machines. In Shannon, Claude E. and MacCarthy, J., editors, *Automata Studies*, pages 129–153. Princeton University Press, Princeton, NJ
12. Reuter, J. W. (2004). Analysis and comparison of 3 code generation tools. In *Proceedings of the SAE 2004 World Congress*. Society of Automotive Engineers.
13. Robert Bosch GmbH (2007). *Kraftfahrtechnisches Taschenbuch*. Vieweg, 26th edition.
14. Sangiovanni-Vincentelli, A. and Natale, M. D. (2007). Embedded system design for embedded automotive applications. *Computer*, 40(10):42–51.
15. Schulz, S., Rozenblit, J. W., and Buchenrieder, K. (2002). Multilevel testing for design verification of embedded systems. *IEEE Design & Test of Computers*, 19(2):60–69.
16. Whalen, M. W. and Heimdahl, M. P. E. (1999). An approach to automatic code generation for safety-critical systems. In *Proceedings of the 14th IEEE International Conference on Automated Software Engineering*, pages 315–318.
17. Wybo, D. and Putti, D. (1999). A qualitative analysis of automatic code generation tools for automotive powertrain applications. In *Proceedings of the 1999 IEEE International Symposium on Computer Aided Control System Design*, pages 225–230.

Chapter 21
Outlining a Risk-Driven Development Model (RDD)

Mira Kajko-Mattsson and Jaana Nyfjord

Abstract On the surface, agile and risk management process models seem to constitute two contrasting approaches. Risk management follows a heavyweight management approach whereas agile process models advocate a lightweight engineering one. Despite this, a merge between those two is possible. This chapter suggests a Risk-Driven Development model (RDD), a model integrating risk management with agile development process models. It first identifies commonalities in these two process models. It then investigates the state of practice of integrating risk management with software development in 37 software organizations. The results of these two steps provide feedback for outlining the RDD model. The RDD model itself was evaluated by 10 software practitioners.

Keywords Agile process models · Business and engineering levels · Risk definition · Horizontal and vertical risks · Stakeholders · Communication · Pre-implementation phase · Implementation phase

21.1 Introduction

Merging risk management and agile process models sounds like mixing black and white and having difficulties to achieve proper shades of gray. At first glance, these two process models embody much contrast with respect to each other. Current risk management process models assume heavyweight serial development whereas agile process models advocate concurrent engineering flavored with the iterative approach and software craftsmanship [15, 24].

Despite the discrepancies mentioned above, agile software development models claim to be risk-driven [2, 8, 14]. They state that their iterative approach enables continuous attention to risks throughout the whole development project.

M. Kajko-Mattsson (✉)
Department of Computer and Systems Sciences, Stockholm University/Royal Institute of Technology, SE-164 40 Kista, Sweden
e-mail: mira@dsv.su.se

S.-I. Ao et al. (eds.), *Trends in Communication Technologies and Engineering Science*, Lecture Notes in Electrical Engineering 33, DOI 10.1007/978-1-4020-9532-0_21,

In this chapter, we suggest a *Risk-Driven Development* (*RDD*) model, a model integrating risk management with agile development process models. On our way towards integrating these two disciplines, we first investigate and analyze agile processes from a risk management perspective. We then investigate the state of integration practice in 37 software organizations. The goal here is not only to find out the current state of practice of integrating risk management with development process models, but also to identify problems concerning the domain and elicit suggestions for how to optimally integrate these two models.

The remainder of this paper is as follows. Section 21.2 briefly describes the process models and the industrial organizations studied. Section 21.3 identifies major common parts of agile and risk management process models. Section 21.4 reports on the state of integration practice, the integration problems and suggestions for solving these problems. Section 21.5 outlines the RDD model. Section 21.6 describes a process scenario during which risks are managed on an organization-wide level. Finally, Section 21.6 makes conclusions and suggestions for future work.

21.2 Method

The contents presented in this chapter is based on the study of both literature and industrial practice. It is a result of several major research stages. Due to space restrictions, we cannot describe them all. We can only briefly list the process models studied and describe the organizations involved in this research. For further details, we cordially invite our reader to follow [16–22].

To achieve both breath and depth of the risk management domain, we chose risk management publications of renowned industrial and academic institutions, including: (1) international or organizational standards, e.g. [10, 12, 24, 33], (2) academic and/or industrial models, such as [3–6, 31, 32, 35], and (3) various investigations made by individual researchers or research groups, e.g. [9, 11, 13, 23, 26, 34]. We then selected a subset of them for our final analysis. They are the IEEE 1540 Standard for Software Lifecycle Processes – Risk Management [10], the Project Management Body of Knowledge (PMBoK) [25], and the Software Risk Evaluation Method (SRE) [35].

Regarding the agile process models, there is a substantial number of them. To avoid bias towards a specific process model and to achieve objective representation, we studied several models and selected their representative subset. The agile models chosen included eXtreme Programming (XP) and Scrum [2, 29].

We selected these because they were the most widely accepted models [7]. According to [1], they also complement each other. Whereas XP has a definite engineering flavor (pair programming, test driven development, refactoring, continuous integration), Scrum concentrates on the management of software projects [27]. Hence, together they represent a comprehensive agile framework covering both the business and engineering process levels. The result of their synthesis is presented in Fig. 21.1.

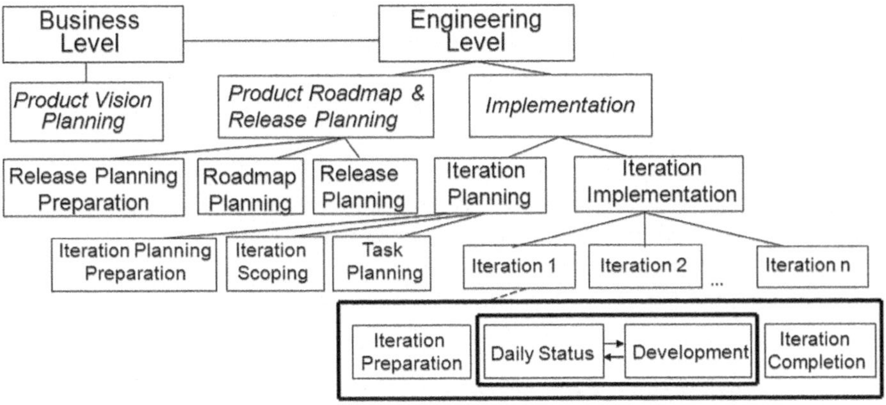

Fig. 21.1 Organizational levels and process phases

Concerning the organizations involved in this research, we were in contact with two groups of organizations. First, we elicited the risk integration practice within 37 different software organizations. Second, after having outlined the RDD model, we evaluated it with 10 software practitioners coming from six Swedish agile organizations.

Regarding the first group of organizations, they manage both traditional and agile development. Out of them, fourteen companies use non-iterative development approaches. Their feedback is still important for outlining the *RDD* model. All the 37 companies are situated in the countries of China, Colombia, Denmark, Finland, Germany, Iran, Mexico, Morocco, Pakistan, Thailand, USA, Spain, and Sweden. They range from small IT consultancies developing business applications to large organizations developing complex software for the medical, defense or space industry. The roles interviewed vary from software engineers to CEOs. Due to confidentiality reasons, we neither name the organizations nor the interviewees.

21.3 Commonalities in Agile and Risk Management Process Models

To be able to identify the commonalities in agile and risk management process models, we used nine comparison criteria in [16]. In this chapter, due to space restrictions, we only use three of them. They are (1) *Risk definition*, (2) *Organizational levels* and (3) *Stakeholders involved*. Below, we briefly we compare and evaluate the risk management models and agile process models using these criteria.

21.3.1 Risk Definition

A definition is a prerequisite for defining risk management process. It helps in understanding the process and facilitates the comparison process. The risk management models studied commonly define risk as an adverse condition or event that will affect the project. In addition, the PMBoK model explicitly recognizes the positive impact of risk as an opportunity [25]. An example of an opportunity may be a development of an innovative, complex and costly feature. In the beginning, its development may imply a serious risk to the organization. If successfully implemented, however, it may turn into a great opportunity to make profit.

Regarding the agile definition of risk, it is similar to those provided by the risk management models. Risk is defined as anything that has not yet happened but may happen and may jeopardize the success of the project . In addition, XP considers risk as opportunity [2].

21.3.2 Organizational Levels and Stakeholders Involved

Software systems are managed by the whole organization. As shown in Fig. 21.1, we distinguish between business and engineering levels managing both pre-implementation and implementation stages.

21.3.2.1 Pre-implementation Stage

The *Pre-Implementation* stage is executed on both *Business* and *Engineering* levels. It covers the phases and their activities during which one makes various preparations for the next-coming implementation. These are *Product Vision Planning,* and *Product Roadmap and Release Planning.*

Organizations generally have an overall picture of their businesses, core values, strategic goals, mission and product portfolios. In the *Product Vision Planning* phase, this overall picture is evolved into a more detailed product vision. At this phase, one focuses on product goals, overall business and product structure, development environment, and return on investment. *Product Vision Planning* guides work in subsequent planning, decision making, and work [2, 29].

The subsequent phase, *Product Roadmap and Release Planning* consists of three sub-phases:*Release Preparation, Product Roadmap Planning* and *Release Planning*. The *Release Preparation* sub-phase involves preparatory activities for the whole phase. Here, one schedules the planning meetings and gathers all the necessary information, including the product vision plan and high-level requirements specification [2, 29].

In the *Product Roadmap and Release Planning* phase, one outlines a high-level roadmap of future product versions. Here, one studies the product vision plan as produced during the *Product Vision Planning* phase, identifies new or revises the old high-level requirements, creates a requirements list (a backlog in Scrum or a set of Stories in XP), and updates the requirements specification [2, 29].

21.3.2.2 Implementation Stage

As illustrated in Fig. 21.1, the*Implementation* stage is executed on the *Engineering* level. It covers phases and activities during which the actual implementation takes place. These are*Iteration Planning* and *Iteration Implementation.*

Iteration Planning is conducted at the start of each iteration. It involves three phases, *Iteration Planning Preparation, Iteration Scoping* and *Task Planning*. Here, the team, product management and other relevant stakeholders meet to plan the implementation work to be conducted in the coming iterations [2, 29].

After having planned the iteration, one steps into the *Iteration Implementation* phase. This phase consists of the sub-phases: *Iteration Preparation, Daily Meetings, Development* and *Iteration Completion*.

The *Iteration Preparation* is required for the subsequent phases. Its objective is to make initial preparations for starting the consecutive iterations. It involves the setting up of an appropriate environment and ensuring that all the pertinent facilities are in place. According to the agile models studied, it is conducted only once, prior to the development start [2, 29].

The *Daily Status* is a meeting that takes place on a day-to-day basis throughout the whole iteration. The agile models studied suggest that the team gathers briefly (usually, for 15 min) to establish project status and to confirm or readjust their commitments, if necessary [2, 29].

The daily meeting is only a quick status meeting. If need arises to discuss additional matters, then one establishes follow-up meetings with the team members concerned. Follow-up meetings usually take place after the daily meeting [2, 29].

In the *Development* sub-phase, the requirements planned for the iteration are implemented and tested. The goal is to deliver an increment of working functionality [2]. Finally, during the *Iteration Completion* sub-phase, one delivers the functionality that has been developed during the iteration, evaluates the iteration, and identifies the improvements to be made in the next iteration [2, 29].

21.3.2.3 Organizational Levels in the Process Models Studied

None of the risk management models places risk management on both the business and engineering level. All the risk management models studied mainly focus on the *Implementation* stage. Regarding the agile models studied, XP clearly takes the engineering standpoint whereas SCRUM manages both business and engineering levels. However, it does not explicitly place any risk management on these levels.

21.3.3 Stakeholders Involved

Stakeholder roles are individual roles or groups of roles who have a stake in or may be impacted by a given activity [10]. Stakeholders can either be internal or external. Internal stakeholders include any managerial or technical roles participating in a project, including project managers, developers, testers, maintainers, product owners, business analysts and managers, quality managers and support personnel [25]. External stakeholders are other roles, such as customers, contractors, suppliers and sponsors [10].

The coverage of stakeholder roles within risk management is very important. It is only then one may be sure that all the risk sources and targets have been identified and scrutinized from all possible perspectives. Designation of roles is a prerequisite for defining risk management process and responsibilities within the process [12, 25].

All the risk management models studied state that the process should support the perspectives of all relevant stakeholder roles, both of the internal and external ones. They however suggest different sets of stakeholder roles. For instance, acquirers are mentioned in IEEE 1540 standard only. Furthermore, the models only identify the roles having responsibilities towards mitigating risks. These roles are commonly referred to as risk owners. Out of them, guidelines for how to manage risks are restricted to the project manager role only.

The agile process models suggest teams as main stakeholder roles. The teams consist of (1) internal roles such as developers, testers, project managers, business analysts and (2) external roles such as customers. However, the agile models do not provide any specific guidelines for the designation of roles and responsibilities for managing risks. The team commonly owns the risk management process [30]. The agile project manager facilitates the process and makes the results visible.

21.4 Process Integration Practice

In this section, we describe the status within the 37 organizations studied. We first describe their status with respect to their practice to define and identify risks, the organizational levels and process phases on which risks are managed. We then list the stakeholders involved in the risk management activities, and list the process integration problems as identified within the organizations studied.

21.4.1 Risk Definition and Identification

All the 37 organizations studied define and identify various risk types. These include *Business, Financial, Project, Process, Planning, Resource, Technical, Organizational, Legal, Partner/Subcontractor, Country, Product* and *Quality* risks. Some organizations further classify risks into (1) *Internal* and *External*, and (2) *Horizontal* and *Vertical. Internal* risks are risks that are encountered locally within the development organization whereas *External* ones come from the external environment. The *Horizontal* risks, as illustrated in Fig. 21.2, spread across the whole software development cycle and organizational levels whereas *Vertical* risks only concern risks for a certain process phase.

21.4.2 Organizational Levels and Stakeholders Involved

Thirty-two out of the 37 studied companies have the *Business* and*Engineering* levels. In the remaining five companies, the interviewees were not familiar with the work conducted on the *Business* level.

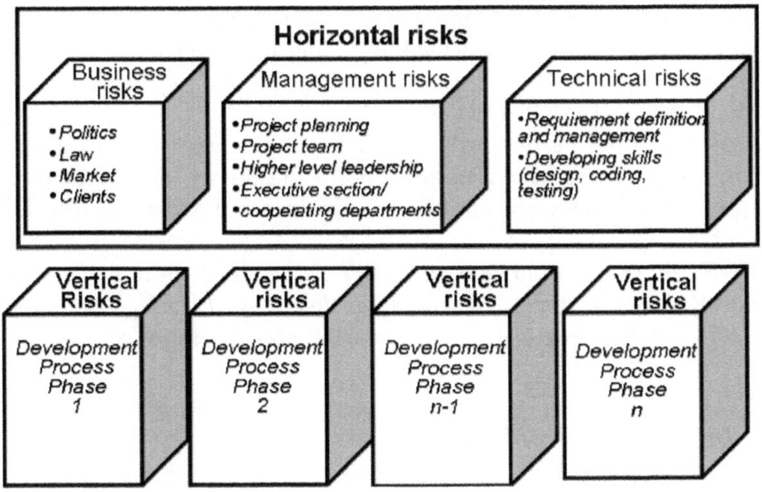

Fig. 21.2 One way of defining and categorizing risks

Twenty-eight out of the 32 companies have a phase corresponding to the *Product Vision Planning* phase during which they manage risks. The risks handled at this stage are primarily business and market risks.

When managing risks in the *Product Vision Planning* phase, the organizations conduct their own risk management processes, mainly by having face-to-face meetings. The stakeholders involved in them are primarily represented by various senior management roles (e.g. *CEO, CIO,* and*CTO)* and the roles coming from the business department, such as sales and product managers.

Concerning risk management on the *Engineering level*, thirty-two companies claim that they conduct risk management using their own organizational risk management process models. They claim that the choice of activities, the types of outcomes and the roles involved in the activities vary depending on the engineering phase.

In the *Product Roadmap Planning* phase, the roles involved are mainly represented by various managers (business, product, project), customer, business analysts and requirement engineers.

Regarding risk management in the *Release Planning* phase, it follows the same organizational risk management process as in previous phases. However, some differences were identified with respect to the roles and the risk management process phases. The roles identified in this phase include release managers, technical leaders, team leaders, senior software engineers and QA. The phases identified are *Risk Identification, Risk Analysis* and *Risk Management Planning.* There is also a shift of the focus on the types of risks managed in this phase. For instance, as stated by one interviewee: "Risks in this phase concern issues such as the stakeholders' satisfaction with the release plan, and not only the business risks".

Regarding risk management in the *Iteration Planning* phase, fourteen companies state that they do not conduct *Iteration Planning* because they use non-iterative development approaches. In the remaining organizations, risk management in the *Iteration Planning* phase is conducted according to the organizational standards. The differences identified concern the roles involved, the risk management activities, and the types of risks focused on.

The roles involved on this level are mainly engineers, represented by system architects, software engineers, testers, system integrators, and other roles. In a majority of the companies having iteration planning, risk management is led by the project manager.

Generally, the activities in the *Iteration Planning* phase cover almost all the risk management phases, including *Risk Identification, Risk Analysis, Risk Management Planning, Risk Monitoring and Control* and *Post-Mortem Analysis.*

21.4.3 Integration Problems

Twelve out of the 37 organizations studied claim to have problems with the process integration. The problems identified are the following:

- Training cost is too high
- Lack of resources to conduct risk management
- Different roles have different views on risk and risk managemen
- Lack of competence
- Work overload for project manager
- Lack of control of external risks
- Lack of coordination
- Lack of integration among the processes within the organization
- Lack of time
- Lack of plan
- Lack of process

One organization points out that although integration is important, the success still depends on the project management. If the project manager can control the integrated process, it is an advantage. However, if the project manager has not enough time to have an overview of the whole process, a separate risk management process led by some other role can be more useful.

21.5 RDD

In this section, we present outline our RDD model. The model is illustrated in Fig. 21.3. Its main constituents are: (1) *Risk definition and identification,*(2) *Organizational levels,* (3) *Stakeholders involved,* (4) *Communication channels*, and finally, (5) *Process aspects*. Below, we briefly describe its constituents and map them on our comparison criteria.

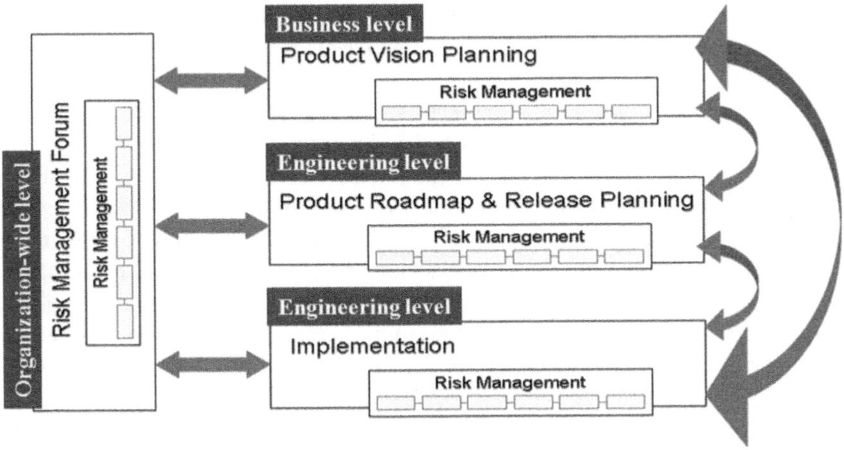

Fig. 21.3 Placing risk management on the organizational levels

21.5.1 Risk Definition and Identification

Risk definition is the driving wheel of the *RDD* model. It is a prerequisite for running an organization-wide software production. *RDD* defines risk as either a loss or a gain (an opportunity). A loss is an unwanted or negative effect whereas a gain is positive or progressive effect. In contrast to other models, *RDD* recognizes both vertical and horizontal risks. It is only in this way, one may embrace risks that span the whole organization and all phases in which risks are managed.

21.5.2 Organizational Levels and Stakeholders Involved

The integrated model covers the entire agile development process. As shown in Fig. 21.3, this involves integration of risk management on two organizational levels: the *Business* and *Engineering* levels. The fact that we organize the RDD into phases and organizational levels does not imply that we advocate their serial execution. Their execution can be conducted concurrently, in any order that suits the process context.

Each phase within the *Business* and *Engineering* levels may undergo a complete risk management process. The risks that cannot be fully managed within one level are channeled to the next level. In general, the following scenarios apply:

- *Product Vision Planning:* This phase involves creating a product vision plan guiding the work carried out in subsequent planning, decision making, and development [28]. Risk management within this phase mainly concerns the identification and analysis of business related risks, such as budget and resource risks.

- *Product Roadmap and Release Planning*: Here, one first creates a high-level roadmap plan for the product releases which one then regularly revisits before the start of each new release [2]. The risk management in the *Product Roadmap and Release Planning* phase comprises risk identification, analysis and action planning. It involves both business and technical risks.
- *Implementation:* Here, the team, product management and other stakeholders plan the work to be conducted in the coming iteration. The plan is then executed to deliver an increment of working product functionality [2]. Risk management in this phase primarily covers the monitoring and controlling of the risks that have been transferred to this phase from previous development phases. New risks are also continuously identified, analyzed and planned for during this phase. Risks in this phase are mainly of technical character.

Most of the risks undergo a complete risk management process within a phase. The ones that do not get mitigated may have to be transferred to the next phase, and/or get reported to the *Risk Management Forum (RMF).* The forum is a function for coordinating risk management across the organization. It manages any risks that will have to be promptly disseminated within the organizations. It consists of a cross-functional group represented by the roles responsible for or concerned with or capable of managing these kinds of organizations-wide risks.

21.5.3 Stakeholders Involved

The RDD model identifies several roles having various responsibilities with respect to risk management and its communication. Generally, however, the risks are owned by the roles in the phase where the risk is originally identified. The roles and responsibilities are:

- *RMF Members* own all the organization-wide risks. Their main task is to supervise and coordinate all the organization-wide risks and make decisions on them. However, they may delegate their management to other roles either within the *Business* or *Engineering* levels or both.
- *Business Manager* is responsible for managing risks at the *Business Level.* This role owns all the risks relevant for this level. However, he may delegate their management to the roles in the *Product Roadmap and Release Planning* or *Implementation* phase. The choice depends on the character of the risk and where in the organization it is most adequately managed. Still however, he keeps the risk ownership till the delegated risks get mitigated.
- *Product Manager* is responsible for all the risks managed in the *Product Roadmap and Release Planning* phase. This role owns all the risks relevant for this phase. However, similarly to the business manager, he may delegate their management to other roles in the organization, if needed. He still keeps the risk ownership till the risks get mitigated.

- *Team Leader* and *Team Members* are responsible for managing risks within the *Implementation* phase. The team leader supervises the risk management. Usually, team members own the risks that concern the development tasks assigned to them. However, the team may also decide to delegate risks to others in the organization depending on the risk and its management needs.

21.5.4 Communication Channels

An effective management of organization-wide risks rests on how risks are communicated within an organization. This is because, in the integrated model, risk management takes place throughout the entire development process. Hence, to warrant effective information flow, one needs define communication channels. As depicted by the double-edged arrows in Fig. 21.3, the integrated model identifies six main two-way *Communication Channels.* The difference between them concerns the different roles in each phase who act as the main senders and receivers of risk information. The communication channels are: (1) *Product Vision Planning – RMF,* (2) *Product Vision Planning – Product Roadmap and Release Planning,* (3) *Product Roadmap and Release Planning – RMF,* (4) *Product Roadmap and Release Planning – Implementation*, (5) *Product Vision Planning – Implementation,* and (6) *Implementation – RMF.*

21.5.5 Process Aspects

Processes are variable and dynamic. Their instances vary due to many different aspects affecting the process design. In this integration model, they are covered by some fundamental aspects of risk management as identified in [16]. Each of these aspects determines the magnitude of risk management required within the integrated process, thus aiding in adapting an instance of an integrated model to the specific situation at hand. The following aspects should be considered:

- *Risk definition is a main prerequisite for identifying risks and to communicate them effectively:* The definition of a risk is of high relevance for an effective communication on risk and hence for making decisions on when to perform risk management in software development.
- *Risk assessment results identify pertinent actions for performing risk-driven development:* The risk classification and assessment techniques are important for knowing when and how to manage risks effectively in subsequent process phases. Hence, guidelines for assessing risks are needed.
- *The product's software lifecycle stage aids in designating appropriate risk management actions:* The portfolio of risk types and risk management activities differs considerably depending on whether a project concerns new development or

maintenance of a legacy system. Hence, the process is adapted to the prevailing risk types and their prioritization.

- *Designation of and coverage of stakeholders, roles and responsibilities are determined by factors such as project risk profile, type and size:* The coverage of stakeholder roles and the degree of their involvement depends on the type of project and its size, risk complexity, scope and criticality.
- *The use of supporting tools and repositories are determined by factors such as project risk profile, type, size and team distribution:* The relevance of tools depends on the organization size, project needs and risk profile.
- *The formality of recording risk information varies by risk profile, project type, size, and team distribution:* Templates provide relevant support for describing and communicating risks [18]. However, the extent of using them and the degree of their formality varies with respect to project type, size, team distribution and risk severity.
- *Product status, life expectancy and business value help determining the amount of risk management process needed:* Risks and their criticality vary by product status, life expectancy and business value. Hence, the amount of attention put into the risk management process varies.
- *Environment and the project's physical context determine the formality of the risk management process:* Large organizations generally need a more conventional risk management process due to the need of more coordination.
- *Organizational maturity and training aids in adopting a risk management program successfully:* Organizational maturity, such as people's attitude towards risks, competency, and capability to perform risk management are of relevance for successful adoption of risk management.
- *Software development need be integrated with other organizational processes:* The development and risk management processes need be integrated to provide useful feedback to the organization, especially in larger organizations. Hence, one needs guidelines for making such integration explicit.

21.6 Epilogue

In this chapter, we have outlined a *Risk-Driven Development* model (*RDD*), a model integrating risk management and agile software development models. The *RDD* is primarily targeted towards process engineers and business developers or other groups involved in process engineering and improvement.

Our model only provides a general skeleton for how to manage risks on an organization-wide basis. Its constituents need to be complemented with flesh and blood. This implies that one need to do more studies for eliciting theory on how to manage risks both within and across the organizational levels and their process phases. Hence, still a lot of challenging and difficult work remains. We are however prepared meet these challenges and continue our work.

References

1. Abrahamsson P. et al., Agile Software Development Methods: Review and Analysis. VTT Electronics/ Oulu University, 2002.
2. Beck K., *Extreme Programming Explained: Embrace Change.*2nd Ed. Addison-Wesley, Upper Sadle River, NJ, 2004.
3. Boehm B., "A Spiral Model of Software Development and Enhancement", *IEEE Computer*, Vol. 21 (5), 1988, pp. 61–72.
4. Boehm B.,"Software Risk Management: Principles and Practices". *IEEE Software*, Vol. 8 (1), 1991, pp. 32–41.
5. Carr M.J. et al., "Taxonomy-Based Risk Identification". SEI Technical Report CMU/SEI-93-TR-006 ESC-TR-93-183, SEI/CMU, Pittsburg, PA, 1993. URL: http://www.sei.cmu.edu/pub/documents/93.reports/pdf/tr06.93.pdf, Accessed March 2007.
6. Charette R., *Software Engineering: Risk Analysis and Management*. McGraw-Hill, New York, NY, 1989.
7. Charette R., The Decision is in: Agile vs Heavy Methodologies (2001). Cutter Consortium, Vol. 2 (19) http://www.cutter.com/content/ project/fulltext/updates/2001/epmu0119.html. Accessed in October 2007.
8. Cohn M., *Agile Estimating and Planning*. Pearson Education, Upper Saddle River, NJ, 2006.
9. Hulett D.T., "Key Characteristics of a Mature Risk Management Process". Proceedings of the European Project Management Conference/PMI Europe 2001, 2001.
10. 1040, "IEEE 1540 Standard for Lifecycle Processes-Risk Management". IEEE, New York, NY, 2001.
11. IEEE Software, " Managing Risk" (special issue). *IEEE Software*, Vol. 14 (3), 1997.
12. Institute of Risk Managment, Association of Insurance and Risk Managers and National Forum for Risk Management in the Public Sector, "A Risk Management Standard". UK, 2002. URL: http://www.theirm.org/ publications/PUstandard.html. Accessed February 2007.
13. Kontio J., "Risk Management in Software Development: a Technology Overview and the Riskit Method". Proceeding of the International Conference on Software Engineering, 1999.
14. Mahnic V. and Drnovscek V., "Agile Development with Scrum". Proceedings of the European University Information Systems Conference, 2005.
15. McBreen P., Software Craftsmanship – The New Imperative, Addison-Wesley, New York, 2002.
16. Nyfjord J. and Kajko-Mattsson M., Commonalities in Risk Management and Agile Process Models, In Proceedings, IEEE International Conference on Software Engineering Applications, IEEE Computer Society Press: Los Alamitos, CA, 2007.
17. Nyfjord J. and Kajko-Mattsson M., Degree of Agility in Pre-Implementation Process Phases, In Proceedings, International Conference on Software Process, Lecture Notes in Computer Science, Springer Verlag, 2008.
18. Nyfjord J. and Kajko-Mattsson M., Communicating Risk Information in Agile and Traditional Environments, In Proceedings, 33rd Euromicro Conference on Software Engineering and Advanced Applications, IEEE, Computer Society Press: Los Alamitos, CA, 2007.
19. Nyfjord J. and Kajko-Mattsson M., Agile Implementation Phase in Two Canadian Organizations, In Proceedings, IEEE Australian Software Engineering Conference, IEEE Computer Society Press: Los Alamitos, CA, 2008.
20. Nyfjord J. and Kajko-Mattsson M., Software Risk Management: Practice contra Standard Models, In Proceedings, International Conference on Research Challenges in Information Science, IEEE, Computer Society Press: Los Alamitos, CA, 2008.

21. Nyfjord J. and Kajko-Mattsson M., Integrating Risk Management with Software Development: State of Practice, In: Proceedings, IAENG International Conference on Software Engineering, BrownWalker Press, Boca Raton, USA, 2008.
22. Nyfjord J. and Kajko-Mattsson M., Outlining a Model Integrating Risk Management and Agile Software Development, In Proceedings, 34rd Euromicro Conference on Software Engineering and Advanced Applications, IEEE, Computer Society Press, Los Alamitos, CA, 2008.
23. Padayachee K., "An Interpretive Study of Software Risk Management Perspectives". Proceedings of the Annual Research Conference of the South African Institute of Computer Scientists and Information Technologists on Enablement Through Technology, 2002.
24. Poppendieck M. and Poppendieck T., Lean Software Development – An Agile Toolkit, Adidison-Wesley, New York, 2003.
25. Project Management Institute, *A Guide to the Project Management Body of Knowledge (PMBoK)*, 3rd Ed. ANSI/PMI 99-001-2004, Project Management Institute, Newton Square, PA, 2004.
26. Ropponen J. and Lyytinen K., "Components of software development risk: how to address them? A project manager survey". *IEEE Transactions on Software Engineering*, Vol. 26(2), 2000, pp. 98–112.
27. Schwaber K., Scrum Development Process. Proc. of Conference on Object-Oriented Programming, Systems, Languages and Applications, 1995.
28. Schwaber K., *The Enterprise and Scrum*. Microsoft Press, Redmond, WA, 2007.
29. ScrumMethodology, Scrum Methodology: Incremental, Iterative Software Development from Agile Processes. Rev. 0.9. Advanced Development Methods Inc., 2003.
30. Sliger M., "Relating PMBoK Practices to Agile Practices" (Part 3 of 4). URL: http://www.stickyminds.com/sitewide.asp?Function=edetail&ObjectType=COL&ObjectId=11133&commex=1#4993. Accessed February 2007.
31. Software Engineering Institute/Carnegie Mellon University, "Risk Management Overview". URL: http://www.sei.cmu.edu/risk/. Accessed February 2007.
32. Sommerville I., *Software Engineering*, 7th Ed. Addison-Wesley, Reading, MA, 2006.
33. Standards Australia and New Zealand, "Australian/New Zealand Standard Risk Management AS/NZS 4360:2004".3rd Ed. Standards Australia/New Zealand, Sydney/Wellington, Australia/New Zealand, 2004.
34. Westfall L., "Software Risk Management". The Westfall Team, 2001. URL:http://www.westfallteam.com/Papers/risk_management_paper.pdf. Accessed February 2007.
35. Williams R. et al., "Software Risk Evaluation (SRE) Method Description (Version 2.0)". Technical Report CMU/SEI-99-TR-029, SEI/CMU, Pittsburg, PA, 1999.

Chapter 22
Mining Frequent Itemsets in Distributed Environment

Using Trie Data Structure

Ebrahim Ansari Chelche, G.H. Dastghaibyfard, M.H. Sadreddini, Morteza Keshtakaran and Hani Kaabi

Abstract Finding association rules is one of the most investigated fields of data mining. Computation and communication are two important factors in distributed association rule mining. In this problem Association rules are generated by first mining of frequent itemsets in distributed data. In this paper we proposed a new distributed trie-based algorithm (DTFIM) to find frequent itemsets. This algorithm is proposed for a multi-computer environment. In second phase we added an idea from FDM algorithm for candidate generation step. Experimental evaluations on different sort of distributed data show the effect of using this algorithm and adopted techniques.

Keywords Frequent itemsets · Distributed environment · Trie data structure · Association rules · Data mining

22.1 Introduction

The association rule mining (ARM) is very important task within the area of data mining [1]. Given a set of transactions, where each transaction is a set of literals (called items), an association rule is an expression of the form $X \rightarrow Y$, where X and Y are sets of items. The intuitive meaning of such a rule is that transactions of the database which contain X tend to contain Y. An example of an association rule is: "30% of transactions that contain beer also contain diapers; 2% of all transactions contain both of these items". Here 30% is called the confidence of the rule, and 2% the support of the rule. The problem is to find all association rules that satisfy user-specified minimum support and minimum confidence constraints. Frequent patterns discovered via mining processes not only themselves are interesting, but also are useful to other data analysis and mining tasks, including associative classification, clustering, cube computation and analysis, and gradient mining and multi-dimensional discriminant analysis [2].

E.A. Chelche (✉)
Department of Computer Science and Engineering, Shiraz University, Shiraz, Iran

S.-I. Ao et al (eds.), *Trends in Communication Technologies and Engineering Science*,
Lecture Notes in Electrical Engineering 33, DOI 10.1007/978-1-4020-9532-0_22,

The main task of every ARM algorithm is to discover the sets of items that frequently appear together, the frequent itemsets. Finding frequent itemsets in transaction databases has been demonstrated to be useful in several business applications [3].

Hipp et al. [4] provides a general survey on efficient mining of association rules in transaction and/or relational databases. AIS [5], SETM [6] and Apriori [7] can be considered as the first generation of association rule mining algorithms. Apriori algorithm is by far the most well-known association rule mining algorithm. AprioriTID [7] is an extension of the basic Apriori approach. Instead of relying on the raw database, AprioriTID internally represents each transaction by current candidates it contains. In AprioriHybrid both Apriori and AprioriTID approaches are combined [7].

Many algorithms have been proposed to find frequent itemsets from a very large database. The number of database scans required for the task has been reduced from a number equal to the size of the largest itemset in Apriori [7], to typically just a single scan in modern ARM algorithms such as Sampling and DIC [8, 9]. Efficient mining of association rules in transaction And/or relational databases has been studied substantially [7–11].

When data is saved in a distributed database, a distributed data mining algorithm is needed to mine association rules. Mining association rules in distributed environment is a distributed problem and must be performed using a distributed algorithm that doesn't need raw data exchange between participating sites. Distributed association rules mining (DARM), has been addressed by some researches and number of distributed algorithms have been proposed [12–16].

Apriori [7] is one of the most popular data mining approaches for finding frequent itemsets from transactional datasets. The Apriori algorithm is the main basis of many other well-known algorithms and implementations. The main challenge faced by the researchers in frequent itemset mining has been to reduce the execution time. One of the best implementation of apriori algorithm is published by Bodon [1]. We use Bodon sequential idea to provide a distributed algorithm. The main reason we adopted Bodon's implementation for parallel computing is because Bodon's implementation using the trie data structure outperforms the other implementations using hash tree [11, 17, 18]. In this algorithm a Trie-based structure has been used and has been constructed in each local site that at the end of each of iterations all local sites synchronized their Tries. This algorithm called DTFIM (Distributed Trie-based frequent itemset mining).

In next step, one of the ideas proposed in FDM [13] has been used to improve our implementation. This part has been embedded to main algorithm. The more skewed distributed database, the more efficiency of the technique is.

The rest of the paper is organized as follows. Section 22.2 introduces related work on frequent itemsets mining. Section 22.3 presents our implementation for DTFIM and revised DTFIM. Section 22.4 presents the experimental results of our implementation on a multi computer environment. Section 22.5 concludes the paper.

22.1.1 Notation and Problem Definition

Let $I = \{i_1, i_2, \ldots, i_n\}$ be the items in a certain domain. An *itemset* is a subset of I. A k-itemset is an itemset with k items from I. A database DB is a list of *transactions* where each transaction T is also a subset of I.

Now assume that there are n sites $S_1, S_2, \ldots, S_n$ in a distributed system, which communicate by message passing. Let $DDB = \{DB^1, DB^2, \ldots, DB^n\}$ be a "horizontal" partition of DB into n parts. We allocate each DB^i to the site S_i.

For any itemset X and transaction T we say T *contains* X if and only if $X \subseteq T$. For any itemset X and any group of transactions A, *Support(X, A)* is the number of transactions in A which contain X. We call *Support(X, DB)* the *global support count* of the itemset X in the database DB and *Support(X, DB^i)*, the *local support count* of X at site i. For a given minimum support threshold s, X is *globally large* (or *globally frequent*) if $Support(X, DB) \geq s \times D$, where D is the number of transactions in database DB; correspondingly X is *locally large* (or *locally frequent*) at site if $Support(X, DB) \geq s \times D_i$, where D_i is the number of transactions in database partition DB^i. In the following, L denotes the globally large itemsets in DB, and $L_{(k)}$ the globally large k-itemsets in L. The essential task of a distributed association rule mining algorithm is to find the globally large itemsets L.

22.2 Previous Work

Since its introduction in 1993 [1], many algorithms with different approaches have been suggested for the ARM problem. Here we review some of the related works that form a basis for our algorithm.

22.2.1 The Apriori Algorithm

The Apriori algorithm is proposed by Agrewal in [7] and is the basis for many other FIM algorithms.

In the first pass, the occurrences of each item is being counted and the items with insufficient support get removed to create $L_{(1)}$, the collection of large 1-itemsets.

A subsequent pass, say pass k, consists of two steps:

- The large *(k–1)*-itemsets collection $L_{(k-1)}$ found in the previous pass is used to generate C_k, the list of candidate k-itemsets; which is a superset of the set of all large k-itemsets. A candidate is generated from every two large *(k–1)*-itemsets which are similar in their first k–2 items. Then, the candidates that have an infrequent subset are removed from the set of candidates.
- The database is scanned and the support count for each candidate itemset in C_k is determined. Removing items with support counts less than the minimum required gives us the large k-itemsets ($L_{(k)}$).

22.2.2 The Trie-Based Apriori

Bodon shows that using efficient data structures and implementation is very important in improving the performance of Apriori algorithm [11]. He proposed a fast Apriori implementation using the trie data structure instead of a hash tree which was used in the classical approaches.

A trie is a rooted, labeled tree. In the FIM setting each label is an item. The root is defined to be at depth 0 and a node at depth d can point to nodes at depth d+1. A pointer is also referred to as edge or link. Each node represents an item sequence that is the concatenation of labels of the edges that are on the path from the root to the node. So a path from root to each node represents an itemset. In this implementation, the value of each node is the support count for the itemset it represents. In Fig. 22.1 a trie is displayed. The path to the node representing itemset {A, B} is shown in blue. The support count for this itemset is 7.

For each transaction record T in the database the trie (containing the candidate itemsets) is recursively traversed and the value of each leaf node will be incremented if T contains the itemset represented by that node. The traverse of the trie is driven by elements of T. At the end, nodes whit a support count less than the required minimum will be pruned.

In the candidate generation phase, we just need to add a leaf node to its left siblings to create new valid candidates, eliminating the need for further processing. In Fig. 22.2 shows a trie structure before and after the new candidates are generated.

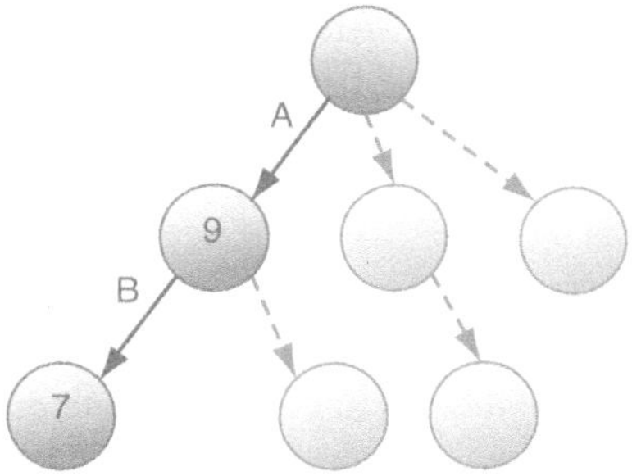

Fig. 22.1. A sample trie

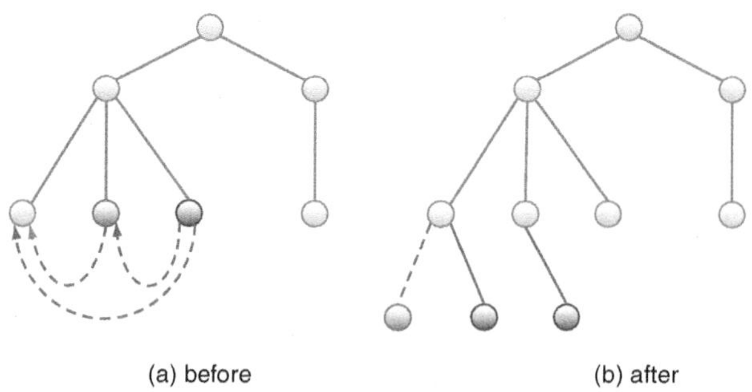

Fig. 22.2 Candidate generation on a trie data structure

22.2.3 The FDM Algorithm

We used an idea proposed by FDM [13] to improve performance on CD [12]. If itemset X is globally frequent, there is at least one site S_i in which all subsets of X are locally frequent. For a proof to this lemma see [13]. Therefore, if we cannot find at least one site S_i in which all subsets of X are locally frequent, X is not globally frequent and we can prune it.

22.3 Our Implementation

The point of this algorithm is that every site keeps a copy of trie locally, and they synchronize their data so that all local trie copies are the same at the end of each stage. After local support is counted, all sites share their support counts and determine the global support counts, in order to remove infrequent itemsets from their local trie.

22.3.1 The DTFIM Algorithm

At the beginning, each site scans its local database independently, and determines the local count of items (1-itemsets). For this purpose, a vector is used to keep count of every item. Each site reads its local transaction records one by one and increase the count of items accordingly. At the end of this stage, sites synchronize their data to determine globally large 1-itemsets ($L_{(1)}$). Using $L_{(1)}$ each site initializes its local trie copy; thus local trie copies are all alike at the end of the pass.

At second pass, the support counts for 2-itemsets shall be calculated. For this purpose a two-dimensional array is created in each site. Like Bodon, we used a triangular structure to better utilize memory. Unlike future passes, for each transaction record T, we determine all 2-itemsets that are subsets of T and increase their count. At the end, the counts are synchronized and global support count for 2-itemsets is calculated, and each site inserts large 2-itemsets into its local trie copy.

From here on, in each pass k ($k \geq 3$) a candidate large k-itemset collection (C_k) is created from the list of large (k–1)-itemsets created in the previous pass ($L_{(k-1)}$). Each site generates its own local copy of C_k, and as the results are uniform at the end of each pass, so will be the list of created candidates. As with the sequential trie-based Apriori, we add every leaf node to its left sibling nodes in the trie.

Now we must prune the infrequent k-itemsets from C_k. To reach this goal, for each transaction record T in local database, each site traverses its trie recursively and finds leaves; if T contains the k-itemset represented by a leaf, the support count on that node is increased.

Before we can prune infrequent itemsets from C_k we have to synchronize local support counts. Each site uses the same depth-first traversal algorithm to find leaves and put their values (local support counts) into a vector, ensuring uniform order in all the sites. Therefore, only the new support count values from the trie are being transferred between sites.

For the pruning step, each site does a depth-first traversal of the trie one more time, updating the value of each leaf node upon traversing it with the value from the global support counts vector and removing the itemset if its support count does not satisfy the minimum required. The output of this step will be L_k and is uniform in every site.

Figure 22.3 shows a simple illustration of DTFIM algorithm.

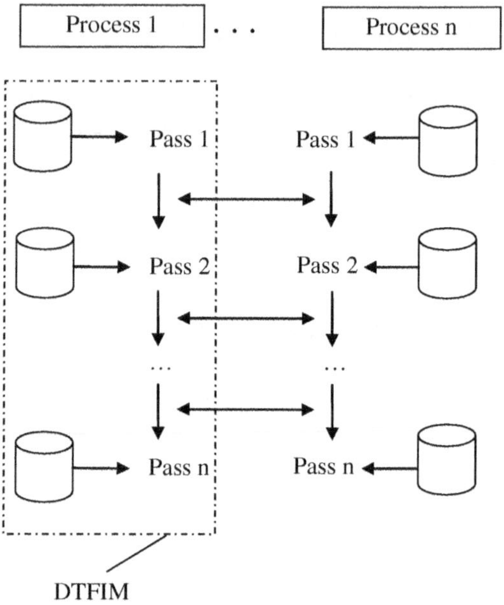

Fig. 22.3 A schema of DTFIM algorithm

22.3.2 *The Revised DTFIM*

In DTFIM algorithm, if we can predict and remove an infrequent itemset X from C_k before each site starts to calculate its local support counts, we save the time needed to process X for every T in transaction records database on every site, just to find out that X is an infrequent itemset later. One technique to predict an itemset X is infrequent is proposed by FDM and discussed in Section 22.2.3. If X is globally frequent, there is at least one site in which all subsets of X are locally frequent.

We revised our algorithm to use this theorem, as it would not impose extra processing time and would give a considerable performance boost in most cases. We added a local frequency indicator bit vector to each trie node, each bit j of this vector signifying whether the corresponding itemset is locally large in site

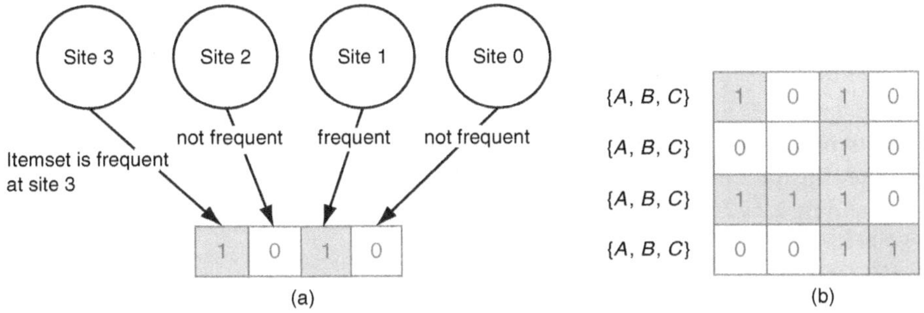

Fig. 22.4 Frequency indicator creation and usage

j or not. Each site updates its own bit at the end of every support counting step and sends it along with local support counts. At the candidate generation step we do not candidate a k-itemset if by examining the local frequency indicators of its subset (k–1)-itemsets we find out there is no site in which all of the subsets are frequent.

Figure 22.4 (a) shows how the local frequency indicator for an itemset is created. Each site contributes its own bit. In Fig. 22.4(b) local frequency indicators for all 3-itemset subsets of an example itemset {A, B, C, D} is displayed. In this case there is site 1 in which all of these subsets are frequent, so {A, B, C, D} is a candidate large itemset.

22.4 Experimental Results

We have implemented all programs in C++ using Visual Studio 2005. The implementations have been tested on a workstation for which Windows XP is running on every node. This workstation consists of eight 3.4 GHz Pentium IV PC with 512 MB of main memory, which are interconnected via 10 M/100 M hub. Parallel message passing software MPICH 2(.Net version) is used here [19]. To empirically evaluate the effect of using proposed technique several tests are performed on the datasets kosarak, accident and T40I10D100K. These datasets are available on FIM repository.[1]

Proposed algorithm is implemented and tested in our environment. The communication and computation are measured with various numbers of nodes and various minimum support values. In every experiment the original dataset is horizontally divided in a number of fragments, each of them is assigned on a node.

In Tables 22.1, 22.2 and 22.3 experimental results for three samples database is shown. Each column shows the result of algorithm for various numbers of sites and

[1] http://fimi.cs.helsinki.fi/data/

every row illustrates one minimum support. "1 site" column represents sequential results.

In Figs. 22.5, 22.6 and 22.7 , there are three experimental result diagram of our implementation in various numbers of sites to show speed up. First Figure show results on *kosarak* database by support threshold equals to 0.00171. And second Figure illustrate the results by database T40I10D100K and support 0.0009. Third Figure, Fig. 22.7, shows result for database accident with support threshold equal to 0.49973.

Table 22. 1 Execution times of DTFIM for database kosarak

Minimum support	1 site	2 sites	4 sites	8 sites
0.04848	16.91	8.61	4.33	2.22
0.00303	22.38	11.31	5.8	3
0.00202	34.81	17.75	9.16	4.98
0.00171	89.25	45.6	24.49	14.13
0.000121	173.75	92.02	49.93	29.25
0.00091	552.137	287.29	165.67	101.7

Table 22.2 Execution times of DTFIM for database T40I10D100K

Minimum support	1 site	2 sites	4 sites	8 sites
0.03	9.52	4.80	2.44	1.28
0.01	37.90	20.03	11.32	6.90
0.009	119.91	63.17	36.73	23.08
0.008	148.62	77.82	45.96	29.23
0.0058	297.91	156.90	94.17	62.01

Table 22.3 Execution times of DTFIM for database accident

Minimum support	1 site	2 sites	4 sites	8 sites
0.61731	21.85	11.22	5.85	3.11
0.49973	33.55	19.53	11.60	7.02
0.41154	119.91	74.47	45.62	27.59

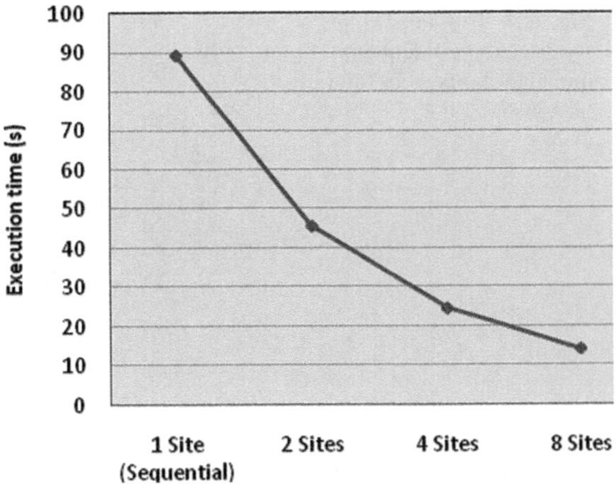

Fig. 22.5 Execution times for database kosarak with minimum support threshold of 0.00171

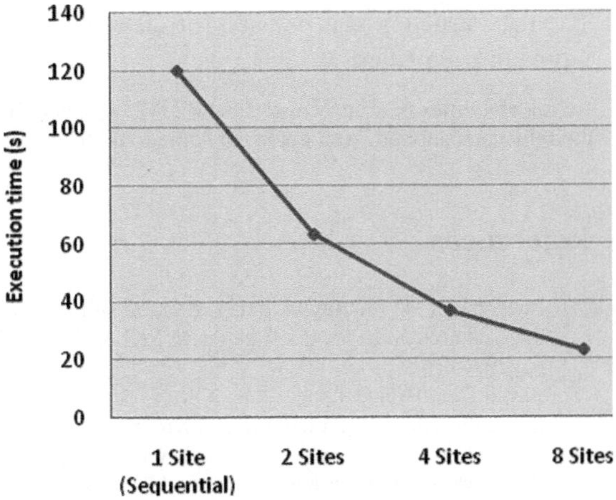

Fig. 22.6 Execution times for database T40I10D100K with minimum support threshold of 0.0009

22.5 Conclusion

Frequent itemsets mining is one of the most important areas of data mining. Bodon presented an implementation that solved frequent itemsets mining problem in most cases faster than other well-known implementations. In this paper, we used the Bodon's ideas for design an algorithm to distributed computing in a no shared memory multi computer environment. The proposed algorithm is revised with some FDM algorithm ideas. In some cases these adopted techniques causes more efficiency. Our experimental results show the efficiency of proposed algorithm. These results show

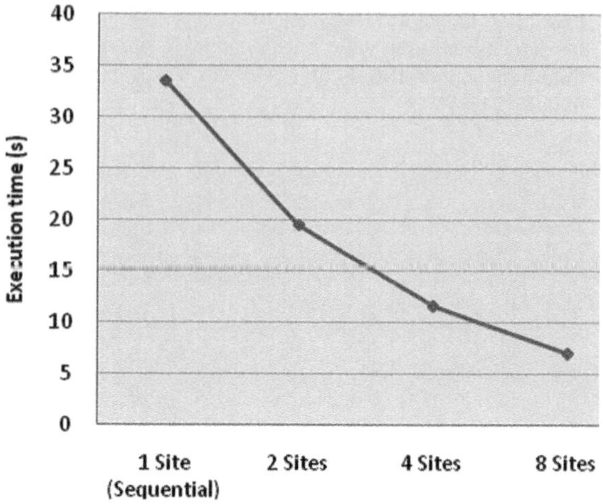

Fig. 22.7 Execution times for database accident with minimum support threshold of 0.49973

Trie data structure can be used for distributed association rule mining not just for sequential algorithms.

Acknowledgments The Authors thank ITRC (Iranian Telecommunication Research Center) for their financial support. And thanks F. Alimardani for her assistance.

References

1. R. Agrawal, T. Imielinski and A. Swami. Mining association rules between sets of items in large databases. In Proc. of the ACM SIG-MOD Conference on Management of Data, 1993, pp. 207–216.
2. Jiawei Han, Hong Cheng, Dong Xin, Xifeng Yan, Frequent pattern mining: current status and future directions, Data Mining and Knowledge Discovery, Vol. 15, 2007, 55–86
3. M.S. Chen, J. Han, and P.S. Yu, "Data mining: An overview from a database perspective, IEEE Transactions on Knowledge and Data Engineering, Vol. 8, No. 6, 1996, pp. 866–883.
4. J. Hipp, Ulrich Güntzer, Gholamreza Nakhaeizadeh, Algorithms for association rule mining – a general survey and comparison, ACM SIGKDD Explorations Newsletter, 2000, Vol. 2, No 1, pages 58–64
5. R. Agrawal, T. Imielinski, A. Swami, Mining association rules between sets of items in large databases, in: Proceedings 1993 ACM SIGMOD Intl. Conf. on Management of Data, Washington, DC, May 1993, pp. 207–216.
6. Maurice Houtsma, Arun Swami, Set-oriented data mining in relational databases, Data & Knowledge Engineering, Vol. 17, No 3, December 1995, Pages 245–262
7. R. Agrawal and R. Srikant, Fast Algorithms for Mining Association Rules, Proceedings of the 20th International Conference on Very Large Data Bases, 1994, pp. 487–499.
8. H. Toivonen, T.M. Vijayaraman, A.P. Buchmann, C. Mohan, and N.L. Sarda, Sampling large databases for association rules, In Proceedings 22nd International Conference on Very Large Data Bases, 1996, pages 134–145.
9. S. Brin, R. Motwani, J.D. Ullman, and S. Tsur. Dynamic itemset counting and implication rules for market basket data. In Proceedings of the 1997 ACM SIGMOD International Conference on Management of Data, Vol. 26(2) of SIGMOD Record, 1997, pp. 255–264.

10. J. Han, J. Pie, Y. Yin and R. Mao. Mining frequent pattern without candidate generation: A frequent-pattern tree approach. Data Mining and Knowledge Discovery, 2003.
11. F. Bodon, "A Fast Apriori Implementation," In B. Goethals and M. J. Zaki, editors, Proceedings of the IEEE ICDM Workshop on Frequent Itemset Mining Implementations, Vol. 90 of CEUR Workshop Proceedings, 2003.
12. R. Agrawal and J. Shafer. Parallel mining of association rules. IEEE Transaction on Knowledge and Data Engineering, Vol. 8, No. 6, 1996, pp. 962–969.
13. D. W. Cheung, and et al., A Fast Distributed Algorithm for Mining Association Rules. In Proc. Parallel and Distributed Information Systems, IEEE CS Press, 1996, pp. 31–42.
14. A. Schuster and R. Wolf, Communication-Efficient Distributed Mining of Association Rules, In Proc. ACM SIGMOD International Conference on Management of Data, ACM Press, 2001, pp. 473–484.
15. A. Schuster, R. Wolf, and D. Trock. A High-Performance Distributed Algorithm for Mining Association Rules, Knowledge And Information Systems (KAIS) Journal, Vol.7, No. 4, 2005.
16. M. Z Ashrafi, D. Taniar and K. Smith, ODAM: an Optimized Distributed Association Rule Mining Algorithm, IEEE Distributed Systems Online, Vol. 5, No. 3, 2004.
17. F. Bodon, "Surprising Results of Trie-based FIM Algorithm," In B. Goethals, M. J. Zaki, and R. Bayardo, editors, Proceedings of the IEEE ICDM Workshop on Frequent Itemset Mining Implementations, Vol. 90 of CEUR Workshop Proceedings, 2004.
18. F. Bodon, A Survey on Frequent Itemset Mining, Technical Report, Budapest University of Technology and Economic, 2006.
19. M. Snir, S. Otto, S. Huss-Lederman, D.Walker, J. Dongarra, MPI: The Complete Reference, The MIT Press, Cambridge, 1996

Chapter 23
A Study on the Inequalities for Fast Similarity Search in Metric Spaces

Tao Ban and Youki Kadobayashi

Abstract Similarity search has also been a heated research topic in the fields of data mining, computer vision, and pattern recognition, where it usually serves as a prominent data processing step. Among the various metric search algorithms proposed in the past few decades, the technique called the Approximating and Eliminating Search Algorithm (AESA) has for 20 years been the baseline method in terms of saved distance computations. In this chapter, we introduce two novel inequalities to estimate the inter-object distances in a metric space. For easy conceivability, the new approximations are defined in the 2D and 3D embeddings of the metric space. To evaluate their efficiency for metric search, we will incorporate them into two indexing algorithms. They are also adaptable to generic metric spaces if appropriate modification can be made to the metric. Experiments showed that the search efficiency of PAESAs, with the new inequalities incorporated into the AESA search algorithm, was much better than the classical AESA, especially for high dimensional datasets.

Keywords Similarity search · Approximating and eliminating search algorithm · Metric spaces · Inequalities · Inter-object distances · Indexing algorithms

We are faced today with a rapidly increasing amount of published information. It would be hard to survive this information explosion without the search engines that are readily accessible on the web, PC, and various database servers. In these systems, similarity search techniques are always involved to retrieve relevant information from a large amount of data with limited computational resources. Meanwhile, similarity search has also been a heated research topic in the fields of data mining, computer vision, and pattern recognition, where it usually serves as a prominent data processing step.

One popular approach for similarity search is mapping data objects into vectors and then conducting a search in the feature space. This method is computationally efficient because the geometrical properties of the feature space and

---

T. Ban (✉)
National Institute of Information and Communications Technology, Tokyo, Japan
e-mail: bantao@nict.go.jp

S.-I. Ao et al. (eds.), *Trends in Communication Technologies and Engineering Science*, Lecture Notes in Electrical Engineering 33, DOI 10.1007/978-1-4020-9532-0_23,

the vectorial presentation of the data help to speed up the search. However, it also introduces the undesirable element of indirection into the process, especially for sophisticated applications where finding the vectorial presentation of the data is itself an open question. A more direct approach is to define a distance function directly between objects and then build indexing structures based on the distance information. Researches have shown that if the distance function satisfies some basic conditions, namely, non-negativity, identity, symmetry, and triangle inequality, efficient algorithms can be formulated for fast similarity search, [6, 10, 14, 20]. Such a distance function, usually called a metric, together with the data domain, are commonly known as a metric space. Since defining a metric between objects can be accomplished more intuitively than mapping objects to feature vectors, the metric space model has found numerous applications in the processing of complex data such as multimedia objects, texts, protein sequences, images, etc.

Among the various metric search algorithms proposed in the past few decades, the technique called the Approximating and Eliminating Search Algorithm (AESA), Vidal [17], has for 20 years been the baseline method in terms of saved distance computations. The key to the use of AESA to avoid unnecessary distance computations is the extensive application of the triangle-inequality to approximate the distances from the query object to the indexed objects and prune objects that do not satisfy the query condition. It is interesting to observe that the triangular inequality is defined in a 1D embedding space: objects are embedded in the space defined by the distance from the reference object, whereas all information perpendicular to this dimension is ignored. Assuming that the indexed objects can be embedded into a Euclidean space with a finite dimension, it can be expected that more accurate approximations of inter-object metrics can be obtained in a higher dimensional embedding space. The higher the dimension of the embedding space, the better the projected distances approaches the metrics. Following this idea, we introduce two novel inequalities to estimate the inter-object distances in a metric space. For easy conceivability, the new approximations are defined in the 2D and 3D embeddings of the metric space. To evaluate their efficiency for metric search, we will incorporate them into two indexing algorithms. The first algorithm is AESA, and we aim to estimate a new baseline for similarity search algorithms in metric spaces. The second algorithm is the Linear AESA (LAESA), Mico et al. [13], which alleviates the quadratic storage and side computation costs of AESA and thus can have a much broader range of application.

This chapter is organized as follows. Section 23.1 briefly reviews the problem of metric search, together with some related works on metric search. Section 23.2 specifies the two inequalities which can give better approximations for the inter-object distances in a metric space. Section 23.3 discusses the implementation of AESA and Linear AESA incorporating the new inequalities. Section 23.4 reports the numerical results. Section 23.5 gives some concluding remarks on this chapter.

23.1 Previous Works on Metric Search

Before specifying the related metric search algorithms, we will first review the properties of a metric space and commonly used metric queries.

23.1.1 Similarity Search in Metric Spaces

Let $\mathbb{D}$ be the domain of objects, $d : \mathbb{D} \times \mathbb{D} \rightarrow \mathcal{R}$ a distance function on $\mathbb{D}$. The tuple $\mathcal{M} = (\mathbb{D}, d)$ is called a *metric space*, if $\forall \boldsymbol{u}, \boldsymbol{v}, \boldsymbol{z} \in \mathbb{D}$, the following conditions hold, [3].

$$d(\boldsymbol{u}, \boldsymbol{v}) \geq 0 \qquad \text{non} - \text{negativity} \qquad (23.1)$$

$$d(\boldsymbol{u}, \boldsymbol{v}) = 0 \Leftrightarrow \boldsymbol{u} = \boldsymbol{v} \qquad \text{identity} \qquad (23.2)$$

$$d(\boldsymbol{u}, \boldsymbol{v}) = d(\boldsymbol{v}, \boldsymbol{u}) \qquad \text{symmetry} \qquad (23.3)$$

$$d(\boldsymbol{u}, \boldsymbol{v}) + d(\boldsymbol{v}, \boldsymbol{z}) \geq d(\boldsymbol{u}, \boldsymbol{z}) \qquad \text{triangular inequality} \qquad (23.4)$$

A metric query is generally defined by a query object $\boldsymbol{q}$ and a proximity condition. For a nearest neighbor (NN) search, the closest object to $\boldsymbol{q}$ is retrieved as the answer to the query. The concept can be generalized to search for k nearest neighbors, and the answer to a kNN search is

$$Q(\boldsymbol{q}, k, \mathbb{U}) = \mathbb{Q} : \{\mathbb{Q} \subseteq \mathbb{U}, |\mathbb{Q}| = k, \\ \forall \boldsymbol{u}_i \in \mathbb{Q}, \boldsymbol{v} \in \mathbb{U} \backslash \mathbb{Q}, d(\boldsymbol{q}, \boldsymbol{u}_i) \leq d(\boldsymbol{q}, \boldsymbol{v})\}. \qquad (23.5)$$

For simplicity, we confine our discussion mainly to the NN search, which is the most widely explored metric search type. All of the discussions can be easily extended to a kNN search algorithm by maintaining a list of the k candidates seen so far and using the largest distance among the k candidates to eliminate the faraway objects. Discussions on other kinds of queries, e.g. range query, point query, reverse NN query, similarity join, and a combination of the enumerated search types, can be found in Samet [14] and Zezula et al. [20].

23.1.2 AESA

During the search, AESA makes extensive use of the following property which is directly derived from the triangular inequality in (23.4).

Theorem 1 *(1D lower bound Vidal [17]) Let $\mathcal{M} = (\mathbb{D}, d)$ be a metric space. For all $\boldsymbol{u}, \boldsymbol{p}, \boldsymbol{q} \in \mathbb{D}$, the following inequality holds:*

$$d_{1\text{D}}(\boldsymbol{q}, \boldsymbol{u}|\boldsymbol{p}) = |d(\boldsymbol{q}, \boldsymbol{p}) - d(\boldsymbol{u}, \boldsymbol{p})| \leq d(\boldsymbol{q}, \boldsymbol{u}), \qquad (23.6)$$

Let $\mathbb{P}$ be the set of pivots whose distances from $\boldsymbol{q}$ and $\boldsymbol{u}$ are known, we can get a lower bound $d_{\mathrm{ID}}(\boldsymbol{q}, \boldsymbol{u}|\mathbb{P})$ on $d(\boldsymbol{q}, \boldsymbol{u})$ as

$$d(\boldsymbol{q}, \boldsymbol{u}) \geq d_{\mathrm{ID}}(\boldsymbol{q}, \boldsymbol{u}|\mathbb{P}) = \max_{\boldsymbol{p}_i \in \mathbb{P}} |d(\boldsymbol{q}, \boldsymbol{p}_i) - d(\boldsymbol{u}, \boldsymbol{p}_i)|. \tag{23.7}$$

Since this inequality helps to prune some of the objects from further consideration, it is generally called a pruning rule, Fukunaga and Narendra [9]. To apply the pruning rule, the distances from the object $\boldsymbol{p}$ to the indexed object $\boldsymbol{u}$ and the query object $\boldsymbol{q}$ need to be known beforehand. Hence, object $\boldsymbol{p}$ is called a reference point or a pivot.

At the preprocessing step, AESA computes all $N \times N$ inter-object distances for the N objects in the indexed object set $\mathbb{U}$ and stores them in a matrix. At query time, AESA starts by first initializing the pivot set $\mathbb{P} = \emptyset$, $\mathbb{U} = \mathbb{U}_0 \setminus \mathbb{P}$, and $d_{\mathrm{ID}}(\boldsymbol{q}, \boldsymbol{u}) = 0$ for all $\boldsymbol{u} \in \mathbb{U}$. At each iterative step, a new pivot $\boldsymbol{p}$ is selected from $\mathbb{U}$ as the one with the minimal lower bound, and then its distance from $\boldsymbol{q}$ is computed. The algorithm then updates the lower bounds for the remaining objects in $\mathbb{U}$ and elimiates the ones with lower bounds greater than the distance from $\boldsymbol{q}$ to the NN candidate, $\boldsymbol{v}$. AESA terminates once there are no more objects left in $\mathbb{U}$. $\boldsymbol{v}$ is updated if necessary when the distance computation is invoked.

23.1.3 LAESA

The main drawback of the AESA approach is the quadratic storage requirement and side computation overhead. LAESA alleviates these drawbacks by choosing a fixed number, M, of pivots whose distances from all other objects are computed and stored in advance. Thus, for a dataset consisting of N objects, the distance matrix contains $N \times M$ entries rather than $N \times N$ for AESA. The LAESA search algorithm is very similar to that of AESA, especially when the distances from $\boldsymbol{q}$ to the selected pivots are preferentially computed.

In particular, let $\mathbb{P}$ be the set of selected pivots from $\mathbb{U}_0$, $\mathbb{U} = \mathbb{U}_0 \setminus \mathbb{P}$ the remaining objects. To implement the approximating and eliminating strategy, LAESA first computes the distances between $\boldsymbol{q}$ and all of the pivots $\boldsymbol{p} \in \mathbb{P}$. It then estimates the distance lower bounds for all $\boldsymbol{u}_i \in \mathbb{U}$. This helps to eliminate objects whose lower bounds on distance from $\boldsymbol{q}$ are greater than the distance from $\boldsymbol{q}$ to the current NN candidate. The remaining objects are sequentially compared against $\boldsymbol{q}$ in ascending order of their distance lower bounds. In the case of a distance computation, the NN candidate is updated if necessary.

23.1.4 Other Metric Search Algorithms

Besides the mentioned AESA and LAESA which seek to reduce the distance computations as many as possible, there have been many proposals of advanced indexing structures that try to hit a balance between the distance computations, side computations, and storage cost. Reprensentative methods include the metric tree approaches, such as the Vantage Point tree Yianilos [19], Burhard-Keller tree, Burkhard and Keller [4], Generalized Hyperplane tree, Uhlmann [16], Geometric Near-neighbor Access tree, Brin [3], and M-tree, Ciaccia et al. [7], and similarity hashing methods, such as the D-index, Dohnal et al. [8], and its decedents. In these algorithms, objects are organized into locally adjacent groups so that the pruning rules can be applied to multiple objects at one time and therefore side computations can be greatly reduced.

23.2 New Lower Bounds

In general, any successful metric search should pay attention to two key points: the indexing structure used to organize the indexed objects and the techniques to avoid unnecessary distance computations. Hence we can analyze the design of metric search algorithms from two different points of views.

For a metric space, nothing beyond the inter-object distance is available. In this sense, the distance matrix possesses all the information that we can learn from the metric space model. AESA tries to make use of all the information in the distance matrix to reduce distance computations, whereas other algorithms, e.g. the metric trees, only use some sketched information about the matrix. In respect of the indexing structure design, this may be the reason why it is so hard for other methods to beat AESA on the reduction of distance computations.

On the other hand, it might be possible to improve AESA's search efficiency with respect to the pruning technique. As suggested in Ban and Kadobayashi [2], with a slightly more strict assumption of the metric space model, we can effectively tighten the lower bound produced by (23.7), to avoid more unnecessary distance computations. In the following, we first discuss a Euclidean case and then generalize the discussion to metric spaces.

23.2.1 Embedding of the Metric Space

In a multi-dimensional Euclidean space, a 2D embedding space can be defined by a triple of non-identical points $\boldsymbol{o}$, $\boldsymbol{p}$, and $\boldsymbol{u}$, as shown in Fig. 23.1. Let $\overrightarrow{\boldsymbol{op}}$ be a coordinate axis, with $\boldsymbol{o}$ being the origin and $\boldsymbol{p}$ defining the positive direction. Let the projection of $\boldsymbol{u}$ on the axis be $\boldsymbol{u}'$. Then by the law of cosines, the projected distance between $\boldsymbol{o}$ and $\boldsymbol{u}$ along $\overrightarrow{\boldsymbol{op}}$ is:

$$\bar{d}(\boldsymbol{o}, \boldsymbol{p}; \boldsymbol{u}) = \cos\theta d(\boldsymbol{o}, \boldsymbol{u}) = \frac{d^2(\boldsymbol{o}, \boldsymbol{u}) + d^2(\boldsymbol{o}, \boldsymbol{p}) - d^2(\boldsymbol{p}, \boldsymbol{u})}{2d(\boldsymbol{o}, \boldsymbol{p})}. \tag{23.8}$$

We can see that the projected distance $\bar{d}(\boldsymbol{o}, \boldsymbol{p}; \boldsymbol{u})$ can be computed using no information other than the inter-object distances, thus it might be possible to apply the lower bound derived from (23.8) to metric space models. Then it is interesting to

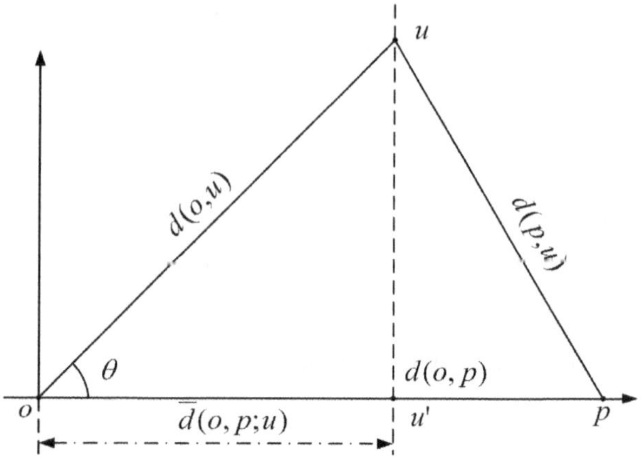

Fig. 23.1 The 2D Euclidean embedding space

ask what circumstance might allow a metric space model to give rise to a configuration of points, $\{\boldsymbol{x}_i\}$, in a Euclidean space, where the associated Euclidean distance $L_2(\boldsymbol{x}_i, \boldsymbol{x}_j) \equiv d(\boldsymbol{u}_i, \boldsymbol{u}_j)$ for all $\boldsymbol{u}_i, \boldsymbol{u}_j \in \mathbb{U}$. We have the following theorem as the answer to this question.

Theorem 2 *(Mardia et al. [12].) Let $\mathcal{M} = (\mathbb{D}, d)$ be a metric space, $\mathbb{U} = \{\boldsymbol{u}_i, i = 1, \ldots, N\} \subset \mathbb{D}$. Define the squared inter-object distance matrix as $\boldsymbol{B}^{N\times N}$, where $[\boldsymbol{B}]_{ij} = d^2(\boldsymbol{u}_i, \boldsymbol{u}_j)$, and the gram matrix $\boldsymbol{G}$ as*

$$\boldsymbol{G} = -\frac{1}{2}\left(\boldsymbol{I} - \frac{1}{N}\boldsymbol{1}\boldsymbol{1}^T\right)\boldsymbol{B}\left(\boldsymbol{I} - \frac{1}{N}\boldsymbol{1}\boldsymbol{1}^T\right), \tag{23.9}$$

where $\boldsymbol{1} = \{1, \ldots, 1\}$ is the n-ary all ones vector and $\boldsymbol{I}$ the identity matrix. If $\boldsymbol{G}$ is positive semi-definite, then there exists a configuration $\boldsymbol{x}_i$, $i = 1, \ldots, N$, in a Euclidean space with a dimension up to $(N-1)$ that satisfies

$$L_2(\boldsymbol{x}_i, \boldsymbol{x}_j) \equiv d(\boldsymbol{u}_i, \boldsymbol{u}_j),\ (i, j = 1, \ldots, N). \tag{23.10}$$

We call a metric space model that always produces a positive semi-definite gram matrix positive semi-definite. Fortunately, most of the commonly used metric space models are positive semi-definite so that the lower bounds derived from this property can be applied. Note that the concept of the positive semi-definite metric spaces is in accordance with positive semi-definite kernels, Shawe-Taylor and Cristianini [15], which have recently been extensively studied in the field of machine learning. In fact, a positive semi-definite kernel always corresponds to a positive semi-definite metric space. Hence, the techniques explored here are readily applicable to most modern applications.

23.2.2 New Bounds on the Unknown Distances

By Theorem 2,the following theorem defines a novel lower bound on the distance between two objects given their distances to two pivots are known.

Theorem 3 *(2D lower bound, Ban and Kadobayashi [1].) Let $\mathcal{M} = (\mathbb{D}, d)$ be a positive semi-definite metric space. Then for all $\boldsymbol{o}, \boldsymbol{p}, \boldsymbol{u}, \boldsymbol{q} \in \mathbb{D}$, $\boldsymbol{o} \neq \boldsymbol{p}$, the following inequality holds:*

$$d(\boldsymbol{q}, \boldsymbol{u}) \geq d_{2\mathrm{D}}(\boldsymbol{p}, \boldsymbol{u}|\boldsymbol{o}, \boldsymbol{p}) = |\bar{d}(\boldsymbol{o}, \boldsymbol{p}; \boldsymbol{q}) - \bar{d}(\boldsymbol{o}, \boldsymbol{p}; \boldsymbol{u})|. \tag{23.11}$$

It follows directly from the basic properties of the Euclidean space. As shown in Fig. 23.2, since $\overrightarrow{\boldsymbol{q}'\boldsymbol{u}'}$ is the projection of $\overrightarrow{\boldsymbol{q}\boldsymbol{u}}$ along $\overrightarrow{\boldsymbol{o}\boldsymbol{p}}$, the equation holds if and only if $\overrightarrow{\boldsymbol{q}\boldsymbol{u}}$ parallels $\overrightarrow{\boldsymbol{o}\boldsymbol{p}}$. In Theorem 3, only the projected distances along $\overrightarrow{\boldsymbol{o}\boldsymbol{p}}$ is considered. The following theorem employs the additional information perpendicular to $\overrightarrow{\boldsymbol{o}\boldsymbol{p}}$ for a tighter lower bound.

Theorem 4 *(3D lower bound, Ban and Kadobayashi [1].) Let $\mathcal{M} = (\mathbb{D}, d)$ be a positive semi-definite metric space, $\boldsymbol{o}, \boldsymbol{p}, \boldsymbol{u}, \boldsymbol{q} \in \mathbb{D}$, $\boldsymbol{o} \neq \boldsymbol{p}$. Let $\boldsymbol{u}'$ and $\boldsymbol{q}'$ be the projections of $\boldsymbol{u}$ and $\boldsymbol{q}$ on $\overrightarrow{\boldsymbol{o}\boldsymbol{p}}$, respectively (see Fig. 23.2). Then $d(\boldsymbol{q}, \boldsymbol{u})$ is lower bounded by*

$$d(\boldsymbol{q}, \boldsymbol{u}) \geq d_{3\mathrm{D}}(\boldsymbol{p}, \boldsymbol{u}|\boldsymbol{o}, \boldsymbol{p}) = \sqrt{d^2(\boldsymbol{u}, \boldsymbol{q}|\boldsymbol{o}, \boldsymbol{p}) + (d(\boldsymbol{q}, \boldsymbol{q}') - d(\boldsymbol{u}, \boldsymbol{u}'))^2}. \tag{23.12}$$

Generally, for a metric search algorithm, the tighter the incorporated distance lower bound, the better the pruning ability to avoid unnecessary distance computations. It is obvious that the 3D lower bound is tighter than its 2D counterpart. For a positive semi-definite metric space, it can be proven that the 3D lower bound is also tighter than the 1D lower bound in Theorem 1, as stated in the following theorem.

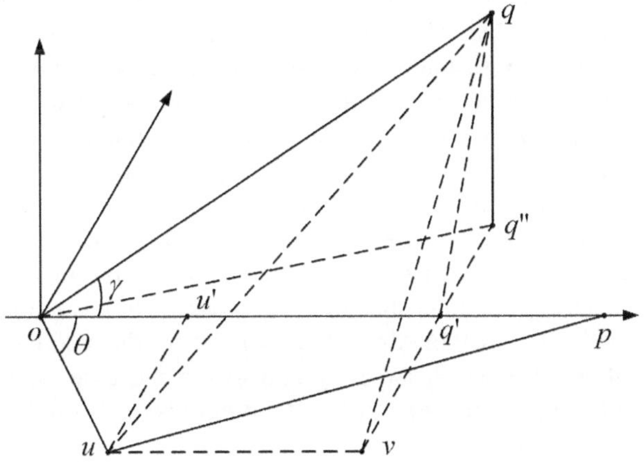

Fig. 23.2 The 3D Euclidean embedding space

Theorem 5 *(Ban and Kadobayashi [1].) Let* $\mathcal{M} = (\mathbb{D}, d)$ *be a positive semi-definite metric space. For all* $\boldsymbol{o}, \boldsymbol{p}, \boldsymbol{u}, \boldsymbol{q} \in \mathbb{D}, \boldsymbol{o} \neq \boldsymbol{p}$*, the following inequality holds:*

$$d_{3\mathrm{D}}(\boldsymbol{q}, \boldsymbol{u}|\boldsymbol{o}, \boldsymbol{p}) \geq \max\big(d_{1\mathrm{D}}(\boldsymbol{q}, \boldsymbol{u}|\boldsymbol{o}), d_{1\mathrm{D}}(\boldsymbol{q}, \boldsymbol{u}|\boldsymbol{p})\big). \tag{23.13}$$

23.2.3 Extending to Generic Metric Spaces

To apply the inequalities to search algorithms in generic metric spaces, consideration should be given to the case when a metric leads to a matrix $\boldsymbol{G}$ that is not positive semi-definite. According to Lingoes [11], the solution for this problem is adding a constant to the squared metrics in $\boldsymbol{B}$, except for the self-distances $(\boldsymbol{u}_i, \boldsymbol{u}_i)$. The new square metrics $\{d_{ij}'^2\}$ in the form of $d_{ij}'^2 = d_{ij}^2 + C(1 - \delta^{ij})$, where C is a constant and δ^{ij} the Kronecker delta, makes $\boldsymbol{G}$ positive semi-definite. This has been referred to as the additive constant problem. The smallest value of C that can turn $\boldsymbol{G}$ into a positive semi-definite is $-2\lambda_n$, where λ_n is the smallest eigenvalue of $\boldsymbol{G}$, Lingoes [11].

23.3 Search Algorithms

In this section, we specify the search algorithms of AESA and LAESA with the new lower bounds incorporated. We call the new distance bounds *projection based distance bounds*, and term the algorithms as PAESA (Projection-based AESA) and LPAESA (Linear Projection-based AESA), respectively.

23.3.1 PAESA

As specified in Algorithm 23.1, PAESA basically follows the procedure of AESA, Vidal [17, 18]. The major difference between PAESA and AESA is the application of the new distance lower bounds for approximation and elimination. In AESA, for each pivot $\boldsymbol{p}$ selected from the dataset, Theorem 1 is applied only once to update the associated lower bound $d_{1\mathrm{D}}(\boldsymbol{q}, \boldsymbol{u}_i)$ on $d(\boldsymbol{q}, \boldsymbol{u}_i)$. However, for PAESA, when a pivot is added to the pivot set, multiple lower bounds for $d(\boldsymbol{q}, \boldsymbol{u}_i)$ can be obtained from Theorem 3 or 4. Thus, given M pivots, the lower bounds $d_{2\mathrm{D}}(\boldsymbol{q}, \boldsymbol{u}_i)$ or $d_{3\mathrm{D}}(\boldsymbol{q}, \boldsymbol{u}_i)$ for $d(\boldsymbol{q}, \boldsymbol{u}_i)$, are selected as the maximum from the $M(M + 1)/2$ approximations. Hence, the lower bounds produced by PAESA will generally be much tighter than that of AESA and hence more distance computations can be avoided. Line 9 of Algorithm 23.1 applies the inequality to estimate the lower bound of the distance between $\boldsymbol{q}$ and $\boldsymbol{u}_i$. In the case where Theorem 3 is applied, we term the algorithm PAESA2D, otherwise it is dubbed PAESA3D if Theorem 4 is employed.

23.3.2 LPAESA

Like AESA, PAESA also requires a matrix storing $N \times N$ inter-object distances, which becomes impractical for large datasets. According to LPAESA, we can alleviate this drawback by choosing a fixed number of M pivots, whose distances from all the objects are computed and stored beforehand.

Algorithm 23.2 specifies the search algorithm for LPAESA where a set of M pivots, $\mathbb{P}$, are previously selected from the dataset. During the search, the query object is first compared against the pivots and then the distances from the pivots are used to compute the lower bounds for the non-pivots. After that, the non-pivots are visited in ascending order of the lower bounds until no object in the set can be the NN of the query object. The NN candidate is updated if necessary when a distance computation is invoked. LPAESA is called LPAESA2D or LPAESA3D according to the employed lower bound.

Alg. 23.1 The PAESA search algorithm.

PAESA_search$(q, U_0 = \{u_1, \ldots, u_N\})$
  $P \leftarrow \emptyset$; $U \leftarrow U_0$; // initialization
  $D(u_i) \leftarrow 0$, for $u_i \in U$; // lower bound
  $d_n \leftarrow \infty$; // distance to the nearest neighbor
  while $U \neq \emptyset$ **do**
    $p \leftarrow \arg\min_{u_i \in U} D(u_i)$; $U \leftarrow U \setminus \{p\}$; // select pivot
    if $d(q, p) < d_n$ **then** // distance evaluation
      $d_n \leftarrow d(q, p)$; $v \leftarrow p$; // update the candidate
    for $u_i \in U$ **do** // approximation
      $D_{\text{low}}(u_i) \leftarrow \max_{p_j \in P}$ ***appro***(q, u_i, p, p_j);
      $D(u_i) \leftarrow \max(D(u_i), D_{\text{low}}(u_i))$;
      if $D(u_i) \geq d_n$ **then** $U \leftarrow U \setminus \{u_i\}$; // elimination
    $P \leftarrow P \cup \{p\}$; // update pivot set
  return v; // return the nearest neighbor

Alg. 23.2 The LPAESA search algorithm.

LPAESA_search$(q, U_0 = \{u_1, \ldots, u_{N-M}\}, P = \{p_1, \ldots, p_M\})$
  $U \leftarrow U_0 \setminus P$; // set of non-pivots
  $d_n \leftarrow \infty$; // distance to the nearest neighbor
  for $p_i \in P$ **do**
    if $d(q, p_i) < d_n$ **then** // distance evaluation
      $d_n \leftarrow d(q, p_i)$; $b \leftarrow p_i$; // update the candidate
  for $u_i \in U$ **do** //approximation
    $D_{\text{low}}(u_i) \leftarrow \max_{p_j, p_k \in P}$ ***appro***(q, u_i, p_j, p_k);
    if $(D_{\text{low}}(u_i) \geq d_n)$ then $U \leftarrow U \setminus \{u_i\}$; // elimination
  $U_s \leftarrow sort(U, \{D_{\text{low}}(u_i)\})$; // ascending ordering
  for $i = 1$ to $N - M$ **do**
    if $(D_{\text{low}}(u_i) \geq d_n)$ **then break**; // no possible NN
    if $d(q, u_i) < d_n$ **then** // distance evaluation
      $d_n \leftarrow d(q, u_i)$; $b \leftarrow u_i$; // update the candidate
  return b; // return the nearest neighbor

23.3.3 Pivot Selection

To speed up the search, LPAESA requires a distance matrix consisting of the computed distances between the pivots in the pivot set and all of the objects in the dataset. As pointed out in Bustos et al. [5], the way pivots are selected affects the search performance of a metric search algorithm. Among the pivot selection heuristics studied in Bustos et al. [5], the incremental selection strategy shows the best performance for real world metric spaces, both in terms of approximating accuracy and computation cost. In the experiment, we employed an incremental selection algorithm, as specified in Algorithm 23.3, to select the pivot set. By defining the distance from an object $\boldsymbol{u}$ to a set of objects $\mathbb{P}$ as the minimal distance from $\boldsymbol{u}$ to $\boldsymbol{p}_i \in \mathbb{P}$, the idea can be easily stated as follows. First initialize the pivot set with a randomly selected object, then sequentially select the next pivot as the most separated object from the pivot set.

```
Alg. 23.3 Incremental pivot selection algorithm.
PIVOT_selection(U, M)
   P ← {p1 ∈ U}; // random select the first pivot
   for i = 2 to M do // select the most faraway object
      d_i(u_j, P) = min_{p_k∈P} d(u_j, p_k), for u_j ∈ U;
      p_i ← arg max_{u_j∈U\P} d_i(u_j, P);
      P ← P ∪ {p_i}; // update pivot set
   return P; // return the pivot set
```

23.4 Experiments

Inthis section we present the experimental results of PAESA and LPAESA on a series of simulations. The performance of the indexing structures was measured by the reduction in the distance calculations. Because AESA and LAESA, which provide the best effectivity in distance computation reduction, have long been the baseline of metric search algorithms, we only refer to these two algorithms. Other metric search algorithms—typically with less storage cost—generally will require more distance computations. In the experiments, the indexed objects and the query objects were independently drawn from uniform distributions in c-dimensional unit hypercubes. The coordinates of the data points were never used directly in the algorithms. If not otherwise noted, the number of objects in the datasets was 10,000, and the number of queries was 100. The results were averaged over 10 runs.

We first show the results of the pivot selection algorithms for LAESA, LPAESA2D, and LPAESA3D. Figure 23.3 shows the curves of the number of distance calculations as a function of pivot set size for a 10D dataset. Evident saddle points can be found in all the curves. That is, the number of distance calculations first drops because the approximation accuracy increases as pivots are added. However, after a certain threshold, the number of distance calculations goes up since computations against the pivots increase considerably. The saddle point is therefore chosen

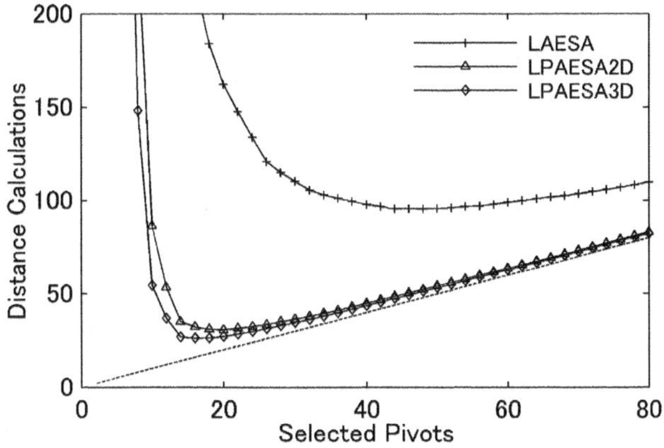

Fig. 23.3 Number of distance computations as a function of number of pivots in 10D

as the result of the pivot selection algorithm. Experiments showed that, for independently generated datasets with the same distribution, the optimal number of pivots is quite stable. It can also be seen that the curves for LPAESA2D and LPAESA3D are very close to the dashed line, which stands for the number of pivots. This is because the distance lower bounds given by LPAESA2D and LPAESA3D are so tight that most of the non-pivot objects are pruned without distance computations.

Figure 23.4 reports the results of pivot selection for datasets up to 20 dimension. It shows that the number of pivots selected by LAESA was exponential in the dimension of the dataset. LPAESA2D and LPAESA3D needed comparable numbers of pivots for all the cases. For datasets with over 5 dimensions, the LPAESAs selected considerably fewer pivots than LAESA. For datasets in 10D, LAESA, LPAESA2D, and LPAESA3D selected 44, 15, and 12 Pivots, respectively, and for 20 dimensions, they selected 1002, 132, and 118 pivots.

The aim of the second experiment was to compare the performance of the referenced metric search algorithms in different dimensions. The algorithms were tested

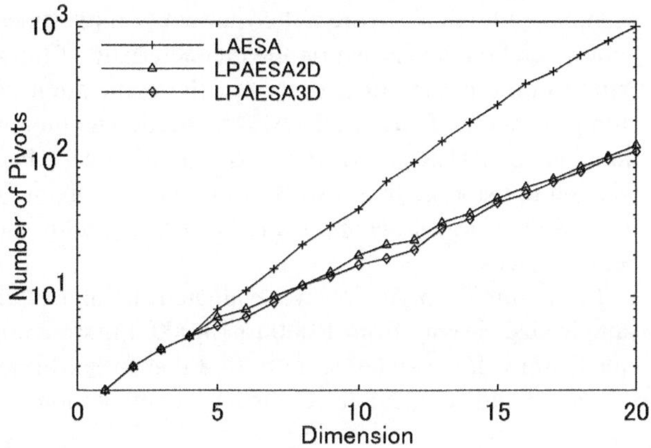

Fig. 23.4 Pivot selection results on different dimensions

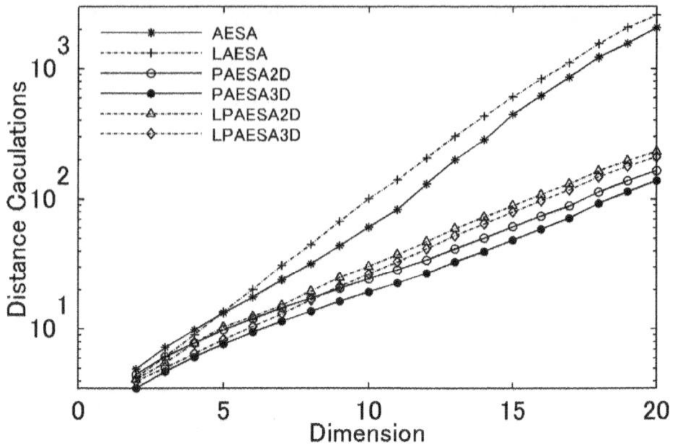

Fig. 23.5 Average number of distance computations as a function of dimension

on datasets up to 20 dimensions. Figure 23.5 shows the number of distance calculations during NN search. It can be seen that the search difficulty increases exponentially with the number of dimensions because of the so called curse of dimensionality. The performance comparison is clear enough: For a dataset with a dimension below 5, since the selected pivots can produce very tight lower bounds, the pivot set based methods are more preferable than the full matrix based methods, especially when the storage cost is considered. When the search difficulty increases as the dimension goes up, accurate lower bounds can only be obtained by engaging more pivots. For datasets over 8D, we have the following performance order:

$$\begin{aligned} LAESA \lessdot AESA \lessdot LPAESA2D \\ \lessdot LPAESA3D \lessdot PAESA2D \lessdot PAESA3D, \end{aligned} \tag{23.14}$$

where '$\lessdot$' stands for an ascending order in terms of the avoided distance computations. Since Theorem 4 produced tighter lower bounds than Theorem 3, PAESA3D needed slightly fewer distance computations than PAESA2D, and LPAESA3D needed fewer than LPAESA2D. However, the differences were not very significant. If the computation cost to estimate the distance lower bounds is also taken into account—note that Theorem 4 costs two or three times more than Theorem 3—the choice can be made based on the characteristic of the specific application. Another thing to note is the effectiveness of the application of pivot sets. With much less storage cost than PAESAs, LPAESAs needed no more than twice the distance computations of PAESAs. For the most intensive case in 20D, LPAESAs saved about 99% on the storage than PAESAs at a cost of about 50% more distance computations. So for large scale applications, or for systems short on storage, LPAESAs are better choices.

In the third experiment, we evaluated the algorithms on 10D datasets with the sample size varying from 1,000 to 13,000. Figure 23.6 shows the required distance calculations. It was interesting to find that for all these algorithms fewer objects in the dataset did not mean fewer distance computations required to retrieve the nearest

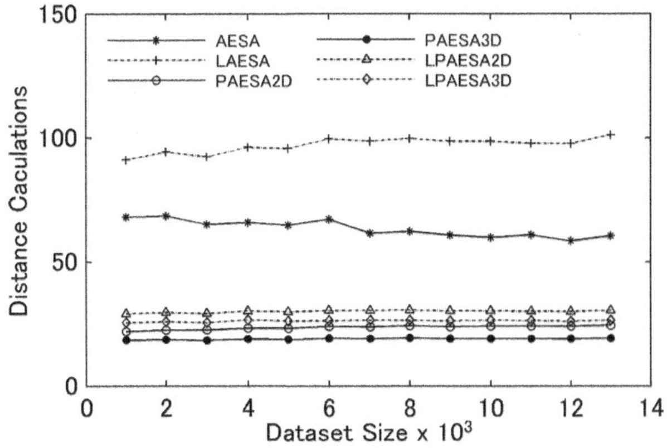

Fig. 23.6 Average number of distance computations as a function of sample size

neighbor. On the contrary, each of the algorithms maintained a stable number of distance computations as the sample size increased. Note that for AESA and the PAESAs, there was no influence from the selected number of pivots. This property is especially useful for large scale datasets. For all the cases, the performance order in (23.14) was preserved.

In the fourth experiment, we extended the algorithms to find the k nearest neighbors by introducing an ordered list of the k nearest objects to the query object and updating it when the distance computation was invoked. The number of pivots were fixed for LAESA and the LPAESAs. Figure 23.7 shows the curves of the distance calculations against k, which changed from 1 to 100. The experiments were carried on 10D datasets. We can learn from the figure that as k increased, the search difficulty increased accordingly. For each of the algorithms, the number of distance calculations was in linear relation to the retrieved number of nearest neighbors. We have to note that this linearity in increased distance computations is only appli-

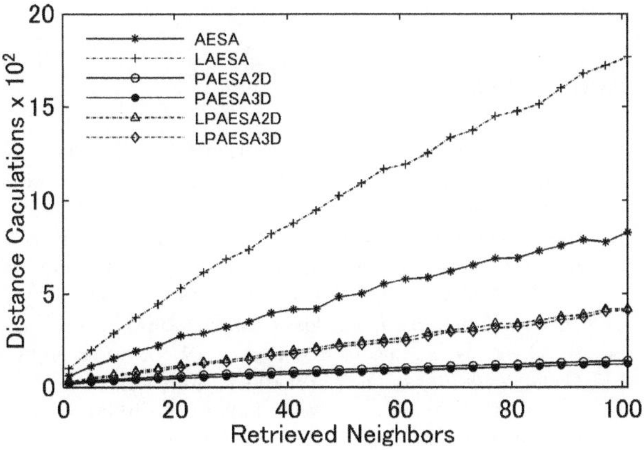

Fig. 23.7 Average number of distance computations as a function of retrieved neighbors

cable for small k values. Since the number of pivots were fixed for LAESA and LPAESAs in the experiments, the results were in favor of PAESA and AESA. Better results could be produced by LAESA and LPAESAs if more pivots were selected for large k values. A more adaptive strategy could be selecting a large enough pivot set beforehand, and then selectively using a portion of the pivot set according to the user defined k parameter.

23.5 Concluding Remarks

In this chapter, we have discussed two new inequalities for approximating the distances in a metric space. The inequalities were derived from the geometrical properties of the embedding space and are applicable to positive semi-definite metric space models. They are also adaptable to generic metric spaces if appropriate modification can be made to the metric. Experiments showed that the search efficiency of PAESAs, with the new inequalities incorporated into the AESA search algorithm, was much better than the classical AESA, especially for high dimensional datasets. When the dimension exceeded 10, the two types of PAESAs both needed fewer than half of AESA's distance computations to retrieve the nearest neighbor. For 20D datasets, the PAESAs only needed about 10% of the distance computations of AESA. For better storage and side-computation efficiency, we had also applied the new inequalities to the LAESA algorithm. Results showed that LPAESAs can not only outperform LAESA in saving up unnecessary distance computations, but further alleviate the storage cost as well.

For most of the experiments, we have discovered the performance order in (23.14). Hence, we can reach the conclusion that, for applications where the distance computations are the most essential computational cost, the application of PAESA is the best choice. The LPAESAs could be better choices if the storage and side computation costs are taken into consideration. Whether to apply the 3D lower bound or 2D lower bound depends on the comparative cost of the side computations against distance evaluations.

References

1. Ban, T. and Kadobayashi, Y. (2008). On tighter inequalities for efficient similarity search in metric spaces. *IAENG International Journal of Computer Science*, 35(3): 392–402.
2. Ban, T. and Kadobayashi, Y. (2007). New prune rules for similarity search. *The 11th IASTED International Conference on Artificial Intelligence and Soft Computing*, pp. 120–126.
3. Brin, S. (1995). Near neighbor search in large metric spaces. *The 21st International Conference on Very Large Data Bases*, pp. 574–584.
4. Burkhard, W. A. and Keller, R. M. (1973). Some approaches to best-match file searching. *Communications of the ACM*, 16(4):230–236.
5. Bustos, B., Navarro, G., and Chávez, E. (2003). Pivot selection techniques for proximity searching in metric spaces. *Pattern Recognition Letters*, 24:2357–2366.

6. Chávez, E., Navarro, G., and Marroquin, J. L. (2001). Searching in Metric Spaces. *ACM Computing Surveys*, 33 (3): 273–321.
7. Ciaccia, P., Patella, M., and Zezula, P. (1997). M-tree: an efficient access method for similarity search in metric spaces. *The 23rd International Conference on Very Large Data Bases*, pp. 426–435.
8. Dohnal, V., Gennaro, C., Savino, P., and Zezula, P. (2003). D-Index: distance searching index for metric data sets. *Multimedia Tools and Applications*, 21(1):9–33.
9. Fukunaga, L. and Narendra, P. M. (1975). A branch and bound algorithm for computing *k*-nearest neighbors. *IEEE Transactions on Computers*, 24(7):750–753.
10. Gisli, H. R. and Samet, H. (2003). Index-driven similarity search in metric spaces. *ACM Transactions on Database Systems*, 28:517–580.
11. Lingoes, J. C. (1971). Some boundary conditions for a monotone analysis of symmetric matrices. *Psychometrika*, 36:406–407.
12. Mardia, K. V., Kent, J. T., and Bibby, J. M. (1979). *Multivariate Analysis*, London: Academic Press.
13. Micó, M. L., Oncina, J., and Vidal, E. (1994). A new version of the nearest-neighbor approximating and eliminating search algorithm (AESA) with linear preprocessing time and memory requirements. *Pattern Recognition Letters*, 15(1):9–17.
14. Samet, H. (2005). *Foundations of Multidimensional and Metric Data Structures*, Morgan Kauf-mann Publishers Inc., San Francisco, CA, USA.
15. Shawe-Taylor, J. and Cristianini, N. (2004). *Kernel Methods for Pattern Analysis*, Cambridge University Press, Cambridge, England.
16. Uhlmann, J. K. (1991). Satisfying general proximity/similarity queries with metric trees. *Information Processing Letters*, 40(4):175–179.
17. Vidal, E. (1986). An algorithm for finding nearest neighbors in (approximately) constant average time. *Pattern Recognition Letters*, 4:145–157.
18. Vidal, E. (1994). New formulation and improvements of the nearest-neighbor approximating and eliminating search algorithm (AESA). *Pattern Recognition Letters*, 15(1):1–7.
19. Yianilos, P. N. (1993). Data structures and algorithms for nearest neighbor search in general metric spaces. *ACM-SIAM Symposium on Discrete Algorithms*, pp. 311–321.
20. Zezula, P., Amato, G., Dohnal, V., and Batko, M. (2006). *Similarity Search—The Metric Space Approach*, Springer.

Chapter 24
Discovering Knowledge of Association Using Coherent Rules

Alex Tze Hiang Sim, Samar Zutshi, Maria Indrawan and Bala Srinivasan

Abstract Mining of association rules is of interest to data miners. Typically, before association rules are mined, a user needs to determine a support threshold in order to obtain only the frequent item sets. Having users to determine a support threshold attracts a number of issues. We propose an association rule mining framework that does not require a pre-set support threshold. The framework is developed based on implication of propositional logic. The experiments show that our approach is able to identify meaningful association rules.

Keywords Association rules · Propositional logic

24.1 Introduction

Association Rule Mining (ARM) is a learning technique used to discover knowledge from transaction records. Each rule discovered has its importance measured against many interest measures [1] such as *support* and *confidence*. Typically, ARM technique necessitates a cut-off support threshold to be predefined to separate frequent patterns from the infrequent ones. Two item sets are said to be *associated* if they occur together frequently above a minimum support threshold value.

There are major disadvantages to having a predefined threshold. Firstly, some rules are inevitably lost if the support threshold is set inaccurately. In addition, it is usually not possible to remove the support threshold in order to find infrequent items because ARM relies on a downward closure property of *support*, which necessitates a threshold to search for frequent item sets. That is, if an item set passes a minimum support requirement then all its subsets also pass this requirement. This minimum support threshold value is used as the basis for pruning, without which mining rules are not feasible due to the exponential search space. In summary, in

---

A.T.H. Sim (✉)
Monash University, Melbourne, Australia

S.-I. Ao et al. (eds.), *Trends in Communication Technologies and Engineering Science*, Lecture Notes in Electrical Engineering 33, DOI 10.1007/978-1-4020-9532-0_24,

traditional ARM, a minimum support threshold is needed, and should be determined accurately in order to produce useful rules for users.

To overcome the above limitation, we investigate the possibility of developing a new association rule mining framework that works without having to determine a support threshold. We base our framework on the notion of implication of propositional logic. We explain our proposed model in detail in Section 24.3 after a discussion of previous work is presented in Section 24.2. Coherent rules discovered have its interestingness defined and quantified using our proposed measure of interest in Section 24.4. We design an algorithm to mine coherent rules without a pre-set support threshold in Section 24.5. Experiments based on an implementation of the framework and a discussion of the results is presented in Section 24.6. Finally, conclusion is made in Section 24.7.

24.2 Previous Work

Recently, mining infrequent rules have started to gain momentum as many have begun to accept that rules based on infrequently occurring items are also important because it represents knowledge not found in frequent rules, and these infrequent rules are often interesting [2–6]. Association rules among infrequent items have been relatively ignored by association rule mining algorithms mainly due to the problem of the large search space and the consequent explosion of total number of association rules reported [2–8]. Some of these reported rules may in fact be based on noise in the data. However, there have been some attempts towards finding infrequent association rules, such as [9], where a generalised association rules mining framework using correlation is proposed. Correlation is measured by Pearson's Goodness of Fit Chi Square measure. However, this chi-square measure suffers from the limitation of measuring the association rules inaccurately at small expected values, if one of the expected values is lower than the value five [9]. In practice, this is often being observed. This limit the use of a Chi Square based framework. In addition, the authors' algorithm relies on a modified support hence, is not really suitable to find infrequent rules except the ones that are above a threshold. Reference [10] finds independent rules measured by interest (leverage) and below a minimum support threshold. Authors in [10] also use the measure in [11], which is derived from correlation, and necessitates a minimum confidence threshold. Mining below a minimum support threshold has the same problem as mining above a maximum support threshold in the sense that the threshold needs to be accurately pre-set. In addition, the measure used in [11] inherits the drawbacks of a correlation measure in [9]. Reference [12] filters uninteresting rules using leverage as a measure. References [13, 14] finds rules using measure such as leverage or lift; these can be performed without other thresholds in place. Since rules are found independently from a minimum support threshold, theoretically all infrequent rules may be found. The measure of leverage is non directional. It does not show if a rule antecedent has an effect on the rule consequence. It only denotes the number of co-occurrences of both antecedent and consequence item set that is above the case if both are independent to each other [15, 16].

There is relatively little research on finding association rules that are both infrequent and interesting. Two fundamental constraints are (i) the selection of the measure used and (ii) the use of this measure to search for infrequent and interesting rule directly without post-processing the found rules. The measure should justify the search time in discovering rules. Such a measure must possess properties that can be used to search for infrequent association rules directly. Otherwise, the measure might be theoretically interesting but of limited practical use.

24.3 Coherent Rules Framework

The current section discusses the proposed theoretical framework for coherent rules. The salient features of the framework are, informally, (i) a novel, strong definition of association based on the notions of implications from propositional logic, (ii) the taking into account of frequency-based measures without requiring arbitrary thresholds and (iii) the use of mutually reinforcing rule pairs. These features are addressed in detail below.

In our study on the definition of an *association*, we found that it is defined in many ways which can be referred to different types of relationships among item sets. A typical definition of association is *co-occurrence* [17]. Association can also be generalized into *correlation* [9] or *dependence* rule [18]. Each definition has their merits. For the purpose of our model, we define association using implication of propositional logic in that an implication must be supported by its *inverse.*[1] Such association rules mined has implications stronger than the typical association rules based on single co-occurrences.

We study the frequency of occurrences between two item sets and rather than relying on a pre-set support threshold. To illustrate our proposed framework, consider Table 24.1 that contains relations between a rule antecedent (LHS), 'A' and a rule consequence (RHS), 'C' as an association rule. The rule antecedent 'A' consists of a combination of items, called an item set X. An item set X may exist, represented by X, or absence, represented by $\neg X$. Similarly, the rule consequence 'C' may contain existence or absence of item set Y. They are represented as Y and $\neg Y$. The frequency of occurrence of item sets X and Y is represented by $Q1$, X and $\neg Y$ by $Q2$, $\neg X$ and Y by $Q3$, finally, $\neg X$ and $\neg Y$ by $Q4$. The total of occurrence of Y is represented by $C1$, the occurrence of $\neg Y$ is given by $C2$, where $C2 = m\text{-}C1$. The same representation applied to X and $\neg X$ with the statistics $A1$ and $A2$.

Association rules:

1. $X \Rightarrow Y$ is mapped to propositional logic implication $p \rightarrow q$ if and only if $Q1 > Q2$, $Q1 > Q3$, and $Q1 > Q4$.

[1] Both inverse and contrapositive have the same number of co-occurrences in transaction records.

Table 24.1 Frequency of occurrences among item set X and Y

		A rule consequence (RHS), C		
		Y	$\neg Y$	Total
A rule antecedent (LHS), A	X	$Q1$	$Q2$	$A1$
	$\neg X$	$Q3$	$Q4$	$A2$
	Total	$C1$	$C2$	m

2. $X \Rightarrow \neg Y$ is mapped to propositional logic implication $p \rightarrow \neg q$ if and only if $Q2 > Q1$, $Q2 > Q3$, and $Q2 > Q4$.
3. $\neg X \Rightarrow Y$ is mapped to propositional logic implication $\neg p \rightarrow q$ if and only if $Q3 > Q1$, $Q3 > Q2$, and $Q3 > Q4$.
4. $\neg X \Rightarrow \neg Y$ is mapped to propositional logic implication $\neg p \rightarrow \neg q$ if and only if $Q4 > Q1$, $Q4 > Q2$, and $Q4 > Q3$.

Having mapped, each association rule (1–4) is called a *pseudo implication*. By pseudo implication, we mean that it approximates a real implication (according to propositional logic). It is not a real implication yet because there are fundamental differences – pseudo implication is judged true or false based on comparison of supports, which has a range of integer values. On the other hand, an implication is based on binary values. The former still depends on the frequencies of co-occurrences between item sets (supports) in a transaction records, whereas the latter does not and is based on truth value. We again mapped pseudo implication into specific modes of implication called equivalents. Each equivalent would follow the same truth values of the respective relations in logic. For example, in equivalents, the negation and the inverse-negation of an implication is always false. That is, to map association rules $X \Rightarrow Y$ to logic equivalent $X \equiv Y$, we need to check if the support value on its negation $X \Rightarrow \neg Y$ and inverse-negation $\neg X \Rightarrow Y$ are lower than other support values.

Coherent rules are a pair of item sets, X and Y represented using a pair of rules $X \Rightarrow Y$, $\neg X \Rightarrow \neg Y$, where following mappings and conditions must be met,

1. $X \Rightarrow Y$ is mapped to logic equivalent $p \equiv q$ if and only if, $Q1 > Q2$, $Q1 > Q3$, $Q4 > Q2$, and $Q4 > Q3$.
2. $X \Rightarrow \neg Y$ is mapped to logic equivalent $p \equiv \neg q$ if and only if, $Q2 > Q1$, $Q2 > Q4$, $Q3 > Q1$, and $Q3 > Q4$.
3. $\neg X \Rightarrow Y$ is mapped to logic equivalent $\neg p \equiv q$ if and only if, $Q2 > Q1$, $Q2 > Q4$, $Q3 > Q1$, and $Q3 > Q4$.
4. $\neg X \Rightarrow \neg Y$ is mapped to logic equivalent $\neg p \equiv \neg q$ if and only if, $Q1 > Q2$, $Q1 > Q3$, $Q4 > Q2$, and $Q4 > Q3$.

Each association rule having mapped to its corresponding logic equivalent is called pseudo implication of equivalent. Coherent rules $X \Rightarrow Y$, $\neg X \Rightarrow \neg Y$ can be decoupled into two pseudo implication of equivalents. It implies that item set X and item set Y always occurr or do not occurr together. Thus, coherent rules discover strong association rules in transaction records due to its implication of equivalent properties.

Suppose, $I = \{i_1, i_2, \ldots, i_n\}$ be a set of n items. And, $T = \{t_1, t_2, \ldots, t_m\}$ be a set of m transaction records. A task-relevant transaction record t_j holds a subset of items such that $t_j \subseteq I$. Let I_A and I_C be two sets of items, where, $I_A \subset I$, $I_C \subset I$, and $I_A \cap I_C = \emptyset$. Let P be a power-set function. We are interested in coherent rules between two item sets X and Y, where $X \in P(I_A)$, $Y \in P(I_C)$, $X \neq \emptyset$, and $Y \neq \emptyset$. We can pair any element of the power-sets I_A and I_C together to denote a possible coherent rule pair, which is termed as candidate coherent rules,

$$X..Y \tag{24.1}$$

$X..Y$ is a coherent rule pair, if it meets all the conditions of a coherent rule pair, $Q1>Q2$, $Q1>Q3$, $Q4>Q2$, and $Q4>Q3$.

24.4 Coherent Rules Measure of Interest

Coherent rules' interestingness can be denoted by the degree of usefulness in knowing item set X to predict on item set Y. We propose to augment the well-known measure of association, lambda (λ), in statistics [19, 20]. We highlight that the interest measure λ by default, is not measuring the strength of association between two item sets. We need to augment measure λ to measure the two item sets for coherent rules $X \Rightarrow Y$, $\neg X \Rightarrow \neg Y$.

Measure λ quantifies the association between two (nominal) variables, based on a concept called Proportional Reduction in Error (PRE). According to this concept, a variable is used to predict the existence of another. And, if this prediction performs better than guessing the second variable independent of the first variable, then these two variables are deemed to be related to each other. Otherwise, the second variable can be guessed without the need to know the first variable. That is, the concept compares two predictions together, between knowing an independent variable and not knowing it. Note that we use the term *variable* in the sense of, e.g. [21] where, informally, each variable contains many categories and each category corresponds to an item set. Measure λ quantifies the strength of the association between two variables into a value between zero and one and is defined as [19, 20].

In one example, the original measure λ is used to measure the interestingness between two variables A and B, where $A=\{A_1, A_2\}$ and $B=\{B_1, B_2, B_3\}$. Each of the variables contains two and three categories separately. (The variable A is specifically known as an independent variable and the variable B is a dependent variable.) We show a contingency table on the variables in Table 24.2.

The interestingness of between variable A and B is given as '0.273'. The interestingness on the categories (or item sets), for examples, $A_1 \Rightarrow B_1$ and $A_1 \Rightarrow B_2$ are not known.

Table 24.2 Contingency table of variables A and B

	A_1	A_2	Total
B_1	300	600	900
B_2	600	100	700
B_3	300	100	400
Total	1,200	800	2,000

Table 24.3 Contingency table of categories A_1 and B_1

	A_1	A_2	Total
B_1	300	600	900
B_2	600	100	700
Total	1,200	800	2,000

We use a procedure to augment the measure λ into quantifying the categories hence the analogous item sets. Instead of quantifying the variables itself, we quantify the interestingness between two specific 'categories' of a variable. And, for each category under measure, two statistics on observing a category and not observing a category are taken as the 'sub categories'. For example, based on contingency table in Table 24.2, the category B_1 under A_1 is valued at 300. The statistics that *not* category B_1 under A_1 is 1200 – 300 = 900. Hence, the sub category of B_1 takes the values 300 and 900 separately. Based on this procedure, we can construct many contingency tables for each possible categories (for example, see Table 24.3).

We can now use the measure λ to quantify between categories (e.g. A_1 and B_1).

We generalise that let I_A and I_C be two item sets such that $I_A \subset I$ and $I_C \subset I$. For each element of the power-set of item set I_A such that $X \in P(I_A)$, its binary states is given by θ such that $X = \{\theta_X, \theta_{\neg X}\}$. θ_X is the item set X observed and $\theta_{\neg X}$ is the item set X not observed. The measure λ quantifies the interestingness on two item sets X and Y asymmetrically. Hence, we can use the measure λ to quantify interestingness of $X \Rightarrow Y$, $\neg X \Rightarrow \neg Y$.

Measure λ however gives positive strength value over some candidate coherent rules $X..Y$ even they are not coherent rules. We write a measure of interest for coherent rules that only gives positive value for coherent rules and zero otherwise by conditioning the measure λ with conditions of coherent rules,

$$H(X..Y) = \begin{cases} \lambda, & Q1 > Q2, Q1 > Q3, Q4 > Q2, Q4 > Q3 \\ 0, & otherwise \end{cases} \tag{24.2}$$

where,

$$\lambda = \frac{\min(Q1 - Q3, Q2 - Q4) - \min(Q1, Q2) - \min(Q3, Q4)}{\min(Q1 + Q3, Q2 + Q4)} \tag{24.3}$$

Hence, all coherent rules have strength value of lambda. (Coherent rules have strength value above zero because measure λ gives positive values whenever subsets of the coherent rules conditions are met, (I) $Q1>Q2$, $Q4>Q3$ (II) $Q2>Q1$, $Q3>Q4$. Conditions of measure λ are shown in Appendix A of [22]).

24.4.1 Property of Coherent Rules Measure of Interest H

We highlight an important property of coherent rules measure of interest *H* that affect the discovery of coherent rules from transaction records that contain *n* number of unique items. Coherent rules measure of interest *H* originated from measure λ that produces values in ratio. Each value is an interval level of measurement with an arbitrary zero. For example, the positive value given by the measure *H*(*X*..*Y*) is a ratio over the statistics of the total number of item set *Y* or $\neg Y$. A different denominator value results in a different "arbitrary zero". Therefore, it is meaningless to compare the strength value across coherent rules that having a different item set *Y* because they are of different scale. It is necessary to fix an item set *Y* in mining for coherent rules. Consequently, coherent rules mined can be compared for their interestingness over the same item set. We explain our search strategies to discover coherent rules in the next section.

24.5 Mining Coherent Rules

This section covers an algorithm for the discovery of coherent rules. Suppose, *X*..*Y* is a candidate coherent rules pair, and *XE*..*Y* is another candidate coherent rules pair such that item set $X \subset XE$.[2] If *X*..*Y* does not met an anti-monotone constraint, then further generation of candidate coherent rules *XE*..*Y* is not necessary. As a result, we can avoid exhaustively generate all candidate coherent rules and validate them to be coherent rules. For example, we have item sets $I_A = \{a, b, c, d, e\}$, if $X = \{c, d\}$ does not meet the condition, then further generating of candidate coherent rules with $X = \{a, c, d\}$ and $X = \{b, c, d\}$ can be avoided.

The statistical conditions ($Q1 > Q3$) within property of coherent rules inherits this anti-monotone property. If *X*..*Y* does not have this condition met, then it is not coherent rules together with its *XE*..*Y*. The algorithm does not require a minimum support threshold in advance. The algorithm proceeds to systematically explore the power-set of I_A, but does not need to generate and validate the complete power-set as that would be infeasible. The feasibility of the algorithm is ensured in two ways. Firstly, if *X*..*Y* does not meet the anti-monotone properties, then *XE*..*Y* is not generated (see Lines 4.1.5–4.1.6 in Algorithm 1). Secondly, as a logical consequence, if the cardinality of the item set *X* of *X*..*Y* that does not meet the anti-monotone property consists only of a single item, then this item can be removed from item set I_A

[2]We write the union of item sets *X* and *E* ($X \cup E$) as *XE*.

(see Line 4.1.6.2). Clearly, such a removal cuts down the cardinality of the power set being explored by a factor of 2.

Algorithm discoverCR (candidateCoherentRules *R*, items I_A, itemIndex PV_{X1}, itemIndex PV_{X2}, itemIndex PV_Y, itemIndex PV_{Max}, subItems *T*, orderedSet<index> *RA*, RuleSet *CR*)
//Initial//

1. If $PV_{X1} > 1$
   1.1 $PV_{X2} := PV_{X1}$, $PV_{X1} := 1$
2. Else
   2.1 $PV_{X2} := PV_{Max}$
3. End if

//Generating candidate coherent rules by enumerating item set X//

4. While ($PV_{X1} < PV_{X2}$)
   4.1 If (PV_{X1} != PV_Y)
      4.1.1 $RA \leftarrow$ concatenate(PV_{X1}, RA)
      4.1.2 X $\leftarrow \{i_L : L \in RA\}$
      4.1.3 Let *R* be the set of candidate coherent rules corresponding to (*X*, *Y*) such that $R=(X \Rightarrow Y, \neg X \Rightarrow \neg Y)$

      //START of Anti-monotone Check for Efficient Generations//

      4.1.4 Compute *Q1, Q2, Q3, Q4* based on scans into *T*
      4.1.5 If (*Q1*>*Q3*)
         4.1.5.1 $\forall r \in R$ compute H_r and store it
            4.1.5.1.1 If (H_r >0)
               4.1.5.1.1.1 CR = CR $\cup$ R
            4.1.5.1.2 End
      4.1.6 Else
         4.1.6.1 *itemToRemove* = {*X*: *X* is the item set of some $r \in$ R and = |X| = 1}
         4.1.6.2 $I_A = I_A$ - *itemToRemove*
      4.1.7 End

      //End of Conditions for Efficient Generations//

      4.1.8 If (PV_{X1}>1)
         4.1.8.1 (*R*, I_A, PV_{X1}, PV_{X2}, PV_Y, PV_{Max}, *RA*) = discoverCR(*R*, I_A, PV_{X1}, PV_{X2}, PV_Y, PV_{Max}, *RA*)
      4.1.9 End
      4.1.10 *RA*$\leftarrow$ (*RA* - PV_{X1}) //remove an item from the buffer of item set *X*//
   4.2 End
   4.3 PV_{X1}:= PV_{X1} + 1 //increase the first pointer value//
5. End

Algorithm 1 Discover Coherent Rules

24.6 Experiments and Discussions

In this section, we report and analyse the experimental results from two experiments. In the first, we show that our association rule mining framework can find infrequent association rules that may be difficult to find in traditional association rule mining. The zoo data set [23] is used in this experiment. The second experiment shows that our proposed framework finds smaller number of rules compared to the traditional association rule mining algorithm that finds too many uninteresting rules. The experiment for this purpose is conducted in the mushroom data set [24]. In both zoo and mushroom dataset, we use the *classes* as the consequences in order to find strong association rules from data.

24.6.1 Zoo Dataset

Zoo dataset is a collection of animal characteristics and their categories in a zoo. This dataset is chosen because animal characteristics in each category are very well known. As a result, it is easier to verify the correctness and interestingness of rules mined. Zoo dataset contains seven categories of animals including *mammal* and *amphibian*. While mammals such as elephants, buffalos, and goats are frequently observed in this zoo, amphibians such as frogs and toads are relatively rare. We run our algorithm 1 without setting a minimum support threshold to obtain all coherent rules and report only the most interesting coherent rules and each rule contains not more than four items. We report the results in Table 24.4.

A total of 11 rules are found on mammals. All rules have strength of 1.0 out of 1.0. For example, all mammals has no feathers but milk and backbones therefore feathers (0), milk (1), and backbones (1) are reported associated with mammals (1).

We found these rules describe mammals correctly. In fact, the first and the shortest rule, milk ⇒ mammals(1) describes a fundamental characteristic of mammals explicitly. From literature review, the second rule may be deemed redundant in comparison with the first rule because inclusion of an additional item set, feather(0), which cannot further increase the strength of rule. The strength of the first rule is

Table 24.4 Rules describing *mammals*

No.	Antecedent item set		Consequence item set
1	milk(1)	⇒	mammals(1)
2	feathers(0),milk(1)	⇒	mammals(1)
3	milk(1),backbones(1)	⇒	mammals(1)
4	feathers(0),milk(1),backbones(1)	⇒	mammals(1)
5	milk(1),breathes(1)	⇒	mammals(1)
6	feathers(0),milk(1),breathes(1)	⇒	mammals(1)
7	milk(1),backbones(1),breathes(1)	⇒	mammals(1)
8	milk(1),venomous(0)	⇒	mammals(1)
9	feathers(0),milk(1),venomous(0)	⇒	mammals(1)
10	milk(1),backbones(1),venomous(0)	⇒	mammals(1)
11	milk(1),breathes(1),venomous(0)	⇒	mammals(1)

already at its maximum at 1.0; any further inclusion of items may be redundant. Such a consideration however is application dependent. We could use both items, feathers(0) and milk(1) to describe mammals more comprehensively at the same strength of 1.0. That is, mammals do not have feathers but milk. If we discard feathers(0), we loss this item as a descriptive.

We run the search for amphibians, and found a total of 136 rules. Again, we could not find any incorrect rules. These rules have strength 1.0. While studying at these rules, we are surprised by the fact that amphibians like frogs have teeth! We confirm this via answer.com, and this is indeed correct.

Comparing the two experiments, there is a large difference in their total number of occurrence in transaction records. 41% of transaction records contain mammals, in comparison, only 4% of transaction records contains amphibians. That is, the search for amphibians is a search for infrequent association rules, which is often missed by most association rule mining techniques that demand a minimum support threshold. If we set minimum support threshold to be higher than 4% and use a typical association rule mining technique, we lose rules describing amphibians. In comparison, our technique does not necessitate a minimum support threshold, it finds all necessary rules.

24.6.2 Mushroom Dataset

In our next experiment, we run our search algorithm on mushroom dataset, which contains 8,124 transactions and 119 items. We list these rules in Table 24.5(a) and (b).

Table 24.5 (a) Rules describing *edible* mushroom (b) Rules describing *poisonous* mushroom

(a) No.	Antecedent Item Set		Consequence Item Set
1	odor.almond	⇒	*Edible*
2	odor.almond, stalk-color-below-ring.orange	⇒	*Edible*
(b) No.	Antecedent Item Set		Consequence Item Set
1	cap-color.green, odor.spicy, gill-attachment.free	⇒	*Poisonous*
2	cap-color.green, odor.spicy, gill-attachment.free, stalk-color-below-ring.orange	⇒	*Poisonous*
3	cap-color.green, gill-attachment.free, stalk-color-below-ring.cinnamon	⇒	*Poisonous*
4	cap-color.green, gill-attachment.free, stalk-color-below-ring.orange, stalk-color-below-ring.cinnamon	⇒	*Poisonous*

We leave the correctness of these results to domain experts since we are not experts. The strengths of these rules are around 0.77 out of 1.0, this suggests that there may exist some exceptional cases besides these strongest rules.

In comparison, a typical association rule mining technique such as *a priori* reports more than 100 thousands of rules with confidence value at 100%. Some of these rules are not interesting, and one way to filter these are to select high confidence rules with positive leverage values. Rules with positive leverage are rules that are dependent on each other. However, after filtering high confident rules with positive leverage, we are still left with more than 100 thousands of rules for this dataset. These rules include our six rules. We conclude from these observations that our approach produces rules that are concise and easier to apply.

Based on both the experimental results, we conclude that coherent rules are useful to discover knowledge of associations (both frequent and infrequent) from dataset.

24.7 Conclusion

We have presented a framework to mine association rules without minimum support threshold. The framework employs a novel, strong definition of association based on logical equivalence from propositional logic to avoid using a cut-off support threshold. The experimental results show that implication of propositional logic is a good alternative for the definition of association.

The stronger definition of association also results in the discovery of knowledge that is vital from transaction records represented by coherent rules. These are a pair of rules that can be mapped to a pair of logical equivalence of the propositional logic, which means that the rules reinforce each other. While coherent rules found are important, the interestingness of these rule pairs is further quantified using coherent rules measure of interest H. All coherent rules have positive values for its interestingness and imply that the item set X of a coherent rule pair is useful to predict its consequence item set Y, and is better than a guess without the former.

Rules based on this definition may be searched and discovered within feasible time. This can be done by our proposed strategy to exploit the anti-monotone property inherit in the statistical conditions ($Q1 > Q3$) within property of coherent rules. The experimental results show both infrequent and frequent association rules are discovered. They are also smaller in numbers thus easier to apply.

References

1. X.-H. Huynh, F. Guillet, and H. Briand, "Evaluating Interestingness Measures with Linear Correlation Graph," in *Advances in Applied Artificial Intelligence*, vol. 4031: Springer Berlin / Heidelberg, 2006, pp. 312–321.
2. B. Liu, W. Hsu, and Y. Ma, "Mining association rules with multiple minimum supports," pp. 337–341, 1999.
3. E. Cohen, M. Datar, S. Fujiwara, A. Gionis, P. Indyk, R. Motwani, J.D. Ullman, and C. Yang, "Finding Interesting Associations without Support Pruning," 2000, pp. 489–500.

4. Y. Huang, H. Xiong, S. Shekhar, and J. Pei, "Mining confident co-location rules without a support threshold," 2003, pp. 497–501.
5. Y.S. Koh and N. Rountree, "Finding Sporadic Rules Using Apriori-Inverse," in *Lecture Notes in Computer Science: Advances in Knowledge Discovery and Data Mining*, vol. 3518: Springer Berlin / Heidelberg, 2005, pp. 97–106.
6. Y.S. Koh, N. Rountree, and R. O'Keefe, "Mining Interesting Imperfectly Sporadic Rules," in *Advances in Knowledge Discovery and Data Mining*, vol. 3918: Springer Berlin / Heidelberg, 2006, pp. 473–482.
7. R. Castelo, A. Feelders, and A. Siebes, "MAMBO: Discovering Association Rules Based on Conditional Independencies," in *Advances in Intelligent Data Analysis: 4th International Conference Proceedings*, vol. 2189: Springer Berlin / Heidelberg, 2001, pp. 289–298.
8. W.-Y. Lin, M.-C. Tseng, and J.-H. Su, "A Confidence-Lift Support Specification for Interesting Associations Mining," 2002, pp. 148–158.
9. S. Brin, R. Motwani, and C. Silverstein, "Beyond market baskets: generalizing association rules to correlations," 1997, vol. 26, pp. 265–276.
10. L. Zhou and S. Yau, "Efficient association rule mining among both frequent and infrequent items," *Comput. Math. Appl.*, vol. 54, pp. 737–749, 2007.
11. X. Wu, C. Zhang, and S. Zhang, "Mining Both Positive and Negative Association Rules," pp. 658–665, 2002.
12. G. Piatetsky-Shapiro, "Discovery, analysis, and presentation of strong rules," 1991, pp. 229–248.
13. S. Zhang and G.I. Webb, "Further Pruning for Efficient Association Rule Discovery," *Lecture Notes in Computer Science*, vol. 2256, pp. 205–216, 2001.
14. G.I. Webb and S. Zhang, "k-Optimal Rule Discovery," *Data Mining and Knowledge Discovery*, vol. 10(1), pp. 39–79, 2005.
15. N. Rescher, *Conditionals*. Cambridge, Massachusetts: The MIT Press, 2007.
16. C. Cornelis, P. Yan, X. Zhang, and G. Chen, "Mining Positive and Negative Association Rules from Large Databases," Bangkok, Thailand, June 2006, pp. 1–6.
17. R. Agrawal, T. Imielinski, and A. Swami, "Mining association rules between sets of items in large databases," *SIGMOD Rec.*, vol. 22, pp. 207–216, 1993.
18. C. Silverstein, S. Brin, and R. Motwani, "Beyond Market Baskets: Generalizing Association Rules to Dependence Rules," *Data Min. Knowl. Discov.*, vol. 2, pp. 39–68, 1998.
19. F.J. Kviz, "Interpretating Proportional Reduction in Error Measures as Percentage of Variation Explained," *The Sociological Quarterly*, vol. 22, pp. 413–420, 1981.
20. J.A. Rafter, M.L. Abell, and J.P. Braselton, *Statistics with Maple*. Amsterdam, Boston: Academic Press, 2003.
21. J. Hubert M. Blalock, *Social Statistics*. New York: McGraw-Hill Book Company, 1972.
22. A.T.H. Sim, M. Indrawan, and B. Srinivasan, "Mining Infrequent and Interesting Rules from Transaction Records," in *Advances on Artificial Intelligence, Knowledge Engineering, and Data Bases (AIKED 2008)*, Cambridge, UK, 2008, pp. 515–520.
23. "Zoo dataset," 2007. Available: http://magix.fri.uni-lj.si/orange/doc/datasets/zoo.htm.
24. "Frequent Itemset Mining Dataset Repository." Available: http://fimi.cs.helsinki.fi/data/.

Chapter 25
Particle Swarm Optimization with Diversive Curiosity and Its Identification

Hong Zhang and Masumi Ishikawa

Abstract Particle Swarm Optimization (PSO) is a stochastic and population-based adaptive optimization algorithm. Although the optimized PSO models have good search performance with moderate computational cost and accuracy, they still tend to be trapped in local minima (premature convergence) in solving multimodal optimization problems. To overcome this difficulty, we propose a new method, Particle Swarm Optimization with Diversive Curiosity (PSO/DC). A key idea of the proposed method is to introduce a mechanism of diversive curiosity into PSO for preventing premature convergence, and for managing the exploration-exploitation trade-off. Diversive curiosity is represented by an internal indicator that detects marginal improvement of a swarm of particles, and forces them to continually exploring an optimal solution to a given optimization problem. Owing to the internal indicator representing the mechanism of diversity curiosity, PSO/DC can successfully prevent premature convergence, and manage the exploration-exploitation trade-off. Empirically, PSO/DC is very effective in enhancing the search performance of PSO.

Keywords Fault tolerance · WDM-EPON · Differentiated services · PFEBR

25.1 Introduction

Particle Swarm Optimization (PSO) is a stochastic and population-based adaptive optimization algorithm proposed by Kennedy and Eberhart motivated by social behavior of animals [5, 13]. In recent years, this technique has been widely applied to various disciplines in science and engineering such as applications to large-scale, highly nonlinear, and multimodal optimization problems [6, 9, 17, 20, 21].

Similar to other search methods such as Reinforcement Learning (RL) [23, 24] and Genetic Algorithms (GAs) [8, 10, 16, 27], a trade-off between exploration and exploitation during PSO runs for efficiently solving various optimization problems

H. Zhang (✉)
Kyushu Institute of Technology, 2-4 Hibikino, Wakamatsu, Kitakyushu 808-0196, Japan
e-mail: zhang@brain.kyutech.ac.jp

S.-I. Ao et al. (eds.), *Trends in Communication Technologies and Engineering Science*, Lecture Notes in Electrical Engineering 33, DOI 10.1007/978-1-4020-9532-0_25,

is an important issue [3]. An appropriate trade-off can activate swarm of particles in search to avoid premature convergence and to increase accuracy and efficiency for finding an optimal solution to a given optimization problem [14, 31].

Considerable attention has been paid to this issue, and a number of algorithms have been proposed such as non-global best neighborhoods for promoting exploration and local search for promoting exploitation [19, 22, 26]. Although these endeavors are effective in solving multimodal optimization problems, they suffer from heavy computational cost in managing the exploration-exploitation trade-off.

For obtaining still better search performance in PSO, we proposed Evolutionary Particle Swarm Optimization (EPSO), which used Real-coded Genetic Algorithm (RGA) to optimize PSO models with on-line evolutionary computation [29, 32]. Since a temporally cumulative fitness function is performed for effectually evaluating the performance of PSO, it is expected to suppress stochastic disturbance in dynamic evaluation, and EPSO greatly contributes to model selection in PSO without prior knowledge. Our experiments in a solving 2-dimensional multimodal optimization problem demonstrated that the search performance by PSO models optimized by EPSO is superior to that by the original PSO [29, 32].

Although the optimized PSO models have good search performance with moderate computational cost and accuracy, they still tend to be trapped in local minima (premature convergence) in solving multimodal optimization problems. This is because that the performance and efficiency of PSO heavily depends on the trade-off between exploration and exploitation.

To overcome this difficulty, we propose a new method, Particle Swarm Optimization with Diversive Curiosity (PSO/DC) [31]. Diversive curiosity here is a concept in psychology: tendency of seeking stimulus/sensation in humans and animals. A key idea of the proposed method is to introduce a mechanism of diversive curiosity into PSO for preventing premature convergence, and for managing the exploration-exploitation trade-off. Diversive curiosity is represented by an internal indicator that detects marginal improvement of a swarm of particles, and forces them to continually exploring an optimal solution to a given optimization problem.

It is obvious that the concept of diversive curiosity introduced in PSO/DC is different from that in EPSO in managing the exploration-exploitation trade-off. The internal indicator in PSO/DC is simple with no extra computational cost. It has only two adjustable parameters, i.e., duration of judgment and a tolerance parameter. The former is used for memorizing the change of activity of a swarm of particles, and the latter is used for detecting premature convergence during search.

Similar to the internal indicator in PSO/DC, Adaptive Particle Swarm Optimization (APSO) was proposed [11]. In APSO, a fixed-$gBest$-value method was used for detecting dynamic changes of environment, which monitors the changes of the $gBest$ value and the second-best $gBest$ value during 20 iterations. Since the method checks whether the changes happen for the fixed duration or not, it is not only different from the formulation of the internal indicator mentioned in Section 25.3.2, but also it does not represent boredom for the behavior of particles during search. A characteristic common to APSO and PSO/DC is randomization of the entire swarm of particles for ensuring exploration.

Psychology asserts that diversive curiosity leads to exploration, but also creates anxiety [15]. Needless to say, anxiety affects the efficiency of exploration. It is necessary to verify that the internal indicator in PSO/DC really represents diversive curiosity. It is also necessary to provide a method for parameter selection in the internal indicator. In addition to these, the effectiveness of PSO/DC in swarm intelligence should be demonstrated.

In computer experiments, we use a 2-dimensional multimodal optimization problem. For practical efficacy of PSO/DC, we propose to estimate the appropriate range for the parameter values in the internal indicator. We also investigate the characteristics of PSO/DC, and the trade-off between exploration and exploitation in PSO/DC.

The rest of the paper is organized as follows. Section 25.2 briefly addresses an algorithm of EPSO, and two temporally cumulative fitness functions applied to evaluation of the performance of PSO. Section 25.3 describes a concept of curiosity, an internal indicator for diversive curiosity, and an algorithm of PSO/DC. Section 25.4 discusses the experimental results of computer experiments, analyzes the characteristics of PSO/DC, and compares the search performance with those by other methods such as the original PSO, EPSO and RGA/E. Finally, Section 25.5 gives conclusions.

25.2 Overview of EPSO

25.2.1 Basic EPSO

The PSO is formulated by particles with position and velocity as follows.

$$\boldsymbol{x}^i_{k+1} = \boldsymbol{x}^i_k + \boldsymbol{v}^i_{k+1} \tag{25.1a}$$

$$\boldsymbol{v}^i_{k+1} = c_0 \boldsymbol{v}^i_k + c_1 \boldsymbol{r}_1 \otimes (\boldsymbol{x}^i_l - \boldsymbol{x}^i_k) + c_2 \boldsymbol{r}_2 \otimes (\boldsymbol{x}_g - \boldsymbol{x}^i_k) \tag{25.1b}$$

where c_0 is an inertial factor, c_1 is an individual confidence factor, c_2 is a swarm confidence factor, $\boldsymbol{r}_1, \boldsymbol{r}_2 \in \Re^n$ are random vectors each component of which is uniformly distributed on [0,1], and $\otimes$ is an element-wise operator for vector multiplication. $\boldsymbol{x}^i_l (= arg \max_{k=1,2,\cdots} \{g(\boldsymbol{x}^i_k)\}$, where $g(\boldsymbol{x}^i_k)$ is the fitness value of the ith particle at time k), is the local best position of the ith particle, `lbest`, up to now, and $\boldsymbol{x}_g (= arg \max_{i=1,2,\cdots} \{g(\boldsymbol{x}^i_l)\})$ is the global best position of the swarm of particles, `gbest`, respectively.

EPSO is an algorithm with online evolutionary computation, which provides a new paradigm for meta-optimization in model selection [29, 32]. Figure 25.1 illustrates the flowchart of EPSO.

The procedure of EPSO is composed of two parts. One is an outer loop in which Real-coded Genetic Algorithm with Elitism strategy (RGA/E) [28] is applied to

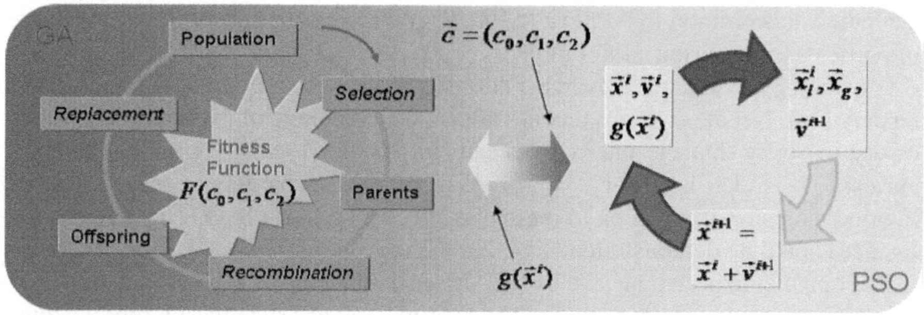

Fig. 25.1 The flowchart of EPSO

solving real-valued optimization problems. While the PSO finds an optimal solution to a given optimization problem, RGA/E is used to optimize the values of parameters in the PSO. The other is an inner loop in which PSO runs. The PSO with the values of parameters created by RGA/E is expected to achieve higher fitness than the original PSO.

As the genetic operations of RGA/E, specifically, roulette wheel selection, BLX-α crossover [7], random mutation, and rank algorithm are used for optimization.

25.2.2 Fitness Functions

To effectually obtain PSO models with better search performance, we use the following fitness functions in model selection [30, 31]. The first one is a temporally cumulative fitness function of the best particle,

$$F_1(c_0, c_1, c_2) = \sum_{k=1}^{K} g(\boldsymbol{x}_k^b) \,|_{c_0, c_1, c_2} \tag{25.2}$$

where $x_k^b(= arg\,max_{i=1}^P\{g(\boldsymbol{x}_k^i)\}$, P: the number of particles) is the position of the best particle at time k, and K is the maximum number of iterations.

The second one is a temporally cumulative fitness function of the entire swarm, which is expressed by

$$F_2(c_0, c_1, c_2) = \sum_{k=1}^{K} \bar{g}_k \,|_{c_0, c_1, c_2} \tag{25.3}$$

where, $\bar{g}_k = \sum_{i=1}^{P} g(\boldsymbol{x}_k^i)/P$ is the average of fitness values over the entire swarm at time k.

It is obvious that the fitness functions, F_1 and F_2, stress distinctive character of same swarm of particles in search, respectively. For understanding the relationship

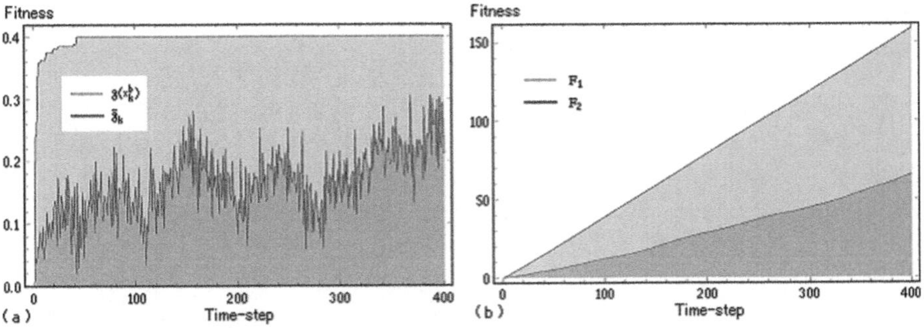

Fig. 25.2 Comparison of two fitness functions. (**a**) $g(\boldsymbol{x}_k^b)$ and $\bar{g}_k$, (**b**) F_1 and F_2

between them, Fig. 25.2 illustrates instantaneous fitness functions, $g(\boldsymbol{x}_k^b)$ and $\bar{g}_k$, and the corresponding cumulative fitness functions, F_1 and F_2. It is to be noted that F_1 and F_2 are approximately straight except near the origin. This suggests that both fitness functions, F_1 and F_2, are suitable for evaluating the performance of PSO.

Since the cumulative fitness functions, F_1 or F_2, are the sum of instantaneous fitness functions, $g(x_k^b)$ or $\bar{g}_k$, over time, their variance is inversely proportional to the interval of summation. Accordingly, both fitness functions can suppress stochastic perturbation in dynamic evaluation, and effectually estimate PSO models with better search performance.

25.2.3 Convergence Speed

To evaluate the search ability of the entire swarm of particles for finding an optimal solution or near-optimal solutions, we define the maximum time-step, k_{max}, as an indicator for convergence speed,

$$\forall k \geq k_{max}, \quad g(\boldsymbol{x}_k^b) - \bar{g}_k \leq \tau, \tag{25.4}$$

where τ is a positive tolerance parameter.

The smaller the maximum time-step, k_{max}, is, the faster the convergence speed of the swarm of particles is.

25.3 PSO/DC

25.3.1 Curiosity

Curiosity is a concept in psychology representing instinct for seeking stimulus/sensation in humans and animals. Berlyne categorized it as *diversive curiosity* and *specific curiosity* [1]. Diversive curiosity signifies instinct to seek novelty, to

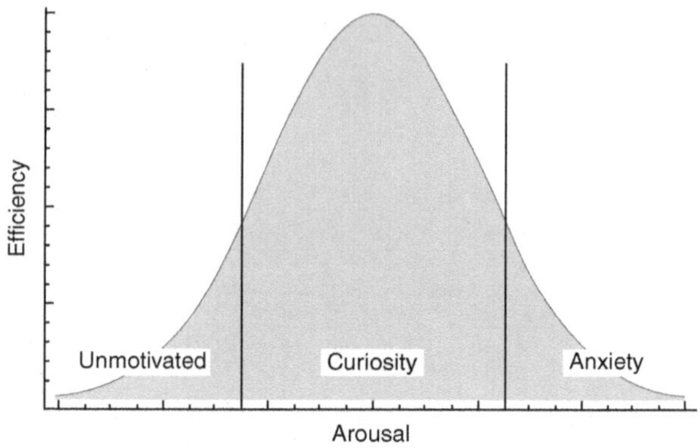

Fig. 25.3 The zone of curiosity

take risks, and to search for adventure. Specific curiosity signifies instinct to investigate a specific object for its full understanding.

According to Berlyne and his colleague Day [4], the diversive curiosity is aroused by external stimuli with complexity, novelty, uncertainty and conflict. The level of stimulation plays an essential role. If it is too low, it causes relaxation; If it is too high, it causes anxiety; If it is moderate, it causes excitement. Figure 25.3 illustrates a hypothesis of "zone of curiosity" in psychology representing intensive exploration. A central issue here is how to realize it by an engineering technique [12, 18, 25].

25.3.2 Internal Indicator

Loewenstein pointed out that "diversive curiosity occupies a critical position at the crossroad of cognition and motivation" [15]. We consider that "cognition" is the action of exploitation, and "motivation" is the intention for exploration. According to this, we will clarify the relation between them.

We define the following internal indicator for monitoring activity of a swarm of particles in search [31].

$$y_k(L, \varepsilon) = max\left(\varepsilon - \sum_{l=1}^{L} \frac{|g(\boldsymbol{x}_k^b) - g(\boldsymbol{x}_{k-l}^b)|}{L}, 0\right) \tag{25.5}$$

where L is the duration of judgment, and ε is a positive tolerance parameter for premature convergence and boredom.

Equation (25.5) indicates the state change of the best particle among the swarm in search process, i.e., when the output value of the internal indicator, y_k, is zero, the fitness value for $\boldsymbol{x}_k^b$ is still significantly changing, and when the output value of the internal indicator, y_k, exceeds zero, the fitness value for $\boldsymbol{x}_k^b$ is not changing significantly. The former means that the swarm of particles is exploring (cognition). The later means that the swarm of particles has lost interest, i.e., boredom, in $\boldsymbol{x}_k^b$.

Equation (25.5) can be interpreted as a mechanism of diversive curiosity by an engineering technique.

When y_k becomes positive, the internal indicator sends information to all particles to reinitialize their locations and velocities for exploring other solutions in search space. The action of continually exploring in search space can be interpreted as a mechanism of diversive curiosity.

In summary, we have clarified the relation between the internal indicator and the diversity curiosity, that between the internal indicator and exploration, and that between the diversive curiosity and exploration. According to this, we may say that Equation (25.5) plays a role of diversive curiosity.

25.3.3 Procedure of PSO/DC

The internal indicator detects whether the swarm of particles continues to change or not, and constantly makes them active to explore an optimal solution in search space.

The procedure of PSO/DC is implemented as follows.

```
Begin
 Set the number of maximum search, K;
 Set k=0, d=-1; Set gbest set to empty;
 While k<=K Do
   If k=0 or d=1 Then initialize swarm;
   Else
     For each particle
       Calculate position and velocity;
     End For
     Update each local best particle; Update global best
        particle;
   End If
   Calculate the value, y_k, of internal indicator;
   If y_k<=0 Then d=y_k;
   Else d=1 Do add global best to gbest set;
   End If
   k=k+1;
 End While
 Select the best result from the gbest set;
End
```

It is to be noted that the optimized PSO models by EPSO described in Section 25.2 have better search performance. Owing to the combination of EPSO and diversive curiosity, PSO/DC attains a good trade-off between exploration and exploitation for efficiently solving a given optimization problem.

25.4 Computer Experiments

25.4.1 Experimental Conditions

Computer experiments are carried out for investigating the characteristics and the search performance of PSO/DC for solving the given 2-dimensional multimodal optimization problem in Fig. 25.4. The search space is limited to 60×60, and the fitness value of the optimal solution, $g(\boldsymbol{x}_g)$, is about 0.4. Table 25.1 gives the major parameters in EPSO for estimating appropriate values of parameters in PSO.

25.4.2 Results of EPSO

All experiments were carried out with 20 trials. Table 25.2 shows the resulting parameter values of PSO models optimized by EPSO with two fitness functions.[1] Comparing with these results, we observed that the EPSO generates 3 types or 4 types of PSO models which all can solve the given problem well, regardless of the appearance frequency within top 20 models. These results indicate that although the values of the inertial factor, c_0, and the individual confidence factor, c_1, could be zero, the value of the swarm confidence factor, c_2, is always non-zero in these models.

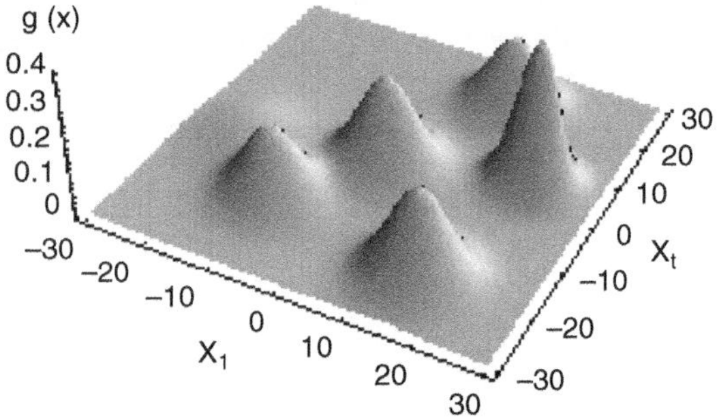

Fig. 25.4 An optimization problem

Table 25.1 The major parameters in EPSO

Parameters	Value	Parameters	Value
The number of individuals, M	100	The number of superior individuals,	
The number of generation, G	20	s_n	1
The number of iterations, K	400	Roulette wheel selection	–
The number of particle, P	10	Probability of random mutation, p_m	1.0
The maximum velocity, v_{max}	30	Probability of BLX-2.0 crossover, p_c	1.0

[1] Computing environment: Intel(R) Xeon(TM); CPU 3.40 GHz; Memory 2.00 GB RAM; Computing tool: Mathematica 5.2; Computing time: about 3 min.

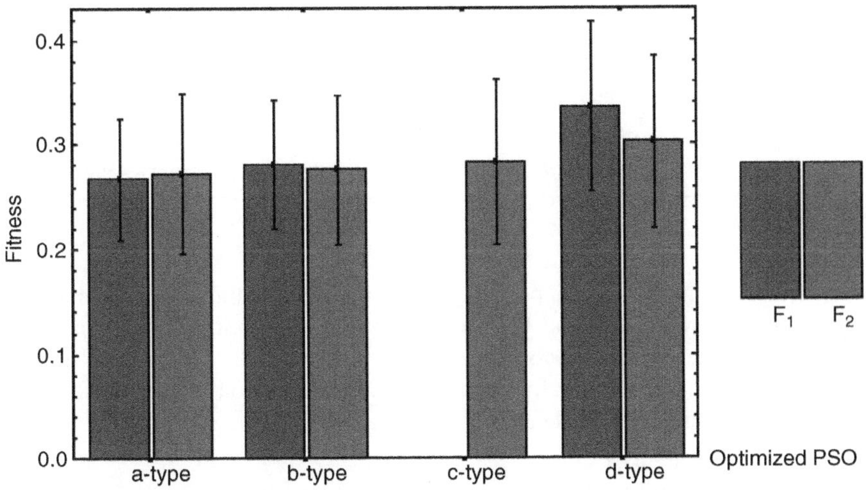

Fig. 25.5 The mean and standard deviation of fitness values for each optimized PSO model

Figure 25.5 gives the mean and the standard deviation regarding the fitness values of the GBEST for each model in Table 25.2. By comparing with the average of fitness values for each optimized PSO model, we observed that the optimized PSO model in d-type has better search performance than that by other ones.

Table 25.3 gives the performance index of the convergence speed, k_{max}, for representing the search activity of the entire swarm corresponding to each model in Table 25.2.

We observed that the maximum time-step, k_{max}, created by the fitness function F_2 is smaller than that by the fitness function F_1. This means that the resulting PSO models generated by the fitness function F_2 converges on a solution faster than that by the fitness function F_1. The ratio of the maximum time-step by the fitness function F_2 to that by the fitness function F_1 is about 2.2 $\sim$ 29.0%; in the

Table 25.2 Estimated parameter values in PSO and the appearance frequency of the resulting models (within top 20 models). PSO model in a-type: $c_0 = 0$, $c_1 = 0$, $c_2 \neq 0$; b-type: $c_0 = 0$, $c_1 \neq 0$, $c_2 \neq 0$; c-type: $c_0 \neq 0$, $c_1 = 0$, $c_2 \neq 0$; d-type: $c_0 \neq 0$, $c_1 \neq 0$, $c_2 \neq 0$

Fitness function	Optimized PSO	Parameter c_0	c_1	c_2	Appearance frequency (%)
F_1	a-type	0	0	3.26 ± 1.35	45
	b-type	0	1.26 ± 0.90	3.33 ± 1.07	30
	c-type	–	–	–	0
	d-type	0.70 ± 0.30	0.64 ± 0.36	2.86 ± 1.84	25
F_2	a-type	0	0	2.00 ± 0.52	40
	b-type	0	0.37 ± 0.39	2.00 ± 0.07	30
	c-type	0.15 ± 0.00	0	1.34 ± 0.25	20
	d-type	0.16 ± 0.01	0.75 ± 0.32	1.19 ± 1.16	10

Table 25.3 The maximum time-step, k_{max}, for each optimized PSO model ($\tau = 0.03$)

Fitness	Optimized PSO			
Function	a-type	b-type	c-type	d-type
F_1	21.2 ± 10.2	$325. \pm 69.2$	–	81.5 ± 30.3
F_2	6.15 ± 1.46	7.30 ± 2.84	5.65 ± 0.58	6.05 ± 0.94

experiments. In the meantime, because it is just easy to be trapped in local minima, the search performance of the PSO models optimized by the fitness function F_2 is inferior to that by the fitness function F_1. These results indicate the different characteristics of two temporally cumulative fitness functions in optimization, and give the hint to create PSO models for designer.

25.4.3 Results of PSO/DC

According to the above results of EPSO, the optimized PSO model in d-type is adopted in PSO/DC for obtaining superior search performance to efficiently solve the given optimization problem.

For estimating appropriate range for the parameter values in the internal indicator, and investigating the search performance of PSO/DC, we change the values of parameters in the indicator, i.e., tolerance parameter, $\varepsilon = 10^{-6}, 10^{-5}, 10^{-4}, 10^{-3}, 10^{-2}$ and duration of judgment, $L = 10, 20, 30, 40, 50, 60, 70, 80, 90$, in the next experiments.

Figure 25.6 indicates the experimental results of PSO/DC for each case. By comparing with the results, the characteristics of PSO/DC can be interpreted as follows.

1. According to Fig. 25.6(a) and (b), when L is longer, particles are hard to be excited (unmotivated) and when L is shorter, adversely they are easy to be excited (anxiety) in the cases of the tolerance parameter $\varepsilon < 10^{-4}$.
2. According to Fig. 25.6(c), with the increment of the duration of judgment, L, the number of reinitializations decrease nonlinearly.
3. By the situation of the fitness values of the best particle are kept to 0.4, the recommended range of the interval of judgment, L, is $30 \sim 50$.
4. By the situation of the ratio of success in each case can be kept to high level ($98 \sim 100\%$), the recommended range of the tolerance parameter, ε, is $10^{-3} \sim 10^{-2}$.
5. The average of fitness values or the ratio of success becomes small, when L or ε is outside of the recommended interval.

These results suggest that the effectiveness of the internal indicator representing the mechanism of diversive curiosity is certificated, even if the obtained experimental data are not so perfect when L is shorter and ε is bigger.

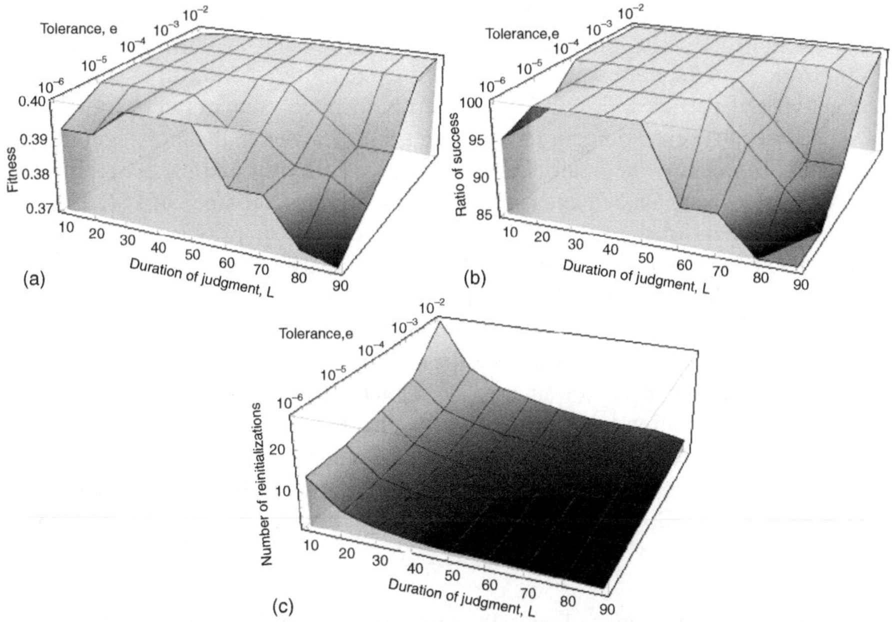

Fig. 25.6 The performance indices of PSO/DC with two adjustable parameters. (**a**) The average of fitness values, (**b**) The ratio of success, (**c**) The number of re-initializations.

25.4.4 Feature of EPSO

For understanding the distinctive feature of PSO/DC in search, Figure 25.7 illustrates a variation of the instantaneous fitness values of the best particle and the entire particles in the limited search period.

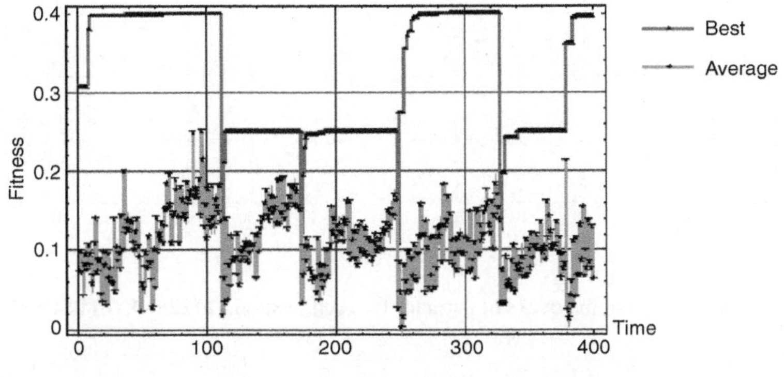

Fig. 25.7 A variation of the fitness values of the best particle and the average of fitness values of the entire particles during search

According to the change status of both fitness values in Fig. 25.7, we observed that the entire swarm has high diversity during search and the number of reinitializations in the search period is four. Since the internal indicator surely detects the activity of particles, PSO/DC successfully avoided premature convergence for several times to escape boredom in this case. It indicates that the increment of reinitialization frequency can greatly improve the possibility for finding an optimal solution in search space, and ensures the superior search performance of PSO/DC.

The increment of the reinitialization frequency also contributes to exploration. For the sake of visual effect, the whole plots of distribution density for the particles corresponding to EPSO and PSO/DC are shown in Fig. 25.8. It is clear that the distribution region of the tracks of particles by PSO/DC in Fig. 25.8(b) is bigger than that by EPSO in Fig. 25.8(a), i.e., these particles are not only around few solutions, but also distributed over other space. This indicates that PSO/DC manages the exploration-exploitation trade-off well, and reflects the contribution of the internal indicator representing diversive curiosity for efficiently solving the given problem.

Figure 25.9 gives the resulting search performance for the original PSO, EPSO, PSO/DC♯,[2] PSO/DC, RGA/E. By comparison with these results, we can confirm the following characteristics.

- The search performance of each PSO method is superior to that by RGA/E. The search performance of PSO/DC (PSO/DC♯) is superior to that by EPSO (the original PSO).
- The search performance of EPSO (the original PSO) is vastly improved by the mechanism of diversive curiosity.

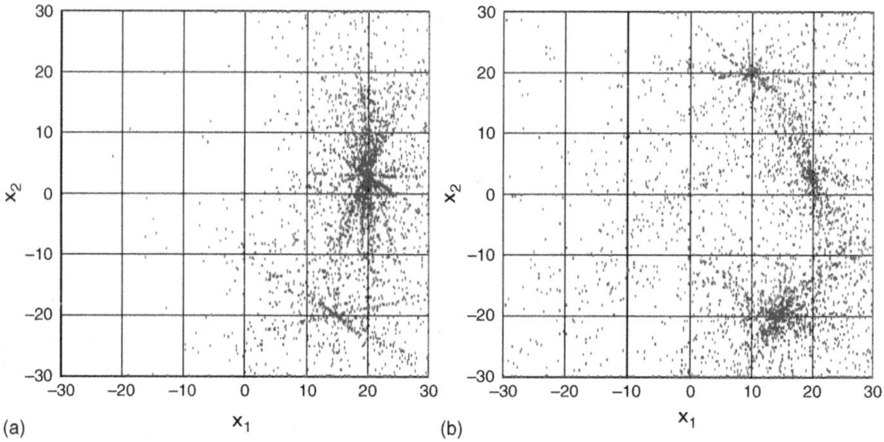

Fig. 25.8 Distribution of the tracks of particles for each method. (**a**) EPSO, (**b**) PSO/DC

[2]It stands for the parameter values of the original PSO is used in PSO/DC.

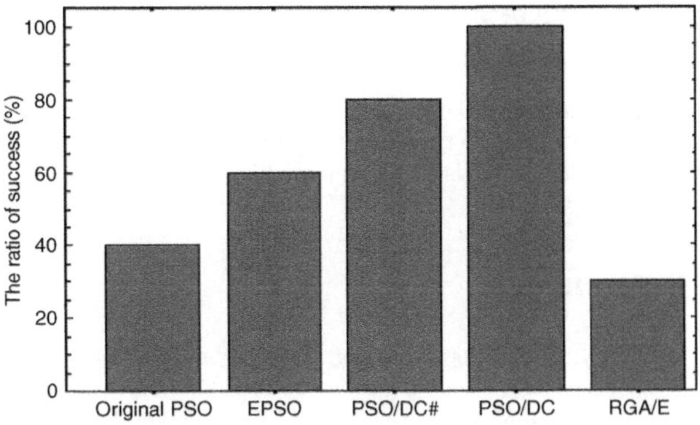

Fig. 25.9 The ratio of success for each method with 20 trials

- The search performance of PSO/DC is inferior to that by PSO/DC, but is superior to that by EPSO.

We can see that the ratio of success finding the optimal solution, $\boldsymbol{x}_g$, corresponding to the given problem vastly improved from 60 to 100%. The result sufficiently indicates that the internal indicator representing the mechanism of diversive curiosity plays an important role in efficiently solving the given optimization problem.

25.5 Conclusions

We proposed Particle Swarm Optimization with Diversive Curiosity, PSO/DC. The key idea of the method is to introduce a mechanism of diversive curiosity into PSO. The mechanism is achieved by an internal indicator which detects premature convergence for improve the search performance of PSO, and sends information to make them active to find an optimal solution corresponding to a given optimization problem. It can be interpreted as the mechanism of diversive curiosity.

Owing to the internal indicator representing the mechanism of diversity curiosity, PSO/DC can successfully prevent premature convergence, and manage the exploration-exploitation trade-off. Applications of PSO/DC to a 2-dimensional multimodal optimization problem well demonstrated its effectiveness. The ratio of success in finding the optimal solution in search space is significantly improved, which reaches 100% with the estimated appropriate values of parameters in the internal indicator.

Empirically, PSO/DC is very effective in enhancing the search performance of PSO. And our experimental results basically accord with the findings called "the zone of curiosity" in psychology. Accordingly, the validity of the internal indicator introduced into PSO/DC was successfully verified. The basis of the PSO in swarm intelligence was further consolidated.

In our experiments, only a 2-dimensional multimodal optimization problem was carried out for demonstrating the effectiveness of PSO/DC. It is left for further study to apply PSO/DC to high-dimensional benchmark problems, and to complex application problems in the real-world.

Acknowledgment This research was supported by a COE program (#J19) granted to Kyushu Institute of Technology by the Ministry of Education, Culture, Sports, Science and Technology (MEXT), Japan. It was also supported by Grant-in-Aid Scientific Research(C)(18500175) from MEXT, Japan.

References

1. Berlyne D (1960) *Conflict, arousal, and curiosity*. McGraw-Hill Book Co. New York, USA
2. Cohen JD, McClure SM, Yu AJ (2007) Should I stay or should I go? How the human brain manages the trade-off between exploitation and exploration. *Philosophical Transactions of the Royal Society*, Part B, 362:933–942
3. Day H (1982) Curiosity and the interested explorer. *Performance and instruction* 21:19–22
4. Eberhart RC, Kennedy J (1995) A new optimizer using particle swarm theory. *Proceedings of the sixth International Symposium on Micro Machine and Human Science* 39–43, Nagoya, Japan.
5. Eberhart RC, Shi Y (2000) Comparing inertia weights and constriction factors in particleswarm optimization. *Proceedings of the 2000 IEEE Congress on Evolutionary Computation* 1:84–88, La Jolla, CA, USA.
6. Eshelman LJ, Schaffer JD (1993) Real-Coded Genetic Algorithms and Interval-Schemata. *Foundations of Genetic Algorithms*, Morgan Kaufman Publishers, San Mateo 2:187–202
7. Goldberg DE (1989) *Genetic Algorithms in Search, Optimization, and Machine Learning*. Addison-Wesley, Boston, USA
8. Gudise VG, Venayagamoorthy GK (2003) Evolving digital circuits using particle swarm. *Proceedings of the International Joint Conference on Neural Networks* Special Issue 1:468–472, Portland, Oregon.
9. Holland JH (1975) *Adaption in natural and artificial systems*. The MIT Press Cambridge, MA, USA
10. Hu X, Eberhart RC (2002) Adaption Particle Swarm Optimization: Detection and Response to Dynamic Systems. *Proceedings of the 2002 Congress on Evolutionary Computation* 2:1666–1670, Honolulu, HI, USA.
11. Kaplan F, Oudeyer P-Y (2006) Curiosity-driven development. *Proceedings of International Workshop on Synergistic Intelligence Dynamics* 1–8 Genova, Italy.
12. Kennedy J, Eberhart RC (1995) Particle swarm optimization. *Proceedings of the 1995 IEEE International Conference on Neural Networks* 1942–1948, Piscataway, NJ, USA.
13. Kennedy J (2006) In Search of the Essential Particle Swarm. *Proceedings of 2006 IEEE Congress on Evolutionary Computations*, 6158–6165, Vancouver, BC, Canada.
14. Loewenstein G (1994) The psychology of curiosity: a review and reinterpretation. *Psychological Bulletin* 116(1):75–98
15. Man KF, Tang KS, Kwong S (1999) *Genetic Algorithms*. Springer-Verlag, London
16. Meissner M, Schmuker M, Schneider G (2006) Optimized Particle Swarm Optimization (OPSO) and its application to artificial neural network training. *BMC Bioinformatics* 7(125)
17. Oudeyer P-Y, Kaplan F, Hafner V (2007) Intrinsic Motivation Systems for Autonomous Mental Development. *IEEE Transactions on Evolutionary Computation* 11(2):265–286
18. Parsopoulos KE, Vrahatis MN (2002) Recent approaches to global optimization problems through Particle Swarm Optimization. *Natural Computing* 1:235-306 Kluwer Academic Publisher, Netherlands.
19. Reyes-Sierra M, Coello CAC (2006) Multi-Objective Particle Swarm Optimizers: A Survey of the State-of-the-Art. *International Journal of Computational Intelligence Research* 2(3):287–308

20. Spina R (2006) Optimisation of injection moulded parts by using ANN-PSO approach. *Journal of Achievements in Materials and Manufacturing Engineering* 15(1–2): 146–152
21. Storn R, Price K (1997) Differential evolution – a simple and efficient heuristic for global optimization over continuous space. *Journal of Global Optimization* 11(4):341–359
22. Sutton RS, Barto AG (1998) *Reinforcement Learning: A Introduction.* The MIT Press, Cambridge, MA, USA
23. Takadama K, Shimohara K (2001) Exploration and Exploitation Trade-off in Multiagent Learning. *Proceedings of the 4th International Conference on Computational Intelligence and Multimedia Applications* (ICCIMA'01), 5 pages Yokusika, Japan.
24. Wohlwill JF (1981) A Conceptual Analysis of Exploratory Behavior: The "specific-diversive" distinction revisited. In Day HI (ed.) *Advances in Intrinsic Motivation and Aesthetics.* Plenum Pub Corp, NY, USA
25. Xiao RB, Xu YC, Amos M (2007) Two hybrid compaction algorithms for the layout optimization problem. *BioSystems* 90(2):560–567
26. Zhang H, Ishikawa M (2004) An Extended Hybrid Genetic Algorithm for Exploring a Large Search Space. *Proceedings of the 2nd International Conference on Autonomous Robots and Agents* (ICARA2004) 244–248 North Palmerston, New Zealand.
27. Zhang H, Ishikawa M (2005) A Hybrid Real–Coded Genetic Algorithm with Local Search. *Proceedings of the 12th International Conference on Neural Information Processing* (ICONIP2005) 732–737, Taipei, Taiwan, R.O.C..
28. Zhang H, Ishikawa M (2007) Evolutionary Particle Swarm Optimization (EPSO) – Estimation of Optimal PSO Parameters by GA. *Proceedings of the International MultiConference of Engineers and Computer Scientists* (IMECS 2007) 1:13-18 Hong Kong China.
29. Zhang H, Ishikawa M (2008a) Designing Particle Swarm Optimization – Performance Comparison of Two Temporally Cumulative Fitness Functions in EPSO. *Proceedings of the 26th IASTED International Conference on Artificial Intelligence and Applications* (AIA 2008) 301–306 Innsbruck Austria.
30. Zhang H, Ishikawa M (2008b) Improving the Performance of Particle Swarm Optimization with Diversive Curiosity. *Proceedings of the International MultiConference of Engineers and Computer Scientists* (IMECS 2008), IAENG 1:1–6 Hong Kong China.
31. Zhang H, Ishikawa M (2008c) Evolutionary Particle Swarm Optimization – Metaoptimization Method with GA for Estimating Optimal PSO Methods. In Castillo O et al. (Eds.) *Trends in Intelligent Systems and Computer Engineering* Lecture Notes in Electrical Engineering, Vol.6, 75–90 Springer, New York.

GPSR Compliance

The European Union's (EU) General Product Safety Regulation (GPSR) is a set of rules that requires consumer products to be safe and our obligations to ensure this.

If you have any concerns about our products, you can contact us on ProductSafety@springernature.com

In case Publisher is established outside the EU, the EU authorized representative is:

Springer Nature Customer Service Center GmbH
Europaplatz 3
69115 Heidelberg, Germany

Batch number: 10761599

Printed by Printforce, the Netherlands